'[Stories] of breathtaking derring-do … A gripping new collection
from Max Hastings that puts you at the heart of the battle … In his
powerful new book, *Soldiers: Great Stories of War and Peace*, he has
collected first-person accounts that illustrate in searing detail and
immediacy all the violence, grief, pathos, black humour and courage
of conflict. In these compelling extracts, a young officer agonises over
his decision to leave a dying comrade, a badly wounded Gurkha gets
back into battle, and a legendary field marshal is executed by his own
side'
Daily Mail

'These accounts show the reality of military life … A pointillist portrait
of enthralling sensitivity … stories tumble from the pages of this book
like gems from a pirate's chest … The anecdotes about female soldiers
in this book are fascinating'
The Times

'An unmissable read … A brilliant, wide-ranging anthology … The
book amply proves Max Hastings's contention that "all generalisations
about soldiers fail" and that "they come in as many sorts and conditions
as does the rest of humanity" … The sheer variety of voices for which
Hastings has found room is impressive … The most famous names of
military history, from Julius Caesar to Erwin Rommel, have their
places, yet some of the most compelling tales are those of ordinary,
often reluctant warriors. For all those who share Hastings's "fascination
with wars and those who have fought through the ages", *Soldiers* is an
unmissable collection'
Sunday Times

SOLDIERS

GREAT STORIES OF WAR AND PEACE

Chosen, Edited and
Introduced by

MAX HASTINGS

WILLIAM
COLLINS

William Collins
An imprint of HarperCollins*Publishers*
1 London Bridge Street
London SE1 9GF

WilliamCollinsBooks.com

HarperCollins*Publishers*
Macken House, 39/40 Mayor Street Upper
Dublin 1, D01 C9W8, Ireland

First published in Great Britain in 2021 by William Collins
This William Collins paperback edition published in 2024

1

A catalogue record for this book is
available from the British Library

ISBN 978-0-00-845426-5

Set in Adobe Garamond Pro
Printed and bound in the UK using 100%
renewable electricity at CPI Group (UK) Ltd

MIX
Paper | Supporting
responsible forestry
FSC™ C007454

This book contains FSC™ certified paper and other controlled
sources to ensure responsible forest management.

For more information visit: www.harpercollins.co.uk/green

For my grandson William,
whose company means so much to me,
and who will not be chivvied to become a soldier,
a war correspondent or even a historian.

Contents

Illustrations

David and Goliath (*Chronicle/Alamy*)
Achilles celebrating the death of Hector (*Alan Carter/Alamy*)
Greek hoplite (© *The Trustees of the British Museum*)
Fourteenth-century warfare at the Siege of Aubenton in France
    (*Album/Alamy*)
The Consequences of War by Peter Paul Rubens (*IanDagnall
    Computing/Alamy*)
The Reduction of Breda by Diego Velázquez (*The Print Collector/
    Alamy*)
The Death of General Wolfe by Benjamin West (*The Yorck Project*)
Marshal Michel Ney painted by François Gérard
Queen Boudicca (*Rapp Halour/Alamy*)
William Waller
Julius Caesar
Joan of Arc (*DEA/M. SEEMULLER/Getty*)
James Graham
William Cobbett
Nadezhda Durova
Peter the Great (*IanDagnall Computing/Alamy*)
James Brudenell (*Hulton Archive/Stringer/Getty*)
Cavalié Mercer
Wellington at Waterloo (*Lifestyle pictures/Alamy*)
An 1830 dragoon (*Author's collection*)
Flanders by Otto Dix
Soldiers sharing a cigarette on D-Day (*Liverpool Echo*)
Leo Tolstoy (*Imagno/Getty*)
Lakshmibai (*Dinodia Photos/Alamy*)
Edward Louis Spears (*Photo 12/Alamy*)
Marcel Proust

Fyodor Dostoevsky (*ITAR-TASS News Agency/Alamy*)

Constance Edwina Lewis (*Historic Collection/Alamy*)

Dorothie Feilding (*Chronicle/Alamy*)

Robert Graves

Adrian Carton de Wiart (*Bettmann/Getty*)

Winston Churchill (*CBW/Alamy*)

Siegfried Sassoon (*Hulton Deutsch/Getty*)

Theodore Roosevelt

Millicent Leveson-Gower (*Chronicle/Alamy*)

A Vietnamese woman holding her child during conflict (*Eddie Adams/AP/Shutterstock*)

Shoshana Johnson (*Sgt. Arledge/US Army/Getty*)

George Orwell

Evelyn Waugh

Eric Linklater (*Ronald Leslie Stewart/Fairfax Media/Getty*)

Erwin Rommel (*Corbis/Getty*)

Enoch Powell

Auberon Waugh (*James Wright/Radio Times/Getty*)

Karl Marlantes

Hiroo Onoda (*The Asahi Shimbun/Getty*)

David Niven (*Mondadori/Getty*)

Patrick Leigh Fermor (*Evening Standard/Getty*)

Michael Howard

Arkady Babchenko

Andrew Milburn

Johnson Beharry (*Sandra Rowse/Alamy*)

Jane Blair (*Christophe Guibbaud/Abaca Press/Alamy*)

Kayla Williams (*courtesy of Kayla Williams*)

SAS troops preparing to abseil from the roof of the Iranian Embassy during the 1980 siege (*Combined Military Services Museum, Maldon, Essex*)

SAS troops storming the Iranian Embassy (© *Crown*)

British soldiers in Afghanistan (*John Moore/Getty*)

Introduction

This is a book for readers who share my lifelong fascination with wars and those who have fought them through the ages. Voltaire wrote that anecdotes are 'the gleanings left over from the vast harvest field of history'. I have gathered stories of men and women which caught my attention first as a schoolboy; then, when I witnessed wars as a correspondent on battlefields around the world; later, when researching as a historian. Any such collection is personal and whimsical, because the range of possible sources is so wide. To do justice to the span of the centuries would demand a library, not a single volume. Think of this as an introductory offer, to tempt you to explore more widely writers and events of which I offer glimpses. Although you will find memorable soldiers' words in what follows, this is neither a book of quotations nor a strategy study. It is, instead, a gathering of tales about people who have borne arms.

Almost all generalisations about soldiers fail; they come in as many sorts and conditions as does the rest of humanity. My choices ring the changes not merely between eras, but also between the heroic, tragic and comic; great moments and slight encounters; the famous and the humble. You will meet Jewish heroes of the Bible, Rome's captain of the gate, Cromwell, Ulysses S. Grant, Napoleon's marshals, George S. Patton, together with less celebrated figures of all ranks and many armies. Garrison duty yields much that is frivolous or absurd, while battlefields are places of life and death. When warriors are not fighting – which is most of the time – they can enjoy the same pleasures as the rest of us. In war, however, they are exposed to extremes of suffering and peril such as most civilians are fortunate enough to escape. That is how soldiers earn their pay, how they justify the stupendous budgets that almost all nations expend on their armed forces.

Some tales are recounted by men who enthusiastically embraced military life, others by those who abhorred it. Where there is comedy, I have tried to choose stories that rise above mere barrack-room humour. The collection emphasizes British experience, because this is my own culture, but also pays homage to Romans, Greeks, Persians, the French, Germans, Russians and Americans. Some stories are those such as every schoolchild learns – or at least, used to learn – of David and Goliath, Sir Philip Sidney dying at Zutphen, Wellington bestriding the ridge of Mont St Jean. There are also tales of less distinguished soldiers who achieved fame as writers, including Cobbett and Dostoevsky, Edward Gibbon and Siegfried Sassoon, Marcel Proust and Evelyn Waugh, George Orwell and George Macdonald Fraser.

There is a bias towards recollections from 1800 onwards, because for the past two centuries writers have expressed themselves in words that strike a more responsive chord with modern readers than do earlier chroniclers. Soldiers have assuredly talked to each other in the same expletive-undeleted language throughout history, from Gideon's and Leonidas' Three Hundreds, through the armies of Henry V and Cromwell, to the twenty-first century. But successive generations compiled very different records: early accounts are, to our eyes, less intimate, more opaque. The two world wars generated the richest reminiscences, because they swept into armies millions of citizens more articulate than most professional soldiers, some being possessed of literary brilliance. Wherever possible, I favour stories recounted by contemporaries, though it is doubtful that the writer of the book of Joshua witnessed the fall of Jericho, if such an event even took place. The biblical quotations are taken from the King James I edition, because the beauty of its language is irresistible.

Dr Johnson quoted a French savant who noted that 'there are many puerilities in war'. Most of those who fight are young, and thus the spirit of armies is distilled from a mingling of soldiers' physical vigour and intellectual immaturity with their licences to kill and opportunities to die. Some familiar stories, such as those of Joan of Arc and Sir John Moore at Corunna, are included. Others, however, such as Michael Strachan's vignette of Col. J. Enoch Powell in 1943 North Africa and Rory Stewart's 2004 experience of Iraq, may come fresh.

There is a rich infusion of tales from the Peninsula, because the campaign yielded so many memoirs. I have omitted or included only brief passages about several remarkable characters – for instance, Marcellin Marbot, Joshua Lawrence Chamberlain, Gustavus Burnaby – whom I explored in my earlier book *Warriors*. Rather than citing the highest authorities, I draw upon writers and narratives that have stirred my own imagination over a lifetime. Modern scholars scorn – for instance – C. V. Wedgwood's writings on the English Civil Wars, those of Cecil Woodham-Smith about the Crimea, and Barbara Tuchman's big books. Would that the dismissive academics wrote a fraction so well! All three women were brilliant narrative historians who have been my faithful companions for sixty years. I likewise include several of Elizabeth Longford's vignettes of Wellington. Her account of the June 1815 Duchess of Richmond's ball before Waterloo, that glittering 'revelry by night', is more vivid than that of any other writer I know, save Byron and Thackeray.

My purpose is to sound notes that ring true about the warrior's condition. The pervasive thesis is that of Lord Tedder, Eisenhower's 1944–5 Deputy Supreme Commander in NW Europe: 'War is organized confusion.' Early pages emphasize the doings of the great, because posterity is informed about them: we possess hundreds of pages of Caesar's shamelessly self-serving writings, almost none from his legionaries. Later passages tell us more about humble folk – for example, Private William Wheeler in the Napoleonic wars: once the poor became widely literate, they too compiled records such as do not survive from earlier times. Some of the contributors loomed large in my own life. Nicholas Tomalin, who wrote a black comic account of a 1966 helicopter ride in Vietnam with an American general, was a close friend and brilliant war correspondent, killed on the Golan Heights in 1973, a few miles from where I was myself watching the Israelis fighting for their lives against the Syrians. Professor Sir Michael Howard OM CH MC, who died in 2019, was a lifelong mentor who profoundly influenced my view of history, and indeed of everything. Other historians whose work is represented here are admired friends such as Rick Atkinson and Antony Beevor; the late John Keegan and Ronald Lewin. My father was editing the old *Strand Magazine* when, in 1949, a reader sent him the priceless bundle of Private Wheeler's letters, found in an attic, which *The Strand* then brought before the world.

A significant portion of the stories below, especially the early ones, appeared in my 1985 anthology the *Oxford Book of Military Anecdotes*. Since that book is long out of print, it would have been perverse to omit gems featured in it. I have, however, re-edited and abridged all the entries for this volume, which emphasizes more recent reading. Men, like it or not, have dominated military affairs throughout most of history, as Margaret MacMillan reminds us in her penetrating study *War*, based on her 2018 Reith Lectures. There were formidable warrior queens and chieftains, but until the twenty-first century only a tiny number of women – for example, the '*abosi*' of Dahomey, some World War II partisans and snipers of the Red Army – killed people with their own hands. Instead, they bore a disproportionate share of war's victimhood. In the past few decades, however, they have played increasingly important roles as protagonists on the world's battlefields, and I include examples of their experiences. An editor who compiles such a collection as this a century hence will be able to draw upon many more women warriors' stories.

The British Army has a longstanding tradition of professing to make war with a light heart, a practice that sometimes bewilders our American allies, wary of soldiers who make too many jokes. Field-Marshal Lord Carver described to me an encounter in June 1944 amid the Normandy bloodbaths, in which he commanded an armoured regiment. He was receiving orders when Brigadier Robert 'Looney' Hinde suddenly broke off and peered intently at the ground beneath him. Carver, under considerable stress, not to say German shellfire, suggested that Hinde should finish the briefing. His eccentric superior responded: 'Shut up, Mike! You can fight a battle every day, but you could go the rest of your life without seeing a caterpillar like this! Somebody find me a matchbox.' It is hard to imagine a modern general daring to indulge himself so.

Soldiers like to clothe their trade in a fancy dress of words that include honour and chivalry, but the history of wars is, in the last analysis, an account of humankind's efforts to kill each other. The reader of this literary march-past should discern a change of pace or mood as the centuries succeed each other, becoming most pronounced in our own, which some are calling the Post-Heroic Age. It has become unfashionable, and surely rightly so, to treat war as a sport to amuse adventurous young men, as did medieval knights and many of Wellington's officers. Modern soldiers' accounts of Iraq and Afghanistan are rich in tales of horror and derring-do, but include far less laughter. This may partly be because 'wars among the people' are especially replete with tragedies, examples of which are recorded here. Also, I believe that the pandemic spread of UAVs – drones; improvised explosive devices – mines; and other technology for remote killing as a substitute for clashes between soldier and soldier makes combat seem less rewarding, more unambiguously terrifying, for even the most adventurous of participants.

A change has also taken place in the chronicling of conflict. Obscenities have punctuated soldiers' speech since the beginning of time, but for two millennia writers omitted them from published narratives. In the past few decades, however, starting around Vietnam time, a consensus has emerged that memoirs and histories lack authenticity unless four-letter words are spelt out, sometimes in the same profusion as semi-literate soldiers use them. I regret this, but bow to its reality.

I have resisted reprising stories that appear in my own works. The only exceptions are the account of Montrose's 1645 triumph at Inverlochy, because my biography of Charles I's lieutenant-general in Scotland is long and deservedly out of print; and my experience on the last day of the 1982 Falklands War, a fond memory, because I came home afterwards unscathed … as many better men and women in all wars do not.

MAX HASTINGS
Chilton Foliat, West Berkshire, February 2021

1.

*Even those who found Religious Studies unrewarding at school never
forget the bloodcurdling Old Testament military history we all
learned. The following incident took place during the Jewish conquest
of Canaan, around 1400 BC, for which the Bible offers the only
evidence.*

Now Jericho was straitly shut up because of the children of Israel; none
went out, and none came in. And the Lord said unto Joshua, See, I
have given into thine hand Jericho, and the king thereof, and the
mighty men of valour. And ye shall compass the city, all ye men of war,
and go round about the city once. Thus shalt thou do six days. And
seven priests shall bear before the ark seven trumpets of rams' horns:
and the seventh day ye shall compass the city seven times, and the
priests shall blow with the trumpets. And it shall come to pass, that
when they make a long blast with the ram's horn, and when ye hear
the sound of the trumpet, all the people shall shout with a great shout;
and the wall of the city shall fall down flat, and the people shall ascend
up every man straight before him.

And Joshua the son of Nun called the priests, and said unto them,
Take up the ark of the covenant, and let seven priests bear seven trumpets of rams' horns before the ark of the Lord. And it came to pass,
when Joshua had spoken unto the people, that the seven priests bearing the seven trumpets of rams' horns passed on before the Lord, and
blew with the trumpets: and the ark of the covenant of the Lord
followed them.

And Joshua had commanded the people, saying, Ye shall not shout,
nor make any noise with your voice, neither shall any word proceed
out of your mouth, until the day I bid you shout. So they did six days.
And on the seventh day, they rose early about the dawning of the day,

8

BIBLE

and compassed the city after the same manner seven times. And at the seventh time, when the priests blew with the trumpets, Joshua said unto the people, Shout; for the Lord hath given you the city. And the city shall be accursed, even it, and all that are therein, to the Lord: only Rahab the harlot shall live, she and all that are with her in the house, because she hid the messengers that we sent. So the people shouted when the priests blew with the trumpets: and when the people heard the sound of the trumpet, and the wall fell down flat, so that the people went up into the city, every man straight before him, and they took the city.

And they utterly destroyed all that was in the city, both man and woman, young and old, and ox, and sheep, and ass, with the edge of the sword. But Joshua had said unto the two men that had spied out the country, Go into the harlot's house, and bring out thence the woman, and all that she hath, as ye sware unto her. And they burnt the city with fire, and all that was therein: only the silver, and the gold, and the vessels of brass and of iron, they put into the treasury of the house of the Lord. And Joshua saved Rahab the harlot alive, and her father's household, and all that she had; and she dwelleth in Israel even unto this day; because she hid the messengers, which Joshua sent to spy out Jericho.

JOSHUA 6

2.

The modern Israeli army regards Gideon, who fought his battles around 1100 BC, as founding father of its special forces.

Then Gideon, and all the people that were with him, rose up early, and pitched beside the well of Harod: so that the host of the Midianites were on the north side of them. And the Lord said unto Gideon, The people that are with thee are too many for me to give the Midianites into their hands, lest Israel vaunt themselves against me, saying, Mine own hand hath saved me. Now therefore go to, proclaim in the ears of the people, saying, Whosoever is fearful and afraid, let him return and depart early from mount Gilead. And there returned of the people twenty and two thousand; and there remained ten thousand.

And the Lord said unto Gideon, The people are yet too many; bring them down unto the water, and I will try them for thee there. So he brought down the people unto the water: and the Lord said unto Gideon, Every one that lappeth of the water with his tongue, as a dog lappeth, him shalt thou set by himself; likewise every one that boweth down upon his knees to drink. And the number of them that lapped, putting their hand to their mouth, were three hundred men: but all the rest of the people bowed down upon their knees to drink water.

And the Lord said unto Gideon, By the three hundred men that lapped will I save you, and deliver the Midianites into thine hand: and let all the other people go every man unto his place. So the people took victuals in their hand, and their trumpets: and he sent all the rest of Israel every man unto his tent. And it came to pass the same night, that the Lord said unto him, Arise, get thee down unto the host; for I have delivered it into thine hand.

And the Midianites and the Amalekites and all the children of the east lay along in the valley like grasshoppers for multitude; and their camels were without number, as the sand by the sea side for multitude. And when Gideon was come, behold, there was a man that told a dream unto his fellow, and said, Behold, I dreamed a dream, and, lo, a cake of barley bread tumbled into the host of Midian, and came unto a tent, and smote it that it fell, and overturned it, that the tent lay along. And his fellow answered and said, this is nothing else save the sword of Gideon the son of Joash, a man of Israel: for into his hand hath God delivered Midian, and all the host.

And it was so, when Gideon heard the telling of the dream, and the interpretation thereof, that he worshipped, and returned into the host of Israel, and said, Arise; for the Lord hath delivered into your hand the host of Midian. And he divided the three hundred men into three companies, and he put a trumpet in every man's hand, with empty pitchers, and lamps within the pitchers. And he said unto them, Look on me, and do likewise: and, behold, when I come to the outside of the camp, it shall be that, as I do, so shall ye do.

When I blow with a trumpet, then blow ye the trumpets also on every side of all the camp, and say, The sword of the Lord, and of Gideon. So Gideon, and the hundred men that were with him, came unto the outside of the camp in the beginning of the middle watch. And the three companies blew the trumpets and brake the pitchers, and held the lamps in their left hands, and the trumpets in their right hands to blow withal: and they cried, The sword of the Lord, and of Gideon. And all the host ran, and cried, and fled. And the three hundred blew the trumpets, and the Lord set every man's sword against his fellow: and the host fled to Beth-shittah in Zererath, and to the border of Abelmeholah.

JUDGES 7

3.

The most famous military anecdote in the Bible, describing an alleged encounter c.1000 BC, which strikes a chord with modern Western commanders struggling to conduct counter-insurgency campaigns in body armour against barefoot adversaries.

And there went a champion out of the camp of the Philistines, named Goliath of Gath, whose height was six cubits and a span. And he had a helmet of brass upon his head, and he was armed with a coat of mail, and the weight of the coat was five thousand shekels of brass. And he had greaves of brass upon his legs, and a target of brass between his shoulders. And the staff of his spear was like a weaver's beam; and his spear's head weighed six hundred shekels of iron, and one bearing a shield went before him. And he stood and cried unto the armies of Israel, Why are ye come to set your battle in array? am not I a Philistine, and ye servants to Saul? choose you a man for you – and let him come down to me. If he be able to fight with me, and to kill me, then will we be your servants – but if I prevail against him, then shall ye be our servants, and serve us.

When Saul and all Israel heard those words of the Philistine, they were dismayed and greatly afraid. Now David was the son of that Ephraimite of Bethlehem-judah, whose name was Jesse. And David was the youngest; and the three eldest followed Saul. But David went and returned from Saul to feed his father's sheep at Bethlehem. And the Philistine drew near morning and evening, and presented himself forty days. And Jesse said unto David his son, Take now for thy brethren an ephah of this parched corn, and these ten loaves, and run to the camp to thy brethren; and carry these ten cheeses unto the captain of their thousand, and look how thy brethren fare.

And David rose up early in the morning, and left the sheep with a keeper – and he came to the trench as the host was going forth to the fight, and shouted for the battle. And David left his carriage in the hand of the keeper of the carriage, and ran into the army, and came and saluted his brethren. And as he talked with them, behold, there came up the champion, the Philistine of Gath, Goliath by name, out of the armies of the Philistines, and spake according to the same words. And all the men of Israel fled from him, and were sore afraid.

And David spake to the men that stood by him, saying, What shall be done to the man that killeth this Philistine, and taketh away the reproach from Israel? For who is this uncircumcized Philistine, that he should defy the armies of the living God? And Eliab his eldest brother heard when he spake unto the men; and Eliab's anger was kindled against David, and he said, Why camest thou down hither? and with whom hast thou left those few sheep in the wilderness? I know thy pride, and the naughtiness of thine heart; for thou art come down that thou mightest see the battle.

And David said, What have I now done? Is there not a cause? And David said to Saul, Let no man's heart fail because of him – thy servant will go and fight the Philistine. And Saul said to David, Thou art not able to go against this Philistine to fight with him; for thou art but a youth, and he a man of war. And David said unto Saul, Thy servant kept his father's sheep, and there came a lion, and a bear, and took a lamb out of the flock – And I went out after him, and smote him, and delivered it out of his mouth; and when he arose against me, I caught him by his beard, and smote him, and slew him. The Lord that delivered me out of the paw of the lion, and out of the paw of the bear, he will deliver me out of the hand of this Philistine.

And Saul armed David with his armour, and he put an helmet of brass upon his head; also he armed him with a coat of mail. And David said unto Saul, I cannot go with these – for I have not proved them. And David put them off him. And he took his staff in his hand, and chose him five smooth stones out of the brook, and put them in a shepherd's bag which he had, even in a scrip – and his sling was in his hand, and he drew near to the Philistine. And the Philistine came on and drew near unto David – and the man that bare the shield went before him. And when the Philistine looked about, and saw David, he disdained him; for he was but a youth, and ruddy, and of a fair countenance. And the Philistine said unto David, Am I a dog, that thou comest to me with staves? And the Philistine cursed David by his gods.

And the Philistine said unto David, Come to me, and I will give thy flesh unto the fowls of the air, and to the beasts of the field. Then said David to the Philistine, Thou comest to me with a sword, and with a spear, and with a shield – but I come to thee in the name of the Lord of hosts, the God of the armies of Israel, whom thou hast defied. This day will the Lord deliver thee into mine hand; and I will smite thee, and take thine head from thee; and I will give the carcases of the host of the Philistines this day unto the fowls of the air, and to the wild beasts of the earth; that all the earth may know that there is a God in Israel.

And it came to pass, when the Philistine arose, and came and drew nigh to meet David, that David hasted, and ran to meet the Philistine. And David put his hand in his bag, and took thence a stone, and slang it, and smote the Philistine in his forehead, that the stone sunk into his forehead, and he fell upon his face to the earth. Therefore David ran, and stood upon the Philistine, and took his sword, and drew it out of the sheath thereof, and slew him, and cut off his head therewith. And when the Philistines saw their champion was dead, they fled. And the men of Israel and of Judah arose, and shouted, and pursued the Philistines, until they came to the valley and to the gates of Ekron.

I SAMUEL 17

4.

Herodotus the Greek (c.484–425 BC) compiled one of the earliest significant records of the past. His Histories constitute a mingling of fact and fable, a compilation of oral traditions. Archaeologists' recent researches on the Eurasian Steppes, and especially the 2019 discovery near Voronezh of a mass grave of warrior women adorned with golden headdresses, lend a gossamer wisp of credibility to this fairy tale.

In the war between the Greeks and the Amazons, the Greeks, after their victory at the river Thermodon, sailed off in three ships with as many Amazons on board as they had succeeded in taking alive. Once at sea, the women murdered their captors, but, as they had no knowledge of boats and were unable to handle either rudder or sail or oar, they soon found themselves, when the men were done for, at the mercy of wind and wave, and were blown to Cremni – the Cliffs – on Lake Maeotis, a place within the territory of the free Scythians. Here they got ashore and made their way inland.

The first thing they fell in with was a herd of horses grazing; these they seized, and, mounting on their backs, rode off in search of loot. The Scythians could not understand what was happening and were at a loss to know where the marauders had come from, as their dress, speech, and nationality were strange to them. Thinking, however, that they were young men, they fought in defence of their property, and discovered from the bodies which came into their possession after the battle that they were women.

The discovery gave a new direction to their plans; they decided to make no further attempt to kill the invaders, but to send out a detachment of their youngest men, about equal in number to the Amazons, with orders to camp near them and take their cue from whatever the Amazons then did: if they pursued them, they were not to fight, but to give ground; then, when the pursuit was abandoned, they were once again to encamp within easy range. The motive behind this policy was the Scythians' desire to get children by the Amazons. The detachment of young men obeyed their orders, and the Amazons, realizing that they meant no harm, did not attempt to molest them, with the result that every day the two camps drew a little closer together. Neither party had anything but their weapons and their horses, and both lived the same sort of life, hunting and plundering.

Towards midday the Amazons used to scatter and go off to some little distance in ones and twos to ease themselves, and the Scythians, when they noticed this, followed suit; until one of them, coming upon an Amazon girl all by herself, began to make advances to her. She, nothing loth, gave him what he wanted, and then told him by signs (being unable to express her meaning in words, as neither understood the other's language) to return on the following day with a friend, making it clear that there must be two men, and that she herself would bring another girl. The young man then left her and told the others what had happened, and on the next day took a friend to the same spot, where he found his Amazon waiting for him and another one with her.

Having learnt of their success, the rest of the young Scythians soon succeeded in getting the Amazons to submit to their wishes. The two camps were then united, and Amazons and Scythians lived together, every man keeping as his wife the woman whose favours he had first enjoyed. The men could not learn the women's language, but the women succeeded in picking up the men's; so when they could understand one another, the Scythians made the following proposal: 'We', they said, 'have parents and property. Let us give up our present way of life and return to live with our people. We will keep you as our wives and not take any others.' The Amazons replied: 'We and the women of your nation could never live together; our ways are too much at variance. We are riders; our business is with the bow and the spear, and we know nothing of women's work; but in your country no woman has anything to do with such things – your women stay at home in their waggons occupied with feminine tasks, and never go out to hunt or for any other purpose.

'We could not possibly agree. If, however, you wish to keep us for your wives and to behave as honourable men, go and get from your parents the share of property which is due to you, and then let us go off and live by ourselves.' The young men agreed to this, and when they came back, each with his portion of the family possessions, the Amazons said: 'We dread the prospect of settling down here, for we have done much damage to the country by our raids, and we have robbed you of your parents. Look now – if you think fit to keep us for your wives, let us get out of the country altogether and settle somewhere on the other side of the Tanais.'

Once again the Scythians agreed, so they crossed the Tanais and travelled east for three days, and then north, for another three, from Lake Maeotis, until they reached the country where they are today, and settled there. Ever since the women of the Sauromatae have kept to their old ways, riding to the hunt on horseback sometimes with, sometimes without, their menfolk, taking part in war and wearing the same clothes as men. The language of these people is the Scythian, but it has always been a corrupt form of it because the Amazons were never able to learn to speak it properly. They have a marriage law which forbids a girl to marry until she has killed an enemy in battle; some of their women, unable to fulfil this condition, grow old and die in spinsterhood.

HERODOTUS

5.

A legend of the sixth century BC, on which Macaulay based the epic poem that nineteenth- and twentieth-century schoolboys learned. Every general in history has urged his soldiers to embrace Horatius's words, 'And how can man die better than facing fearful odds/for the ashes of his Fathers and the temples of his Gods?', though only a small, heroic minority ever agree.

The Tarquins, deposed regal family of Rome, fled to Lars Porsena, king of Clusium. There, with advice and entreaties, they besought him not to suffer them, who were descended from the Etrurians and of the same blood and name, to live in exile and poverty; and advised him not to let this practice of expelling kings from Rome to pass unpunished. Porsena, thinking it would be an honour to the Tuscans that there should be a king at Rome, marched with an army. Never before had such terror seized the Senate, so powerful was the state of Clusium at the time, and so great the renown of Porsena.

Some parts of the city seemed secured by the walls, others by the River Tiber. The Sublician Bridge well-nigh afforded a passage to the enemy, had there not been one man, Horatius Cocles, who happening to be posted on guard at the bridge, when he saw the Janiculum taken by a sudden assault and the enemy pouring down thence at full speed, and that his own party, in terror and confusion, were abandoning their arms and ranks. Laying hold of them one by one, standing in their way and appealing to the faith of gods and men, he declared that their flight would avail them nothing if they deserted their posts.

He charged them to demolish the bridge by sword, by fire, or by any means whatever; declaring that he would stand the shock of the enemy as far as could be done by one man. He then advanced to the first entrance of the bridge, and faced about to engage the foe hand to hand, and by his surprising bravery he terrified the enemy. Two indeed remained with him from a sense of shame: Sp. Lartius and T. Herminius, men eminent for their birth, and renowned for their gallant exploits.

But as they who demolished the bridge called upon them to retire, he obliged them also to withdraw to a place of safety on a small portion of the bridge that was still left. Then casting his stern eyes toward the officers of the Etrurians in a threatening manner, he now challenged them singly, and then reproached them, slaves of haughty tyrants who came to oppress the liberty of others. They hesitated for a time, looking round one at the other, to begin the fight; shame then put the army in motion, and a shout being raised, they hurled weapons from all sides at their single adversary; and when they all stuck in his upraised shield, and he with no less obstinacy kept possession of the bridge, they endeavoured to thrust him down from it by one push, when the crash of the falling bridge was heard, and at the same time a shout of the Romans, raised for joy at having completed their purpose, checked their ardour with sudden panic.

Then said Cocles: 'Holy Father Tiber, I pray thee, receive these arms, and this thy soldier, in thy propitious stream.' Armed as he was, he leaped into the Tiber, and amid showers of darts, swam across safe to his party, having dared an act which is likely to obtain with posterity more fame than credit. The state was grateful for such valour; a statue was erected to him in the Comitium, and as much land given to him as he could plough in one day.

LIVY

6.

Marching upon Greece in 480 BC, the army of Xerxes I (519–465 BC) witnessed an eclipse at the crossing of the Hellespont. The king was satisfied by the assurances of the soothsayers that this augured evil to his Greek enemies, not to himself. But others were less convinced.

The army had not gone far when Pythius the Lydian, in alarm at the sign from heaven, was emboldened to come to Xerxes with a request. 'Master,' he said, 'there is a favour I should like you to grant me – a small thing, indeed, for you to perform, but to me of great importance, should you consent to do so.' Xerxes agreed to grant it and told Pythius to say what he wanted. This generous answer raised Pythius' hopes, and he said, 'My lord, I have five sons, and it happens that every one of them is serving in your army. I am an old man, Sire, and I beg you in pity to release from service the eldest, to care for me and my property. Take the other four – and may you return with your purpose accomplished.'

Xerxes was furious. 'You miserable fellow,' he cried, 'have you the face to mention your son, when I am marching to the war against Greece with my sons, brothers, kinsmen and friends – you, my slave, whose duty it was to come with me with every member of your house, including your wife? Your punishment will be less than your impudence deserves. Yourself and four of your sons are saved; but you shall pay with the life of the fifth, whom you cling to most.' Xerxes at once gave orders to find Pythius' eldest son, cut him in half and lay the two halves one on each side of the road, for the army to march out between them.

HERODOTUS

7.

As Xerxes' 100,000 marched south, the main body of the Greek army retired beyond the Isthmus of Corinth, leaving a rearguard of three hundred at the Pass of Thermopylae, commanded by Leonidas of Sparta. What followed created one of the most influential myths or legends in Western culture.

The Greeks at Thermopylae had their first warning of the death that was coming with the dawn from the seer Megistias, who read their doom in the victims of sacrifice; deserters, too, had come in during the night with news of the Persian movement to take them in the rear. Moreover, just as day was breaking, the pickets had come running down from the hills. A conference was held, at which opinions proved divided, some officers urging that they must on no account abandon their position; others taking the opposite view. The outcome was that the army divided: some men dispersed, setting out for their various homes. Others made ready to stand with Leonidas.

Another account suggests that Leonidas himself dismissed a part of his force, to spare their lives, but thought it unbecoming for his own Spartans to quit the position which they had come to hold. I myself incline to think that he dismissed those whom he realized that had no heart for the fight, were unwilling to accept their share of the peril; meanwhile, honour forbade that he himself should go. And indeed, by remaining at his position he left a great name behind him, and Sparta did not forfeit her wealth, as might otherwise have happened; for at the commencement of the war the Spartans had been told by the oracle which they consulted, that either their city must be laid waste by foreign foes, or one of their kings must be killed.

I believe it was the memory of this prophecy, combined with his desire to lay up for the Spartans a treasure of fame in which no other city should share, that caused Leonidas to dismiss those troops. Moreover, I am strongly supported in this view by the evidence of Megistias, the seer from Acarnania who foretold the coming doom by his inspection of the sacrificial victims. This man was with the army, and was directly ordered by Leonidas to quit Thermopylae, to preserve him. But he refused to go, sending away instead an only son.

In the morning Xerxes poured a libation to the rising sun, then waited till about the time the marketplaces fill, when he began to move forward. As the Persian army launched its assault, the Greeks, knowing that the fight would be their last, pressed forward into the wider part of the pass. Many of the invaders fell; behind them the Persian company commanders plied their whips, driving the men remorselessly on. Many fell into the sea and were drowned, and still more were trampled to death by their friends. No one could count the dead. The Greeks, who knew that the enemy were on their way round by the mountain track and that death was inevitable, fought with reckless desperation. By this time most of their spears were broken, and they were killing Persians with their swords.

Leonidas fell, having fought like a man indeed. Many distinguished Spartans were killed at his side – I have familiarized myself with their names, and indeed those of all the three hundred, because they deserve to be remembered. Amongst the Persian dead, too, were many men of high distinction – for instance, two brothers of Xerxes, Habrocomes and Hyperanthes. There was a bitter struggle over the body of Leonidas; four times the Greeks drove the enemy off, and at last by their valour succeeded in dragging it away. So the contest went on, until the Greeks withdrew into the narrow neck of the pass, behind their walls, and took up a position in a single compact body – all except the Thebans on the little hill at the entrance to the pass, where the stone lion in memory of Leonidas stands today.

Here they resisted to the last, with their swords, if they had them, and, if not, with their hands and teeth, until the Persians, advancing frontally over the ruins of the wall and also closing in from behind, finally overwhelmed them. Of all the Spartans and Thespians who fought so valiantly on that day, the most signal proof of courage was given by the Spartan Dieneces. It is said that before the battle he was told by a native of Trachis that, when the Persians loosed their arrows, these were so numerous that they hid the sun. Dieneces, however, unmoved by the thought of the terrible strength of the Persian army, merely remarked: 'This is pleasant news that the stranger from Trachis brings us: for if the Persians hide the sun, we shall have our battle in the shade.'

After Dieneces the greatest distinction was won by two Spartan brothers, Alpheus and Maron, the sons of Orsiphantus; and of the Thespians a certain Dithyrambous, the son of Harmatides. The dead were buried where they fell. Over them is this inscription: Four thousand here from Pelops' land against three million once did stand. The Spartans have a special epitaph; it runs:

Go tell the Spartans, you who read:
We followed their orders, and are dead.

HERODOTUS

8.

Until the twentieth century, loot was recognized as the legitimate perquisite of every conquering army. An extravagant example was the garnering after the 5 November 333 BC battle at Issus, in Anatolia. King Alexander III of Macedon (356–323 BC) – the Great – had succeeded his father at the age of twenty. Three years later, he gained one of his greatest victories, over Darius of Persia. The spoils, both material and human, almost defied imagination.

Even in his camp, Darius had encumbered himself with riches and paraphernalia, though these were only a foretaste of what lay abandoned at his base in Damascus. The Macedonians had plundered all that was to hand, reserving the royal tent for the man who now deserved it, so that when Alexander returned at midnight, bloodstained and muddied, expressing a wish to wash off his sweat in Darius's bath, they could lead him forwards to his rightful prize, a Companion reminding him that Darius's bath was in future to be known as Alexander's. On the threshold of the royal tent, Alexander stood surprised, struck by a sight which no young man from Pella could ever have imagined to be true: When he saw the bowls, pitchers, tubs and caskets, all of gold, most exquisitely worked and set in a chamber which breathed a marvellous scent of incense and spices, when he passed through into a tent whose size and height were no less remarkable, whose sofas and tables were even laid for his dinner, then he looked long and hard at his companions and remarked: 'This, it would seem, is to be a King.'

But there is more to kingship than its treasures. Alexander was tired; he wanted his bath and dinner; he was limping from a dagger-wound in his thigh which court gossip attributed to a thrust from King Darius himself. And yet he was perturbed by the sound of ladies wailing close to where he stood, and was told that these were Darius's wife, mother and children, weeping for the king whom they believed to be dead. Promptly, Alexander sent a companion, Leonnatus, to reassure them and to tell them, perhaps in Persian, that Darius lived, though his cloak and weapons had been captured in his chariot: Alexander would grant them royal state and the continuing rank of Queen.

The following morning, Alexander summoned Hephaistion and went to visit his royal captives. When they entered her tent, it was said, the Queen Mother did obeisance to Hephaistion, mistaking him for Alexander as he seemed so plainly the taller of the two. Hephaistion recoiled and an attendant corrected her; she stood back, flustered at her mistake. Alexander had the tact to cope with a lady's embarrassment: 'No mistake,' he replied, 'for he too is an Alexander.' Then, he complimented Darius's wife on her six-year-old son and confirmed the ladies' privileges, presenting them with dresses and jewellery and giving them leave to bury their Persian dead; they were to live unmolested in quarters of their own, honour being paid to their beauty. Once more, Alexander had shown himself able to respect feminine nobility; his captives could have been valuable hostages, but he never used them for political bargaining, and not for nine years did he marry Darius's daughter.

Respect for captive royalty had a long history in the ancient East, and Alexander was not the man to betray it; the Persian Queen Mother, especially, came to recognize his kindness. His army were shown this quality in his own inimitable way. Despite his wound, he went round all the other wounded and talked to them; he collected the dead and buried them magnificently with his army arrayed in their full battle-finery; he had a word of congratulation for all whom he himself had seen distinguishing themselves particularly bravely or whose valour he heard from agreed reports: with extra presents of money, he honoured them all according to their deserts. That is the way to lead one's men.

The royal tent and the royal family were not the sum of Alexander's reward. Parmenion was sent to Damascus with orders to capture the treasures; the guards surrendered him 2,600 talents of coin and 500 pounds of silver, unminted as was the Great King's practice. The coin alone equalled one year's revenue from Philip's Macedonia, and sufficed for all debts of army pay and six months' wages; 7,000 valuable pack-animals heaved it back to the main camp. Parmenion further reported that '329 female musicians, 306 different cooks, 13 pastry chefs, 70 wine waiters and 40 scent makers' had been captured. With them came two more personal prizes, the first being the precious casket in which Alexander, after much debate, decided to store his copy of the *Iliad*, the second the Persian lady Barsine, some thirty years old, with an attractive family history. She had been married first to Memnon's brother, then to Memnon himself, and was thus brought up to a Greek way of life.

She was the daughter of the respected Persian satrap Artabazus, who was of royal blood on his mother's side, and she had taken refuge at Philip's Pella some twenty years earlier when her father was exiled from Asia Minor. Barsine met Alexander when he was a boy. 'On Parmenion's advice,' wrote Aristobulus, 'Alexander attached himself to this well-mannered and beautiful noblewoman.' It was a fitting climax to what may have been a childhood friendship, and Alexander retained his first bilingual mistress for the next five years.

ROBIN LANE FOX

9.

Pyrrhus, Greek king of Epirus (319–272 BC), aspired to emulate Alexander the Great, but was unable to translate his considerable military talents into secure political power. In 280 BC, he fought alongside the Tarentines against Rome, and defeated the consul Laevinus at Heraclea. Legend holds that 15,000 men died in the battle, which contributed a lasting memorial of the king to Western figures of speech.

Pyrrhus said to one who was congratulating him on his victory, 'If we are victorious in one more battle with the Romans, we shall be utterly ruined.'

PLUTARCH

10.

Hannibal Barca (247–183 BC), the Carthaginian general who has been called 'the father of strategy', crossed the Alps in 218 BC, his army led by its war elephants.

Wherever they went they rendered the army safe from the enemy, because men unacquainted with such animals were afraid of approaching too nearly. On the ninth day they came to a summit, chiefly through places trackless; and after many mistakes of their way, which were caused either by the treachery of the guides, or, when they were not trusted, by entering valleys at random. For two days they remained encamped on the summit; and rest was given to the soldiers, exhausted with toil and fighting: and several beasts of burden, which had fallen down among the rocks, arrived at the camp. A fall of snow, it being now the season of the setting of the constellation of the Pleiades, caused great fear to the soldiers, already worn out with so many hardships. On the standards being moved forward at daybreak, when the army proceeded slowly over all places entirely blocked up with snow, languor and despair were apparent in the countenances of all.

Hannibal, having advanced before the standards, ordered the soldiers to halt on a certain eminence, whence there was a prospect far and wide. He pointed out to them Italy and the plains of the Po, lying beneath the mountains. He said that 'they were now surmounting not only the ramparts of Italy, but also of the city of Rome; that the rest of the march would be smooth and downhill; that after one, or, at most, a second battle, they would have the citadel and capital of Italy in their power and possession'. The army then began to advance, the enemy not interfering beyond petty scavenging, as opportunity offered. But the journey proved much more difficult than it had been in the ascent, as the declivity of the Alps being generally shorter on the side of Italy is consequently steeper; so that neither those who made the least stumble could prevent themselves from falling, nor, when fallen, remain in the same place, but rolled, both men and beasts of burden, one upon another.

They then came to a rock much more narrow, and formed of such perpendicular ledges, that a light-armed soldier, carefully making the attempt, and clinging with his hands to the bushes and roots around, could with difficulty lower himself down. The ground, very steep by nature, had been broken by a recent landslip into a precipice of nearly a thousand feet in depth. Here when the cavalry had halted, as if at the end of their journey, it was reported to Hannibal, that the rock was impassable. Having gone himself to view the place, it seemed clear to him that he must lead his army round it, by however great a circuit, through the pathless and untrodden regions around.

But this route also proved impracticable; there was a wretched struggle, both on account of the slippery ice not affording any hold to the step, and giving way beneath the foot more readily by reason of the slope; and whether they assisted themselves in rising by their hands or their knees, their supports themselves giving way, they would tumble again; nor were there any stumps or roots near, by pressing against which, a man might with hand or foot support himself; so that they floundered on the smooth ice and amid the melted snow.

The beasts of burden sometimes also cut into this lower ice by merely treading upon it, at others they broke it completely through, by the violence with which they struck their hoofs in their struggling, so that most of them, as if taken in a trap, stuck in the hardened and deeply frozen ice. At length, after the men and beasts had been fatigued to no purpose, a camp was pitched on the summit, the ground being cleared for that purpose with great difficulty, so much snow was there to be dug out and carried away. The soldiers being then set to make a way down the cliff, by which alone a passage could be effected, and it being necessary that they should cut through the rocks, having felled and lopped a number of large trees which grew around, they made a huge pile of timber; and as soon as a strong wind fit for exciting the flames arose, they set fire to it, and, pouring vinegar on the heated stones, they rendered them soft and crumbling. They then opened a way with iron tools through the rock thus heated by the fire, and softened its declivities by gentle windings, so that not only the beasts of burden, but also the elephants could be led down it.

Four days were spent about this rock, the beasts nearly perishing through hunger: for the summits of the mountains are for the most part bare, and if there is any pasture the snows bury it. They then descended into the plains, the country and the dispositions of the inhabitants being now less rugged. In this manner they came to Italy in the fifth month after leaving New Carthage, having crossed the Alps in fifteen days.

11.

*Under the Romans' system, by which each of the two ruling consuls
commanded their army on alternate days, the plebeian Terentius
Varro took the decision to attack Hannibal at Cannae (3 August 216
BC), in south-eastern Italy, against the advice of his patrician
colleague Lucius Aemilius, with disastrous results. Close to 50,000 of
the Romans and their allies are alleged to have been killed. The
victory confirmed Hannibal as one of history's great captains. Among
the vanquished Varro escaped, but his fellow consul was left wounded
on the stricken field.*

Cn. Lentulus, a military tribune, saw, as he rode by, the consul covered
with blood sitting on a boulder. 'Lucius Aemilius,' he said, 'the one
man whom the gods must hold guiltless of this day's disaster, take this
horse while you have still some strength left, and I can lift you into the
saddle and keep by your side to protect you. Do not make this day of
battle still more fatal by a consul's death, there are enough tears and
mourning without that.' The consul replied: 'Long may you live to do
brave deeds, Cornelius, but do not waste in useless pity the few
moments left in which to escape from the hands of the enemy. Go,
announce publicly to the Senate that they must fortify Rome and
make its defence strong before the victorious enemy approaches, and
tell Q. Fabius [the great delayer – 'Cunctator'] privately that I have
ever remembered his precepts in life and in death. Suffer me to breathe
my last among my slaughtered soldiers, let me not have to defend
myself again when I am no longer consul, or appear as the accuser of
my colleague and protect my innocence by throwing the guilt on
another.' During this conversation a crowd of fugitives came suddenly
upon them, followed by the enemy, who, not recognizing the consul,
overwhelmed him in a shower of missiles. Lentulus escaped on horseback in the confusion.

12.

Archimedes, the Greek inventor and mathematician (c. 287–212 BC), was among the inhabitants of Syracuse when the city was besieged by Claudius Marcellus in 213 BC and set his genius to design artillery for the garrison.

Archimedes began to ply his engines, and shot against the land forces of the assailants all sorts of missiles and immense masses of stones, which came down with incredible din and speed; nothing could ward off their weight, but they knocked down in heaps those who stood in their way, and threw their ranks into confusion. At the same time huge beams were suddenly projected over the ships from the walls, which sank some of them with great weights plunging down from on high; others were seized at the prow by iron claws, or beaks like those of cranes, drawn straight up into the air, and then plunged stern foremost into the depths, or were turned round and round by means of enginery within the city, and dashed upon the steep cliffs that jutted out beneath the wall of the city, with great destruction of the fighting men on board, who perished in the wrecks. Frequently, too, a ship would be lifted out of the water into mid-air, whirled hither and thither as it hung there, a dreadful spectacle, until its crew had been thrown out and hurled in all directions, when it would fall empty upon the walls, or slip away from the clutch that had held it.

It chanced that Archimedes was alone, working out some problem with the aid of a diagram, and having fixed his thoughts and eyes upon the matter of his study, he was not aware of the incursion of the Romans or of the capture of the city [in 211 BC]. Suddenly a soldier came upon him and ordered him to accompany him to Marcellus. This Archimedes refused to do until he had worked out his problem, whereupon the soldier flew into a passion, drew his sword, and dispatched him. It is generally agreed that Marcellus was afflicted at his death, and turned away from his slayer as from a polluted person, and sought out the kindred of Archimedes and paid them honour.

13.

After Hannibal's defeat at Zama in 202, he remained at Carthage until in 193 BC he was driven into exile with Antiochus III of Syria. At Ephesus, claim Greek writers quoted by Livy, Hannibal's conqueror Scipio 'Africanus' finally met the Carthaginian.

When Africanus asked who, in Hannibal's opinion, was the greatest general, Hannibal named Alexander, king of the Macedonians, because with a small force he had routed armies innumerable and because he had traversed the most distant regions, even to glimpse which transcended human hopes. To the next request, as to whom he would rank second, Hannibal selected Pyrrhus; saying that he had been first to teach the art of castrametation; besides, no one had chosen his ground or placed his troops more discriminatingly; he possessed also the art of winning men over to him, so that the Italian peoples preferred the lordship of a foreign king to that of the Roman people, so long the masters in that land. When he continued, asking whom Hannibal considered third, without hesitation he named himself. Then Scipio broke into a laugh and said 'What would you say if you had defeated me?' 'Then, beyond doubt,' the Carthaginian replied, 'I should place myself both before Alexander and before Pyrrhus and all other generals.'

LIVY

14.

Spartacus was a Thracian shepherd, born around 111 BC, who served in the Roman army before deserting and becoming a brigand. On his capture, he was sold to a trainer of gladiators, and in 73 BC was confined at Capua, where he persuaded seventy of his comrades to join him in a break for freedom. By the end of his first year of liberty his following had swelled to 90,000, and he controlled most of southern Italy after vanquishing two Roman forces. In 72 BC, he defeated both consuls, and reached the foot of the Apennines. But here his movement collapsed, and in the following year he was beaten and killed by

Crassus on the river Silarus, his surviving followers being crucified.
This is a somewhat partisan Roman account of the slave revolt that
Hollywood has made the most famous in history.

One can tolerate, indeed, even the disgrace of a war against slaves; for
although, by force of circumstances, they are liable to any kind of
treatment, yet they form as it were a class (though an inferior class) of
human beings. But I know not what name to give to the war which
was stirred up at the instigation of Spartacus; for the common soldiers
being slaves and their leaders being gladiators – the former men of the
humblest, the latter men of the worst, class – added insult to the injury
which they inflicted upon Rome.

Spartacus, Crixus and Oenomaus, breaking out of the gladiatorial
school of Lentulus with men of the same occupation, escaped from
Capua.

When, by summoning the slaves to their standard, they had quickly
collected more than 10,000 adherents, these men, who had been orig-
inally content merely to have escaped, soon began to wish to take their
revenge also. The first position which attracted them (a suitable one
for such ravening monsters) was Mt Vesuvius. Being besieged here by
Clodius Glabrus, they slid by means of ropes made of vine-twigs
through a passage in the hollow of the mountain down into its very
depths, and issuing forth by a hidden exit, seized the camp of the
general by a sudden attack which he never expected. They then
attacked other camps, that of Varenius and afterwards that of Thora-
nus; and they ranged over the whole of Campania.

Not content with the plundering of country houses and villages,
they laid waste Nola, Nuceria, Thurii and Metapontum with terrible
destruction. Becoming a regular army by the daily arrival of fresh
forces, they made themselves rude shields of wicker-work and the skins
of animals, and swords and other weapons by melting down the iron
in the slave-prisons. That nothing might be lacking which was proper
to a regular army, cavalry was procured by breaking in herds of horses
which they encountered, and his men brought to their leader the insig-
nia and fasces captured from the praetors. He also celebrated the
obsequies of his officers who had fallen in battle with funerals like

those of Roman generals, and ordered his captives to fight at their pyres, just as though he wished to wipe out all his past dishonour by having become, instead of a gladiator, a giver of gladiatorial shows.

Next, actually attacking generals of consular rank, he inflicted defeat on the army of Lentulus in the Apennines and destroyed the camp of Publius Cassius at Mutina. Elated by these victories he entertained the project – in itself a sufficient disgrace to us – of attacking the city of Rome. At last a combined effort was made, supported by all the resources of the empire, against this gladiator, and Licinius Crassus vindicated the honour of Rome. Routed and put to flight by him, our enemies – I am ashamed to give them this title – took refuge in the furthest extremities of Italy. Here, being cut off in the angle of Bruttium and preparing to escape to Sicily, but being unable to obtain ships, they tried to launch rafts of beams and casks bound together with withies on the swift waters of the straits. Failing in this attempt, they finally made a sally and met a death worthy of men, fighting to the death as became those who were commanded by a gladiator. Spartacus himself fell, as became a general, fighting most bravely in the front rank.

FLORUS

15.

Julius Caesar (100–44 BC), bane of the lives of generations of enforced students of Latin and a general whose memoirs match in vanity those of many modern field-marshals, describes his first brief expedition to Britain in 55 BC.

Summer was now drawing to a close, and winter sets in rather early in these parts. Nevertheless I hurried on preparations for an expedition, knowing that Britain had rendered assistance to the enemy in nearly all my Gallic campaigns. Although it was too late in the year for military operations I thought it would be a great advantage merely to have visited the island, to have seen what kind of people the inhabitants were, and to have learned something about the country with its harbours and landing-places. Of all this the Gauls knew virtually nothing; and even their knowledge is limited to the sea coast. Inter-

views with numerous merchants elicited nothing as to the size of the
island, the names and strength of the native tribes, their military and
civil organisation, or the harbours which might accommodate a large
fleet.

The whole army moved into Artois, where the mainland is nearest
to Britain; and ships were ordered to assemble there from all neighbouring districts, including the fleet which had been built last year for
the campaign against the Veneti [a Gallic people]. However, some
traders revealed our plans to the Britons, and a number of tribes sent
envoys promising hostages and offering their submission. They were
received in audience, promised generous terms, and urged to abide by
their undertaking. They were accompanied on their return journey
by Commius, whom I had appointed ruler of the Atrebates after the
subjugation of that people, and of whose honour, discretion, and
loyalty I had received abundant proof.

Arrangements were now complete, the weather was favourable, and
we cast off just before midnight. The cavalry had been ordered to make
for the northern port, embark there, and follow on; but they were
rather slow about carrying out these instructions, and started too late.
I reached Britain with the leading vessels at about 9 a.m., and saw the
enemy forces standing under arms all along the heights. Precipitous
cliffs tower over the water, making it possible to fire from above directly
on to the beaches. It was clearly no place to attempt a landing, so we
rode at anchor until about 3.30 p.m. awaiting the rest of the fleet.

During this interval I summoned my staff and company commanders, and explained my plans. They were warned that, as tactical
demands, particularly at sea, are always uncertain and subject to rapid
change, they must be ready to act at a moment's notice on the briefest
order from myself. After moving about eight miles up-Channel the
ships were grounded on an open and evenly shelving beach.

The natives, however, realized our intention: their cavalry and war
chariots (a favourite arm of theirs) were sent ahead, while the main
body followed close behind and stood ready to prevent our landing.
In the circumstances, disembarkation was an extraordinarily difficult
business. On account of their large draught the ships could not be
beached; and the troops, besides being ignorant of the locality, had

their hands full: weighted with heavy armour, they had to jump from the ships, stand firm in the surf, and fight at the same time. But the enemy knew their ground: unencumbered, they could hurl their weapons boldly from dry land or shallow water, and gallop their horses which were trained to this kind of work. Our men were terrified: they were inexperienced in this kind of fighting, lacked that dash and drive which always characterized their land battles.

The warships, however, were of a shape unfamiliar to the natives; they were swift, too, and easier to handle than the transports. Therefore, as soon as I grasped the situation I ordered them to go slightly astern, clear of the transports, then full speed ahead, bringing up on the Britons' right flank. From that position they were to open fire and force the enemy back with slings, arrows, and artillery. The Britons were scared by the strange forms of the warships, by the motion of the oars, and by the artillery which they had never seen before: they halted, then fell back a little; but our men still hesitated, mainly because of the deep water.

At this critical moment the standard-bearer of the Tenth Legion, after calling on the gods to bless the legion through his act, shouted: 'Come on, men! Jump, unless you want to betray your standard to the enemy! I, at any rate, shall do my duty to my country and my commander.' He threw himself into the sea and started forward with the Eagle. The rest were not going to disgrace themselves; cheering wildly they leaped down, and when the men in the next ships saw them they too quickly followed their example.

The action was bitterly contested. Our fellows were unable to keep their ranks; nor could they follow their appointed standards, because men from different ships were falling in under the first one they reached, and a good deal of confusion resulted. The Britons, of course, knew all the shallows: standing on dry land, they watched the men disembark in small parties, galloped down, attacked as they struggled through the surf, and surrounded them with superior numbers while others opened fire on the exposed flank of isolated units. I therefore had the warships' boats and scouting vessels filled with troops, so that help could be sent to any point where the men seemed to be in difficulties. When everyone was ashore and formed up, the legions charged:

the enemy was hurled back, but pursuit for any distance was impossible as the cavalry transports had been unable to hold their course and make the island. That was the only thing that deprived us of a decisive victory.

The natives eventually recovered from their panic and sent a delegation to ask for peace, promising to surrender hostages and carry out my instructions. These envoys brought with them Commius, who, it will be remembered, had preceded us to Britain. When he landed, and was actually delivering my message in the character of an ambassador, he had been thrown into prison. Now, after their defeat, the natives sent him back: in asking for peace they laid the blame for this outrage upon the common people and asked me to overlook the incident on the grounds of their ignorance.

I protested against this unprovoked attack which they had launched after sending a mission to the Continent to negotiate a friendly settlement, but agreed to pardon their ignorance and demanded hostages. Some of these were handed over at once, others, they said, would have to be fetched from a distance and would be delivered in a few days. Meanwhile they were ordered to return to their occupations on the land, and chieftains began to arrive from the surrounding districts, commending themselves and their tribes to my protection. Peace was thus concluded.

It happened to be full moon that night; and at such times the Atlantic tides are particularly high, of which we were ignorant. The result was that the warships, which had been beached, became waterlogged: as for the transports riding at anchor, they were dashed one against another, and it was impossible to manoeuvre them. Several ships broke up, and the remainder lost their cables, anchors, and rigging. Consternation naturally seized the troops, for there were no spare ships in which they could return and no means of refitting. It had been generally understood, too, that we should winter in Gaul, and consequently no arrangements had been made for winter food supplies in Britain. The British chieftains at my headquarters sized up the situation and put their heads together. They knew we had no cavalry and were short of grain and shipping; they judged the weakness of our forces from the inconsiderable area of the camp, which was all the smaller because we had brought no heavy equipment; and they decided to renew the offensive. Their aim was to cut us off from food supplies and other material and to prolong the campaign until winter. They were confident that if the present expeditionary force were wiped out or prevented from returning, an invasion of Britain would never again be attempted. Accordingly they renewed their vows of mutual loyalty, slipped away one by one from our camp, and secretly reassembled their forces from the countryside.

I had not yet been informed of their intention; but, in view of the disaster to our shipping and the fact that they had ceased to deliver hostages, I had a suspicion of what might happen, and was prepared for any emergency. Corn was brought in every day from the fields; timber and bronze from the badly damaged vessels were used to repair others; the necessary equipment was ordered from the Continent; and, thanks to the energy and efficiency of the troops, all but twelve ships were made tolerably seaworthy.

One day while these repairs were in progress the Seventh Legion was taking its turn in the harvest field: nothing had occurred as yet to arouse suspicion of an impending attack, for many of the natives were still at work on the land and others were frequent visitors to our camp. Suddenly, however, the sentries on the gates reported an unusually large dust cloud in the direction in which the legion had gone. My suspicions were confirmed – the natives had hatched some new plot.

The battalions on guard duty were detailed to go with me to the scene of action, two others were ordered to relieve them, and the rest to arm and follow on immediately. We had not been marching long before I noticed the Seventh was in difficulties: they were only just managing to hold their ground with their units closely packed and under heavy fire. The enemy had hidden themselves in the woods by night, and attacked while the men were unarmed and busy reaping. We lost a few killed. The rest were in confusion before they could form up, and found themselves hemmed in by cavalry and war chariots.

British charioteers begin by driving all over the field, hurling javelins; and the terror inspired by the horses and the noise of the wheels is usually enough to throw the enemy ranks into disorder. Then they work their way between their own cavalry units, where the warriors jump down and fight on foot. Meanwhile the drivers retire a short distance and station the cars in such a way that their masters, if outnumbered, have an easy means of retreat. In action, therefore, they combine the mobility of cavalry with the staying power of foot-soldiers. Their skill, derived from ceaseless training and practice, may be judged by the fact that they can control their horses at full gallop on the steepest incline, check and turn them in a moment, run along the pole, stand on the yoke, and get back again into the chariot as quick as lightning.

Our troops were unnerved by these tactics, and help reached them in the nick of time: for as we approached the enemy halted, and the legion recovered its morale. The moment, however, was clearly inopportune to precipitate a general engagement; so I advanced no further, and shortly afterwards led the troops back to camp.

There followed several days of bad weather, which confined us to camp besides preventing an enemy attack. But during this interval the Britons sent runners all over the countryside to inform the population that our force was very weak, and that if it could be driven from its base they had every chance of obtaining valuable loot and of securing their freedom once and for all. A strong British force assembled and marched on our camp. It was fairly evident that what had happened before would happen again – even if we routed them, their speed would enable them to get clear of further danger. Nevertheless, there were now available some thirty horses brought over by Commius. So the legions were drawn up in battle formation in front of the camp, and after a brief action the enemy was overwhelmed and fled.

We followed as far as our speed and endurance allowed, killed a large number of them, then burned their dwellings over a wide area, and returned to base. That same day envoys came to sue for peace: they were met with a demand for twice as many hostages as before, and were ordered to bring them over to the Continent, because the equinox was close at hand and the ill-condition of our ships made it inadvisable to postpone the voyage until winter. Taking advantage of fair weather we set sail a little after midnight, and the whole fleet reached the mainland in safety.

16.

*Caesar's April 46 BC triumph over Scipio at Thapsus in modern
Tunisia was a decisive event in the Civil War, confirming the victor as
Rome's dictator – and convincing the superstitious of his impending
doom, by the mishap at his Gallic Triumph.*

Caesar celebrated four Triumphs with a few days' interval between
them; and, after defeating the sons of Pompey, a fifth. These Triumphs
were the Gallic – the first and most magnificent – the Alexandrian, the
Pontic, the African, and lastly the Spanish. Each differed completely
from the others in its presentation. As Caesar rode through the Vela-
brum on the day of his Gallic Triumph, the axle of his triumphal
chariot broke, and he nearly took a toss; but afterwards ascended to
the Capitol between two lines of elephants, forty in all, which acted as
his torch-bearers. In the Pontic Triumph one of the decorated wagons,
instead of a stage-set representing scenes from the war, like the rest,
carried a simple three-word inscription: VENI, VIDI, VICI [I CAME, I SAW,
I CONQUERED].

SUETONIUS

17.

*The history, or perhaps the legend, of Boudicca, Queen of the Iceni:
she was born around AD 30 and was thus thirtyish at the time of her
famous AD 61 rising against the Romans occupying Britain. Like
many of my generation of schoolchildren, I first thrilled to this account
in my early teens.*

When the Icenian King, Prasutagus, died, he made [the Emperor]
Nero co-heir to his realm with his own two daughters. By this device
he hoped to preserve his kingdom and household; but in vain. 'Kingdom and household', writes Tacitus, 'were plundered like prizes of war,
the one by Roman officers, the other by Roman slaves. As a beginning,
his widow Boudicca was flogged and her daughters raped. The Icenian
chiefs were deprived of their hereditary estates as if the Romans had
been given the whole country. The King's own relatives were treated

like slaves.' Even so, perhaps nothing would have happened if resentment had not found a focal point in Boudicca.

'She was very tall,' writes Dio Cassius, 'in appearance terrifying, in the glance of her eye most fierce, and her voice was harsh; a great mass of the tawniest hair fell to her hips; around her neck she wore a large golden necklace; and she wore a tunic of divers colours over which a thick mantle was fastened with a brooch. This was her invariable attire.' Though Dio lived long after Boudicca, he was probably drawing on descriptions by men who had seen her. The fierce warrior-Queen is the embodiment of a people's hate. Her red, streaming hair is a banner to the thousands who follow her war-chariot. Hers is no disciplined army, but a furious, plundering mob whose numbers swell as they press down through Suffolk on their way to unprotected Camulodunum – modern Colchester.

After a two-day siege, the attackers broke in. Flames began to lick the walls of the new houses of the settlers. Choked by rising clouds of smoke, with the barbaric yells of the Britons in their ears, the citizens retreated to the white temple of Claudius, the largest and strongest building. In vain; it too was put to the flames, and the last defenders were burned to death or butchered as they ran. The whole town was given up to loot and murder. 'The British did not take or sell prisoners,' writes Tacitus. 'They could not wait to cut throats, hang, burn and crucify …'

The third chief city of Roman Britain had dissolved in flames before Suetonius had assembled his forces and, on carefully chosen ground, turned to face the massed hordes of Boudicca. Again we have no idea where this battle took place. The Roman historians give us no clue. But they have left vivid word-pictures. 'He [Suetonius] chose a position in a defile with a wood behind him,' says Tacitus. 'There could be no enemy, he knew, except in front, where there was open country without cover for ambushes. Suetonius drew up his regular troops in close order, with the light-armed auxiliaries at their flanks and the cavalry on the wings. On the British side, cavalry and infantry bands seethed over a wide area. Their numbers were unprecedented, and they had confidently brought their wives to see the victory, installing them in cars stationed at the edge of the battlefield.'

The disproportion between the numbers of Britons and Romans is again emphasized by Dio. 'Suetonius Paulinus,' he says, 'could not extend his line the whole length of hers [Boudicca's] for, even if the men had been drawn up only one deep, they would not have reached far enough, so inferior were they in numbers; nor, on the other hand, did he dare join battle in a single compact force, for fear of being surrounded and cut to pieces. He therefore separated his army into three divisions, in order to fight at several points at one and the same time.' One does not envy Suetonius or his legions as they took their stand against the massed ranks of fanatical Britons, maddened by success. He must have been an extremely capable general, since his total force, according to Tacitus, was only 10,000 men. The 2nd Augusta Legion had not moved from Gloucester, and the 9th had been severely mauled. Against the Romans stood a British force outnumbering them by at least ten to one.

In the speech to her troops which Dio puts into the mouth of Boudicca, she pours contempt upon the Romans ... 'men who bathe in warm water, eat artificial dainties, drink unmixed wine, anoint themselves with myrrh, sleep on soft couches with boys for bedfellows – boys past their prime at that – and are slaves to a lyre-player [Nero] and a poor one too. Wherefore may this Mistress Domitia-Nero reign no longer over me or over you men; let the wench sing and lord it over Romans, for they surely deserve to be the slaves of such a woman after having submitted to her so long ...'

'Light-armed troops exchanged missiles with light-armed, heavy-armed were opposed to heavy-armed, cavalry clashed with cavalry, and against the chariots of the barbarians the Roman archers contended,' etc. Tacitus says that the regular troops, keeping to the defile as a natural defence, launched their javelins at the approaching enemy. 'Then, in wedge formations, they burst forward. So did the auxiliary infantry. The cavalry, too, with lances couched, demolished all serious resistance.'

Suetonius's 10,000, fighting shoulder to shoulder in that narrow defile, flung back wave after wave of howling tribesmen, stood their ground, obeyed their commanders, kept their ranks, and gradually, throughout the long day, wore down their enemies. 'Finally, late in the day, the Romans prevailed; and they slew many in battle beside the wagons and the forest, and captured many alive,' says Dio. And Tacitus adds: 'The remaining Britons fled with difficulty, since the ring of wagons blocked the outlets. The Romans did not spare even the women. Baggage animals too, transfixed by weapons, added to the heaps of dead. It was a glorious victory, comparable with bygone triumphs. According to one report, almost 80,000 Britons fell. Our own casualties were about 400 dead and a slightly larger number of wounded. Boudicca poisoned herself.'

DIO CASSIUS, TACITUS AND LEONARD COTTRELL

18.

One of the most famous episodes in Jewish history, a tale of resistance to Roman dominance in BC 67.

On the fall of the town of Jotapata to its Roman besiegers, the Jewish garrison commander Josephus sought refuge in a nearby cave. Here the Romans sent word of their willingness to offer him mercy, which he was eager to accept. But to his dismay, his forty companions preferred the prospect of collective suicide. Josephus argued at great length in his desire to avert mass suicide. But desperation made his hearers deaf; they had long ago devoted themselves to death, and now they were furious with him. Running at him from all directions sword in hand, they reviled him for cowardice, everyone appearing about to strike him. But he called one by name, glared like a general at another, shook hands with a third, pleaded with a fourth till he was ashamed, and distracted by conflicting emotions in his critical situation, he kept all their swords away from his throat, turning like an animal at bay to face each assailant in turn.

In this predicament his resourcefulness did not fail him. Putting his trust in divine protection he staked his life on one last throw. 'You have chosen to die,' he exclaimed; 'well then, let's draw lots and kill each other in turn. Whoever draws the first lot shall be dispatched by number two, and so on down the whole line as luck decides. In this way no one will die by his own hand – it would be unfair when the rest were gone if one man changed his mind and saved his life.' The audience swallowed this bait, and getting his way Josephus drew lots

with the rest. Without hesitation each man in turn offered his throat
for the next man to cut, in the belief that a moment later his
commander would die too. Life was sweet, but not so sweet as death
if Josephus died with them! But Josephus – shall we put it down to
divine providence or just to luck? – was left with one other man. He
did not relish the thought either of being condemned by the lot or, if
he was left till last, of staining his hand with the blood of a fellow Jew.
So he used persuasion, they made a pact, and both remained alive.

JOSEPHUS

19.

*The emperor Vespasian (AD 9–79), who ruled from AD 69–79, was a
stickler for the military virtues.*

He missed no opportunity of tightening discipline: when a young man,
reeking of perfume, came to thank him for a commission he had obtained,
Vespasian turned his head away in disgust and cancelled the order, saying
crushingly: 'I should not have minded so much if it had been garlic.'

SUETONIUS

20.

*Having attempted to check the licence of the Praetorian Guard, on 28
March AD 193 the emperor Pertinax (AD 126–193) was killed by
them in one of history's earliest recorded military coups.*

The Praetorians had violated the sanctity of the throne, by the atrocious murder of Pertinax; they dishonoured the majesty of it, by their
subsequent conduct. The camp was without a leader, for even the
Praefect Laetus, who had excited the tempest, prudently declined the
public indignation. Amidst the wild disorder Sulpicianus, the emper-
or's father-in-law, and governor of the city, who had been sent to the
camp on the first alarm of mutiny, was endeavouring to calm the fury
of the multitude, when he was silenced by the clamorous return of the
murderers, bearing on a lance the head of Pertinax.

Though history has accustomed us to observe every principle and
every passion yielding to the imperious dictates of ambition, it is

scarcely credible that, in these moments of horror, Sulpicianus should
have aspired to ascend a throne polluted with the recent blood of so
near a relation, and so excellent a prince. He had already begun to use
the only effectual argument, and to treat for the Imperial dignity; but
the more prudent of the Praetorians, apprehensive that, in this private
contract, they should not obtain a just price for so valuable a commod-
ity, ran out upon the ramparts; and, with a loud voice, proclaimed that
the Roman world was to be disposed of to the best bidder by public
auction.

This infamous offer, the most insolent excess of military licence,
diffused a universal grief, shame, and indignation throughout the city. It
reached at length the ears of Didius Julianus, a wealthy senator, who,
regardless of the public calamities, was indulging himself in the luxury of
the table. His wife and his daughter, his freedmen and his parasites, easily
convinced him that he deserved the throne, and earnestly conjured him
to embrace so fortunate an opportunity. The vain old man hastened [28
March AD 193] to the Praetorian camp where Sulpicianus was still in
treaty with the guards; and began to bid against him from the foot of the
rampart. The unworthy negotiation was transacted by faithful emissaries,
who passed alternately from one candidate to the other, and acquainted
each of them with the offers of his rival. Sulpicianus had already promised
a donative of five thousand drachm to each soldier, when Julian, eager for
the prize, rose at once to the sum of six thousand two hundred and fifty
drachms, or upwards of two hundred pounds sterling.

The gates of the camp were instantly thrown open to the purchaser;
he was declared emperor, and received an oath of allegiance from the
soldiers, who retained humanity enough to stipulate that he should
pardon and forget the competition of Sulpicianus. It was now incum-
bent on the Praetorians to fulfil the conditions of the sale. They placed
their new sovereign, whom they served and despised, in the centre of
their ranks, surrounded him on every side with their shields, and
conducted him in close order of battle through the deserted streets of
the city. The Senate was commanded to assemble; and those who had
been the distinguished friends of Pertinax, or the personal enemies of
Julian, found it necessary to affect a more than common share of satisfaction at this happy revolution. After Julian had filled the Senate-house

with armed soldiers, he expatiated on the freedom of his election, his own eminent virtues, and his full assurance of the affections of the Senate. The obsequious assembly congratulated their own and the public felicity; engaged their allegiance, and conferred on him all the several branches of the Imperial power.

From the Senate Julian was conducted, by the same military procession, to take possession of the palace. The first objects that struck his eyes were the abandoned trunk of Pertinax and the frugal entertainment prepared for his supper. The one he viewed with indifference; the other with contempt. A magnificent feast was prepared by his order, and he amused himself till a very late hour with dice, and the performances of Pylades, a celebrated dancer. Yet it was observed, that after the crowd of flatterers dispersed, and left him to darkness, solitude, and terrible reflection, he passed a sleepless night; revolving most probably in his mind his own rash folly, the fate of his virtuous predecessor, and the doubtful and dangerous tenure of an empire, which had not been acquired by merit, but purchased by money. [Julianus was in turn murdered by soldiers on 1 June AD 193, after occupying the imperial throne for just sixty-six days.]

EDWARD GIBBON

21.

The name of the Chinese general Sun Tzu (?544–496 AD) survives to this day because of his supposed authorship of The Art of War, *a seminal tract.*

Sun Tzu Wu was a native of the Ch'i State. His *Art of War* brought him to the notice of Ho Lu, King of Wu. Ho Lu said to him, 'I have carefully perused your thirteen chapters. May I submit your theory of managing soldiers to a slight test?'

Sun Tzu replied, 'You may.'

Ho Lu asked, 'May the test be applied to women?'

The answer was again in the affirmative, so arrangements were made to bring 180 ladies out of the palace. Sun Tzu divided them into two companies and placed one of the King's favourite concubines at the head of each. He then bade them all take spears in their hands, and addressed them thus: 'I presume you know the difference between front and back, right hand and left hand?'

The girls replied, 'Yes.'

Sun Tzu went on. 'When I say eyes front, you must look straight ahead. When I say "left turn", you must face towards your left hand. When I say "right turn", you must face towards your right hand. When I say "about turn", you must face right around towards your back.'

Again the girls assented. The words of command having been thus explained, he set up the halberds and battleaxes in order to begin the drill. Then, to the sound of drums, he gave the order 'Right turn', but the girls only burst out laughing. Sun Tzu said, 'If words of commands are not clear and distinct, if orders are not thoroughly understood, then the general is to blame. But if his orders are clear, and the soldiers nevertheless disobey, then it is the fault of their officers.' So saying, he ordered the leaders of the two companies to be beheaded.

Now the King of Wu was watching from the top of a raised pavilion, and when he saw that his favourite concubines were about to be executed, he was greatly alarmed and hurriedly sent down the following message: 'We are now quite satisfied as to our general's ability to handle troops. If we are bereft of these two concubines, our meat and drink will lose their savour. It is our wish that they shall not be beheaded.'

Sun Tzu replied, 'Having once received His Majesty's commission to be general of his forces, there are certain commands of His Majesty which, acting in that capacity, I am unable to accept.'

Accordingly he had the two leaders beheaded and straight away installed the pair next in order as leaders in their place. When this had been done the drum was sounded for the drill once more; and the girls went through all the evolutions, turning to the right or to the left, marching ahead or wheeling back, kneeling or standing, with perfect accuracy and precision, not venturing to utter a sound. Then Sun Tzu sent a messenger to the King saying: 'Your soldiers, Sire, are now properly drilled and disciplined, and ready for Your Majesty's inspection. They can be put to any use that their sovereign may desire; bid them go through fire and water, and they will not disobey.'

But the King replied: 'Let our general cease drilling and return to camp. As for us, we have no wish to come down and inspect the troops.'

Thereupon Sun Tzu said: 'The King is only fond of words and cannot translate them into deeds.'

After that Ho Lu saw that Sun Tzu was one who knew how to handle an army, and finally appointed him general.

SU-MA CH'IEN

22.

The Chinese general Wu Qi (440–381 BC), on the eve of action against the Qin State, anticipated the Roman consul Manlius.

Before the battle had begun, one of his soldiers, a man of matchless daring, sallied forth by himself, captured two heads from the enemy, and returned to camp. Wu Qi had the man instantly executed, whereupon an officer ventured to remonstrate, saying: 'This man was a good soldier, and ought not have been beheaded.' Wu Qi replied: 'I fully believe he was a good soldier, but I had him beheaded because he acted without orders.'

TU MU

23.

An incident in the wars of China, in AD 241, when Sun Pin and T'ien Chi led the armies of the Qi State against those of Wei, whose armies were commanded by P'ang Chuan.

Sun Pin said: 'The Qi State has a reputation for cowardice, and therefore our adversary despises us. Let us turn this circumstance to account.' Accordingly, when the army had crossed the border into Wei territory, he gave orders to show 100,000 fires on the first night, on the next, and the night after only 20,000. P'ang Chuan pursued them hotly, saying to himself, 'I knew these men of Qi were cowards; their numbers have already fallen away by more than half.' In his retreat, Sun Pin came to a narrow defile, which he calculated that his pursuers would reach after dark. Here he had a tree stripped of its bark, and inscribed upon it the words: 'Under this tree shall P'ang Chuan die.' Then, as night began to fall, he placed a strong body of archers in ambush nearby with orders to shoot directly they saw a light. Later on, P'ang

Chuan arrived at the spot, and noticing the tree, struck a light in order to read what was written on it. His body was immediately riddled by a volley of arrows, and his whole army thrown into confusion.

TU MU

24.

Attila the Hun (ruled AD 434–453) on the eve of the Battle of Chalons, AD 451.

In one of his early campaigns, he appeared before his troops with an ancient iron sword in his grasp, which he told them was the god of war whom their ancestors had worshipped. It is certain that the nomadic tribes of Northern Asia, whom Herodotus described under the name of Scythians, from the earliest times worshipped as their god a bare sword. That sword-god was supposed, in Attila's time, to have disappeared from earth; but the Hunnish king now claimed to have received it by special revelation. It was said that a herdsman, who was tracking in the desert a wounded heifer by the drops of blood, found the mysterious sword standing fixed in the ground, as if it had been darted down from Heaven. The herdsman bore it to Attila, who thenceforth was believed by the Huns to wield the Spirit of Death in battle; and the seers prophesied that that sword was to destroy the world.

A Roman, who was on an embassy to the Hunnish camp, recorded in his memoirs Attila's acquisition of this supernatural weapon, and the immense influence over the minds of the barbaric tribes which its possession gave him. In the title which he assumed, we shall see the skill with which he availed himself of the legends and creeds of other nations as well as of his own. He designated himself 'Attila, Descendant of the Great Nimrod. Nurtured in Engaddi. By the Grace of God, King of the Huns, the Goths, the Danes, and the Medes. The Dread of the World.' It was during the retreat from Orleans that a Christian hermit is reported to have approached the Hunnish king, and said to him, 'Thou art the Scourge of God for the chastisement of Christians.' Attila instantly assumed this new title of terror, which thenceforth became the appellation by which he was most widely and most fearfully known.

SIR EDWARD CREASY

25.

*Rodrigo Díaz de Vivar (1043–99), El Campeador – 'The Fighter', or
El Cid – 'The Lord', from the Arabic Sayyidi, passed into Christian
legend for his leadership role in the struggle against the Moorish
infidels in Spain. In reality, he fought almost anybody who stood in
the way of his considerable ambitions, and made local alliances with
the Moors when these were convenient. For most of his career, he acted
the part of a ruthless outlaw. In 1090, his defeat and capture of the
Count of Barcelona put this unfortunate grandee off his food.*

Widespread rumours that the Cid Ruy Díaz was harrying the whole
countryside reached the ears of the Count of Barcelona, who was
highly incensed and viewed this as a personal insult. The Count was a
hasty and foolish man and spoke impulsively: 'The Cid, Rodrigo of
Vivar, has done me great wrongs. In my own palace he gave me offence
by striking my nephew and refusing satisfaction for it. Now he is
ravaging lands under my protection. Since he seeks me out, I shall
demand redress.' Great numbers of Moors and Christians flocked to
join his forces, which went in search of the Cid, the mighty Ruy Díaz
of Vivar. They journeyed three days and two nights and came up with
the Cid in the pine wood of Tévar.

The Count was so confident in his strength that he felt certain of
laying hands on him. The Cid, carrying large quantities of booty,
descended from the mountains to a valley. There he received the message
of Count Ramon. When the Cid heard it he sent word: 'Tell the Count
not to take offence. I am carrying off nothing of his, so let him leave me
to depart in peace.' The Count replied: 'Not so! He shall pay for past
and present injuries here and now. The exile from Castile will learn what
sort of a man he has wronged.' The messenger returned with all speed,
and the Cid realized that there was nothing for it but to fight.

'Knights,' (he said), 'put aside the booty, and make ready quickly to
take up arms. Count Ramon is about to engage us in a great battle. He
has brought with him a vast host of Moors and Christians and is determined to fight. As they are advancing towards us, let us engage them
here. Tighten your saddle-girths and don your armour. The enemy are

coming downhill, all wearing hose [without boots]. They have racing saddles and loose girths, but we shall ride with Galician saddles and wear boots over our hose. Though we number only one hundred knights we must defeat this large army. Before they reach the plain we shall attack with lances.'

All were ready by the time the Cid had finished speaking, mounted and armed. They watched the forces of the Franks ride down, then he ordered the attack. His men were delighted to obey and used their pennoned lances to good effect, striking some and unhorsing others. The Cid won the battle and took Count Ramon prisoner. He took for himself Colada, a sword worth more than a thousand silver marks. He brought the Count as a prisoner to his tent and commanded his faithful vassals to stand guard over him. He then left his tent, in high humour. A great feast was prepared for Don Rodrigo, but Count Ramon showed no relish for it. They brought the dishes and placed them in front of him, but he refused to eat and scorned all they offered. 'I shall not eat a mouthful,' he said, 'for all the wealth of Spain. I had rather die outright since such badly shod fellows have defeated me in battle.'

To that the Cid replied: 'Eat this bread, Count, and drink this wine. If you do as I say you will go free, but if not, you will never see Christendom again.' Count Ramon said: 'You eat, Don Rodrigo, and take your ease, for I would rather starve to death than eat anything.' For three days they tried in vain to persuade him. While they were dividing their great booty, they could not induce him to eat even a morsel.

THE POEM OF THE CID

26.

At the beginning of the thirteenth century the Tatar leader Temuchin, better known to the world as Genghis Khan (1162–1227), created by conquest an empire in Asia such as the world had never seen, which spawned countless legends such as these.

It was a progression that made the savagery and destruction brought upon the world by Attila the Hun look like reasoned acts of statesmanship. Temuchin drowned in blood all the nations that dared to resist him and obliterated towns and villages in his path, leaving no one but the steppe wolves and carrion crows to bury the butchered millions. Within a quarter of a century he subdued all the steppe nations and the kuriltai, the council of the Mongol chiefs, proclaimed him Genghis Khan, the Khan of Khans in 1206. It was a fitting name, although the names Mighty Killer of Man and the Perfect Warrior used by the subjects and neighbours of his empire were nearer the truth.

'The merit of an action,' he cautioned his sons, 'lies in finishing it to the end.' The empire-building could not be considered completed until three powerful states bordering his realm were destroyed. But these were not simple steppe kingdoms with armies unable to match the organization, discipline and massed attacks of Mongol cavalry. The Chin empire, with its well-fortified capital city of Peking, had all the advantages of an ancient civilization over the Mongol nomads, and furthermore the Chinese army was well versed in the use of sophisticated war machines and gunpowder. The Tangut kingdom of Hsia Hsi, on the upper reaches of the Hoang Ho (Yellow River) was also a state of town-dwellers, while Kara Khitai was a powerful Muslim kingdom. Genghis chose Hsia Hsi, the weakest of the three, as his first target.

In 1207 he invaded part of the Tangut kingdom, where the fortifications of the city of Volohai appeared too much for his steppe horsemen. But what he could not get with brute force he won with fox-like cunning. He offered to withdraw if he was given by way of tribute one thousand cats and one thousand swallows. The startled Tangut complied. But instead of withdrawing Genghis set them alight and released them in one great rush of living fire. The hapless cats and birds set the city on fire in hundreds of places and, while the garrison fought the flames, the Mongols breached the walls. By the end of 1211 the Tangut empire was subdued and the golden bit of slavery was forced into its emperor's mouth. The following year Genghis invaded the Chin empire and, after battles of gigantic scale and atrocities of terrifying proportions, the Mongol horsemen overran North China to the Yellow Sea. After a prolonged siege, Genghis took Peking in 1215, but by then his reputation was such that, according to a Persian chronicler, some 60,000 Chinese virgins hurled themselves from the city walls to their deaths rather than fall alive into the hands of his soldiery.

27.

During one of Louis VI of France's (1081–1137) many battles, he found himself in imminent peril of capture.

A soldier of the enemy took hold of the bridle of his horse, crying out, 'The King is taken!' 'No, sir,' replied Louis, raising his battleaxe, with which he hewed down the soldier, 'no, sir, a king is never taken, not even at chess.'

THE PERCY ANECDOTES

28.

The decisive battle of the Crusader wars in Palestine took place on 4 July 1187, when the Saracens trapped an army under King Guy of Jerusalem at the Horns of Hattin, high ground six miles from Tiberias, gaining a victory from which the Christian cause never recovered.

The Moslem attack began soon after daybreak. The Christian infantry had only one thought, water. In a surging mass they tried to break through down the slope towards the lake gleaming far below. They were driven up a hillock, hemmed in by flames and by the enemy. Many of them were slaughtered at once, many others were taken prisoner; and the sight of them as they lay wounded and swollen-mouthed was so painful that five of Raymond's knights went to the Moslem leaders to beg that they might be slain, to end their misery. The horsemen on the hills fought with superb courage. Charge after charge of the Moslem cavalry was driven back with losses, but their own numbers were dwindling. Enfeebled by thirst, their strength began to fail them. Before it was too late, at the King's request, Raymond led his knights to burst through the Moslem lines (which) closed up again behind them. They could not make their way back again to their comrades, so, miserably, they rode … away to Tripoli.

There was no hope left for the Christians; but they still fought on, retiring up the hill to the Horns. The King's red tent was moved to the summit, and his knights gathered round him. Saladin's young son al-Afdal was at his father's side witnessing his first battle. 'When the Frankish King had withdrawn to the hilltop [he relates], his knights made a gallant charge and drove the Moslems back upon my father. I watched his dismay. He changed colour and pulled at his beard, then rushed forward crying: "Let us give the devil the lie!" When I saw the Franks flying I cried out with glee: "We have routed them!" But they charged again and drove our men back again to where my father stood. Again he urged our men forward and again they drove the enemy up the hill. Again I cried out: "We have routed them!" But my father turned to me and said: "Be quiet. We have not beaten them so long as that tent stands there." At that moment the tent was overturned. Then my father dismounted and bowed to the ground, giving thanks to God, with tears of joy.'

The Holy Cross was in the hands of the Infidel. Few of the horses had survived. When the victors reached the hilltop, the knights themselves, the King among them, were lying on the ground … with hardly the strength to hand their swords over and surrender. Their leaders were taken off to the tent that was erected on the battlefield for the Sultan. The only magnate to have been killed outright in the battle was the Bishop of Acre, who contrary to usage had concealed a mail shirt under his vestments. God had justly made an example of him for his lack of faith.

Saladin received King Guy and his brother the Constable Amalric, Reynald de Châtillon and his stepson Humphrey of Toron, the Grand Master of the Temple Gerard de Ridefort, the aged Marquis of Montferrat, the lords of Jebail and Botrun, and many of the lesser barons of the realm. [Saladin] seated the King next to him and, seeing his thirst, handed him a goblet of rose water, iced with the snows of Hermon. Guy drank from it and handed it on to Reynald who was at his side. By the laws of Arab hospitality to give food and drink to a captive meant that his life was safe; so Saladin said quickly to the interpreter, 'Tell the King that he gave that man to drink, not I.' Saladin turned to Reynald and berated him for his crimes. Reynald answered back and Saladin there and then struck off his head; after which he gave orders that none of the other lay knights was to be harmed or offered any indignity. But with the exception of the Grand Master of the Temple, all captives belonging to the Military Orders were sent to execution. The prisoners were sent to Damascus, where the barons were lodged in comfort and the poorer folk sold in the slave market. So many were there that the price of a single prisoner fell to three dinars, and you could buy a whole healthy family, a man, his wife, his three sons and two daughters, for eighty dinars the lot. One Moslem even thought it a good bargain to exchange a prisoner for a pair of sandals.

STEVEN RUNCIMAN

29.

King Richard I – Coeur de Lion (1157–99), a monarch as distinguished for his unfitness to rule as for his prowess in battle – received an arrow wound while besieging the castle of Chalus, south-west of Limoges. Gangrene set in, even as the castle fell and its garrison became prisoners.

The King, perceiving he should not live, ordered Bertram de Gourdon, who had shot the arrow, to be brought into his presence. 'What harm did I ever do thee,' asked the king, 'that thou shouldst kill me?' Bertram replied with great magnanimity and courage, 'You killed with your own hand my father and two of my brothers; and you likewise designed to have killed me. You may satiate your revenge. I should cheerfully suffer all the torments that can be inflicted, were I sure of having delivered the world of a tyrant who filled it with blood and carnage.' This bold and spirited answer struck Richard with remorse. He ordered the prisoner to be presented with one hundred marks, and set at liberty; but one of the courtiers, like a true ruffian, ordered him instead to be flayed alive.

NAVAL AND MILITARY ANECDOTES

30.

Guns, introduced to European warfare in the 1330s, were embraced by King Edward III, who took with him for his French campaigns several of the primitive weapons which would eventually pronounce the doom of chivalry, together with 912 pounds of saltpetre and 846

pounds of sulphur for making powder. Cannon were deployed with
the English army at Crécy on 26 August 1346.

Sound was always an accessory of war–drums, trumpets, bagpipes. Shouting was universal. The Genoese crossbowmen at Crécy raised three resounding shouts as they advanced within shooting range. The English met each of these bellows with silence. They simply waited. We don't know exactly when Edward chose to fire his guns. One account relates that the English 'struck terror into the French Army with five or six pieces of cannon, it being the first time they had seen such thundering machines'. Another states that they fired 'to frighten the Genoese'. A third says they 'shot forth iron bullets by means of fire. They made a sound like thunder.'

Astound. Astonish. Stun. Detonate. All of these words derive from a root meaning thunder. On firing, Edward's guns shot out great tongues of flame followed by roiling clouds of white smoke, an impressive and unique sight to the French knights and their allies. More astounding, more stunning, was the powerful sound of the detonations. If, as one chronicler noted, it scared the horses, it most certainly startled the men as well. It was thunder brought to earth, sound hurled forth as a weapon. Like the crash of thunder from a nearby lightning strike, cannon fire from close range is heard not just with the ears but with the gut, the bones, the nerves. It is feeling more than sound, a sudden expansion of air that delivers a physical blow.

Writers who described early guns almost invariably compared their sound to that of thunder. 'As Nature hath long time had her Thunder and Lightning, so Art now hers,' one observer noted. Shakespeare called them 'mortal engines, whose rude throats th'immortal Jove's dread clamours counterfeit'. Early names for guns referred to their booming sound. In Italian they were *schioppi*, or 'thunderers'. The Dutch had their *donrebusse*, 'thunder gun', in the 1350s.

The term 'gun' had a different origin. The word most likely derived from the Norse woman's name 'Gunnildr', familiarly shortened to 'Gunna'. 'Gonne' first shows up in a 1339 document written in Latin. Geoffrey Chaucer, who served in Edward III's royal administration, initiated the vernacular use of the word in 1384:

As swifte as pellet out of gonne,
When fire is in the poudre ronne.

JACK KELLY

31.

On the field of Crécy King Edward III's eldest son, the Black Prince
(1330–76), won his spurs.

Some French, Germans, and Savoyards had broken through the archers of the prince's battalion, and engaged with the men at arms; upon which the second battalion came to his aid, and it was time, for otherwise he would have been hard pressed. The first division, seeing the danger they were in, sent a knight in great haste to the King of England, who was posted upon an eminence, near a windmill. The knight said, on his arrival, 'Sir, the Earl of Warwick, the Lord Stafford, Lord Reginald Cobham, and the others who are about your son, are vigorously attacked by the French; and they entreat that you would come to their assistance with your battalion, for, if their numbers should increase, they fear he will have too much to do.' The King replied, 'Is my son dead, unhorsed, or so badly wounded that he cannot support himself?' 'Nothing of the sort, thank God,' rejoined the knight; 'but he is in so hot an engagement that he has great need of your help.'

The King answered, 'Now, Sir Thomas, return back to those that sent you, and tell them from me, not to send again for me this day, or expect that I shall come, let what will happen, as long as my son has life; and say, that I command them to let the boy win his spurs; for I am determined, if it please God, that all the glory and honour of this day shall be given to him, and to those into whose care I have intrusted him.' The knight returned to his lords, and related the king's answer, which mightily encouraged them, and made them repent they had ever sent such a message.

When, on this Saturday night, the English heard no more hooting or shouting, nor any more crying out to particular lords or their banners, they looked upon the field as their own, and their enemies as beaten. They made great fires, and lighted torches because of the obscurity of the night. King Edward then came down from his post, who all that day had not put on his helmet, and, with his whole battalion, advanced to the Prince of Wales, whom he embraced in his arms and kissed, and said, 'Sweet son, God give you good perseverance: you are my son, for most loyally have you acquitted yourself this day: you are worthy to be a sovereign.' The prince bowed low and humbled himself, giving all honour to the King his father.

SIR JOHN FROISSART

32.

*Chivalry's finest expression in contemporary eyes was the
1351 Combat of the Thirty.*

An action of the perennial conflict in Brittany, it began with a challenge to single combat issued by Robert de Beaumanoir, a noble Breton on the French side, to his opponent Bramborough of the Anglo-Breton party. When their partisans clamoured to join, a combat of thirty on each side was agreed upon. Terms were arranged, the site was chosen, and after participants heard mass and exchanged courtesies, the fight commenced. With swords, bear-spears, daggers, and axes, they fought savagely until four on the French side and two on the English were slain and a recess was called. Bleeding and exhausted, Beaumanoir called for a drink, eliciting the era's most memorable reply: 'Drink thy blood, Beaumanoir, and thy thirst will pass!' Resuming, the combatants fought until the French side prevailed and every one of the survivors on either side was wounded. Bramborough and eight of his party were killed, the rest taken prisoner and held for ransom.

In the wide discussion the affair aroused, 'some held it as a very poor thing and others as a very swaggering business,' with the admirers dominating. The combat was celebrated in verse, painting, tapestry, and in a memorial stone erected on the site. More than twenty years later Froissart noticed a scarred survivor at the table of Charles V, where he was honoured above all others. He told the ever-inquiring chronicler that he owed his great favour with the King to his having been one of the Thirty. The renown and honour the fight earned

reflected the knight's nostalgic vision of what battle should be. While
he practised the warfare of havoc and pillage, he clung to the image of
himself as Sir Lancelot.

BARBARA TUCHMAN

33.

Many Italian wars of the late fourteenth century were fought by
condottieri – *'contract men' – of whom a considerable number were
English veterans of the campaigns in France. The most celebrated, or
notorious, band was the White Company, led by a former archer
named John Hawkwood, born in Essex about 1323. By 1386, when
he assisted his Paduan allies to victory at the Battle of Brentelle, he
had been a successful mercenary leader for almost
a quarter of a century.*

At Brentelle the Paduans captured, inter alia, two hundred and eleven
courtesans in what might excusably be termed the enemy baggage-
train. These girls they crowned with flowers and provided with
bouquets – it was fittingly midsummer, and a delightful sequel to an
unexpectedly easy battle, the Venetian *condottieri* having been bribed
to retreat. The courtesans were led in procession into Padua and entertained to breakfast by Francesco Carrara in his palace. Life evidently
was not all rape and ruin.

Hawkwood's principal lieutenants were the German, Sterz, whom
he had supplanted, and an aristocratic Englishman, Andrew de
Belmonte, whom the Italians called 'Dubramonte', and credited with
royal blood, because his gentle manners contrasted with those of his
more barbaric followers. The company treasurer was a highly important figure. This was William Turton, to whom, as 'Guglielmo
Toreton', the chamberlain of the Pisan republic paid over the
contracted sums.

The names of many a humble soldier, 'Marco' and 'Marcuccio', the
trumpeters, and others, are preserved in the Pisan accounts. There was
a surprising amount of paperwork and the treasurer needed the help
of a considerable staff. There were negotiations with employing
governments, reports, requests, instructions, and complaints. There

were applications for permission to march through the territory of friendly or neutral states – a recurring necessity in a land that was a jigsaw puzzle of interlocking principalities. Within the company there were the accounts, the pay due to each member according to his grade, and a record of any monies advanced.

The troops were apt to gamble and otherwise fritter away their earnings. If they got into debt, they might pawn or sell their arms and equipment, thereby becoming useless to their employers. Florence, at this same date, was meeting this difficulty with her own mercenaries. The city opened a credit fund in February 1362, making interest-free loans of public money to its embarrassed warriors.

GEOFFREY TREASE

34.

In the ranks of the English army on the eve of Agincourt,
24 October 1415.

And when we were at the last rays of light, and darkness fell between us and them, we still stood in the field and heard our foes, everyone calling as the manner is, for his comrade, servant and friend, dispersed by chance in so great a multitude. Our men began to do the same, but the King ordered silence throughout the whole army, under penalty of the loss of horse and harness in the case of a gentleman, and of the right ear in the case of a yeoman or below, with no hope of pardon, for anyone who might presume to break the king's order. And he at once went in silence to a hamlet nearby, in a place where we had only a few houses; most of us had to rest in gardens and orchards, through a night of pouring rain.

And when our enemies considered the quietness of our men and our silence, they thought that we were struck with fright at our small numbers and contemplated flight during the night; so they established fires and strong watches throughout the fields and routes. And it was said they were so sure of us that they cast dice that night for our king and nobles.

HENRICI QUINTI ANGLIAE REGIS GESTA

35.

On the battlefield next day.

As the morning wore on, Henry's confidence, real or assumed, began to waver. He had hoped that the French would have attacked him soon after dawn and would have exhausted themselves in doing so; but now, three hours after daybreak, they still remained motionless. And he must fight his battle that day, for his men would be incapable of fighting it tomorrow. 'The army,' as one of his chaplains put it, 'was very much wearied with hunger, diseases, and marching, and was not likely to obtain any food in this country. The longer they remained there, so much the more would they be subjected to the effects of debility and exhaustion.'

Henry sought the advice of the more experienced soldiers among his knights and asked them whether or not they thought he should attack as the enemy showed no signs of attacking him. They all agreed that he should, and he appeared to be ready to accept their guidance when three horsemen were seen to move out of the ranks of the French army and ride across towards him.

One of these horsemen, the Sire de Heilly, went up to Henry and, without the polite preamble that custom and etiquette demanded, told him that he had heard it said that his captors accused him of having escaped from English captivity 'in a way unbecoming a knight', and that if anyone dared to repeat the accusation he would challenge him to single combat and prove him a liar. The King replied that there could be no question of personal quarrels being settled at such a time and advised Heilly to return to his companions and tell them to begin their attack. 'I trust in God,' Henry added coldly, 'that if you did disgrace the honour of knighthood in the manner of your escape you will today either be killed or recaptured.'

Heilly refused to deliver the King's message. His companions, he said, were not his servants but subjects of the King of France and they would begin the battle at their own pleasure, not at the will of their enemy. 'Depart, then, to your host,' Henry said dismissively, turning his horse's head. 'And however fast you ride, you may find that we shall be there before you.' A body of eighteen French knights swore that they would get near enough to [King Henry] to strike the crown from his head, or die in the attempt. Whether or not they did so, it is certain that someone struck a fleuron from his crown during the fight and that a battle-axe dented his helmet which can still be seen, to attest the fact, above his tomb in Westminster Abbey.

CHRISTOPHER HIBBERT

36.

The petition of Thomas Hostelle, 1429, is a fine illustration of the ingratitude of princes to their defenders in arms, for it went unanswered.

To the king [Henry VI], our sovereign lord,

Beseecheth meekly your povere [poor] liegeman and humble horatour [petitioner], Thomas Hostelle, that, in consideration of his service done to your noble progenitors of full blessed memory, king Henry the iiiith and King Henry the fifth (whose souls God assoile! [absolve]), being at the siege of Harfleur there smitten with a springbolt through the head, losing his one eye and his cheekbone broken; also at the battle of Agincourt, and afore at the taking of the carracks on the sea, there with a gadde of yrene [iron] his plates smitten in to his body and his hand smitten in sunder, and sore hurt, maimed and wounded, by mean whereof he being sore feebled and debrused, now falle to great age and poverty, greatly endetted, and may not help himself, having not wherewith to be sustained ne relieved, but of men's gracious almesse [alms], and being for his said service never yet recompensed ne rewarded, it please your high and excellent grace, the premises tenderly considered, of your benign pity and grace to relieve and refresh your said povere oratour as it shall please you with your most gracious almesse, at the reverence of God and in work of charity, and he shall devoutly pray for the souls of your said noble progenitors and for your most noble and high estate.

THOMAS HOSTELLE

37.

This decently sceptical view of the Maid of Orléans (140?–31), from a French pen, seems preferable to those of more reverential biographers of the saint.

A gentleman upon the frontiers of Lorraine, whose name was Baudri-court, happened to meet with a young servant wench at an inn in the town of Vaucouleurs, whom he thought a fit person to act the character of a female warrior and a prophetess. Joan of Arc – which was the name of this heroine – whom the vulgar look upon as a shepherdess, was in fact only a tavern girl; 'of a robust make,' as Monstrelet says, 'and who could ride without a saddle, and perform other manly exercises which young maidens are unaccustomed to'. She was made to pass for a young shepherdess of eighteen; and yet it is evident from her confession that she was at that time twenty-seven. She had courage and wit sufficient to engage in this delicate enterprise, which afterward became a heroic one, and suffered herself to be carried before the king at Bourges, where she was examined by matrons, who took care to find her a virgin, and by certain doctors of the university, and some members of the parlia-ment, who all without hesitation declared her inspired. Whether they were really imposed upon, or were crafty enough to adopt the project, the vulgar swallowed the bait, and that was sufficient.

The English were at that time, in 1428, besieging Orléans, Charles's last resource, and were upon the point of making themselves masters of the town, when this amazon in man's dress, directed by able officers, undertook to throw reinforcements into the town. Previous to her attempt she harangued the soldiers, as one sent from God, and inspired them with that enthusiastic courage peculiar to all who imagine they behold the Deity Himself fighting their cause. After this she put herself at their head, delivered Orléans, beat the English, foretold to Charles that she would see him consecrated at Rheims, fulfilled her promise, sword in hand, and assisted at the coronation, holding the standard with which she had so bravely fought.

These rapid victories obtained by a girl, with all the appearances of a miracle, and the king's coronation, which conciliated the public respect to his person, had almost restored the lawful prince, and expelled the foreign pretender, when the instrument of all these wonders, Joan of Arc, was wounded and taken prisoner in 1430, while defending Compiègne. Such a person as the Black Prince would have honoured and respected her courage; but the regent, Bedford, thought it necessary to detract from it, in order to revive the drooping spirits of the English. She had pretended to perform a miracle, and Bedford pretended to believe her a witch. My principal end is always to observe the spirit of the times, since it is that which directs the great events of the world. The university of Paris presented a complaint against Joan, accusing her of heresy and witchcraft. Therefore this university either believed what the regent would have it believe; or if it did not believe it, it was guilty of most infamous baseness.

This heroine, who was worthy of that miracle which she had feigned, was tried at Rouen by Cauchon, bishop of Beauvais, by five other French bishops, and one English bishop, assisted by a Dominican monk, vicar to the Inquisition, and by the doctors to the university; who declared her 'a superstitious prophetess of the devil, a blasphemer against God and His saints, and one who had been guilty of numberless errors against the faith of Christ'. As such she was condemned to perpetual imprisonment, and to fast on bread and water.

She made a reply to her judges, which, in my opinion, is worthy of eternal memory. She was asked why she dared to assist at the consecration of Charles, as his standard-bearer. 'Because,' answered she, 'it is but just that the person who shared in the toil should partake likewise of the honour.' Some time after this, being accused of having again put on men's clothes, which had been left in her way purposely to tempt her, her judges, who certainly had no right to try her, as she was a prisoner of war, declared her a relapsed heretic, in 1431; and without further ceremony condemned to the flames a person who, for the services she had rendered her king, would have had altars erected to her in those heroic times when mankind were wont to decree such honours to their deliverers.

38.

*The fall of Constantinople to the Ottomans on 29 May 1453, as
described by its most celebrated chronicler.*

The preceding night had been strenuously employed: the troops, the
cannon, and the fascines were advanced to the edge of the ditch, which
in many parts presented a smooth and level passage to the breach; and
[21-year-old Sultan Mehmed II's] fourscore galleys almost touched,
with the prows and their scaling ladders, the less defensible walls of the
harbour. Under pain of death, silence was enjoined; but the physical
laws of motion and sound are not obedient to discipline or fear: each
individual might suppress his voice and measure his footsteps; but the
march and labour of thousands must inevitably produce a strange
confusion of dissonant clamours, which reached the ears of the watch-
men of the towers.

At daybreak, without the customary signal of the morning gun, the
Turks assaulted the city by sea and land. The foremost ranks consisted
of the refuse of the host, a voluntary crowd who fought without order
or command; of the feebleness of age or childhood, of peasants and
vagrants, and of all who had joined the camp in the blind hope of
plunder and martyrdom. The common impulse drove them onwards
to the wall; the most audacious to climb were instantly precipitated;
and not a dart, not a bullet, of the Christians, was idly wasted on the
accumulated throng. But their strength and ammunition were
exhausted in this laborious defence: the ditch was filled with the bodies
of the slain; they supported the footsteps of their companions; and of
this devoted vanguard the death was more serviceable than the life.

Under their respective bashaws and sanjaks, the troops of Anatolia
and Romania were successively led to the charge: their progress was
various and doubtful; but, after a conflict of two hours, the [defend-
ing] Greeks still maintained and improved their advantage; and the
voice of the emperor [Constantine XI Palaiologos] was heard, encouraging his soldiers to achieve, by a last effort, the deliverance of their
country. In that fatal moment the Janizaries arose, fresh, vigorous, and
invincible. The sultan himself on horseback, with an iron mace in his

hand, was the spectator and judge of their valour; he was surrounded by ten thousand of his domestic troops, whom he reserved for the decisive occasion; and the tide of battle was directed and impelled by his voice and eye.

His numerous ministers of justice were posted behind the line, to urge, to restrain, and to punish; and if danger was in the front, shame and inevitable death were in the rear, of the fugitives. The cries of fear and of pain were drowned in the martial music of drums, trumpets, and attaballs [Arabian kettledrums]; and experience has proved that the mechanical operation of sounds, by quickening the circulation of the blood and spirits, will act on the human machine more forcibly than the eloquence of reason and honour. From the lines, the galleys, and the bridge, the Ottoman artillery thundered on all sides; and the camp and city, the Greeks and the Turks, were involved in a cloud of smoke, which could only be dispelled by the final deliverance or destruction of the Roman empire.

The single combats of the heroes of history or fable amuse our fancy and engage our affections: the skilful evolutions of war may inform the mind, and improve a necessary, though pernicious, science. But in the uniform and odious pictures of a general assault, all is blood, and horror, and confusion; nor shall I strive, at the distance of three centuries and a thousand miles, to delineate a scene of which there could be no spectators, and of which the actors themselves were incapable of forming any just or adequate idea.

The immediate loss of Constantinople may be ascribed to the bullet or arrow which pierced the gauntlet of [the Genoese commander] John Justiniani. The sight of his blood, and the exquisite pain, appalled the courage of the chief, whose arms and counsels were the firmest rampart of the city. As he withdrew from his station in quest of a surgeon, his flight was perceived and stopped by the indefatigable emperor. 'Your wound,' exclaimed Palaiologos, 'is slight; the danger is pressing: your presence is necessary; and whither will you retire?' – 'I will retire,' said the trembling Genoese, 'by the same road which God has opened to the Turks'; and at these words he hastily passed through one of the breaches of the inner wall. By this pusillanimous act he stained the honours of a military life and the few days which he survived were embittered by his own and the public reproach. His example was imitated by the greatest part of the Latin auxiliaries, and the defence began to slacken when the attack was pressed with redoubled vigour.

The number of the Ottomans was fifty, perhaps a hundred, times superior to that of the Christians; the double walls were reduced by the cannon to a heap of ruins: in a circuit of several miles some places must be found more easy of access, or more feebly guarded; and if the besiegers could penetrate in a single point, the whole city was irrecoverably lost. The first who deserved the sultan's reward was Hassan the Janizary, of gigantic stature and strength. With his scimitar in one hand and his buckler in the other, he ascended the outward fortification: of the thirty Janizaries who were emulous of his valour, eighteen perished in the bold adventure. Hassan and his twelve companions had reached the summit: the giant was precipitated from the rampart: he rose on one knee, and was again oppressed by a shower of darts and stones. But his success had proved that the achievement was possible: the walls and towers were instantly covered with a swarm of Turks; and the Greeks, now driven from the vantage ground, were overwhelmed by increasing multitudes.

Amidst these multitudes, the emperor, who accomplished all the duties of a general and a soldier, was long seen and finally lost. The nobles, who fought round his person, sustained, till their last breath, the honourable names of Palaeologus and Cantacuzene: his mournful exclamation was heard, 'Cannot there be found a Christian to cut off my head?' and his last fear was that of falling alive into the hands of the infidels. The prudent despair of Constantine cast away the purple: amidst the tumult he fell by an unknown hand, and his body was buried under a mountain of the slain. After his death resistance and order were no more: the Greeks fled towards the city; and many were pressed and stifled in the narrow pass of the gate of St Romanus.

The victorious Turks rushed through the breaches of the inner wall; and as they advanced into the streets, they were soon joined by their brethren, who had forced the gate Phenar on the side of the harbour. In the first heat of the pursuit about two thousand Christians were put to the sword; but avarice soon prevailed over cruelty; and the victors acknowledged that they should immediately have given quarter, if the valour of the emperor and his chosen bands had not prepared them for a similar opposition in every part of the capital. It was thus, after a siege of fifty-three days, that Constantinople was irretrievably subdued by the arms of Mahomet the Second. Her empire only had been subverted by the Latins: her religion was trampled in the dust by the Moslem conquerors.

EDWARD GIBBON

39.

How to sack a divisional commander: Tewkesbury, 4 May 1471.

Lord Wenlock not having advanced to the support of the first line, but remaining stationary, contrary to the expectations of the Duke of Somerset [1438–1471], the latter, in a rage, rode up to him, reviled him, and beat his brains out with an axe.

RICHARD BROOKE

40.

Pierre du Terrail, chevalier de Bayard (1473–1524), descended from an ancient family in Dauphiny, was dubbed the flower of French chivalry during the dying years of knight-errantry.

His great-grandfather's father fell at the feet of King John in the battle of Poitiers; his great-grandfather was slain at the battle of Agincourt; his grandfather lost his life in the battle of Montlhéry; and his father was desperately wounded in the battle of Guinegate, commonly called the battle of the Spurs. The chevalier himself had signalized himself from his youth by incredible acts of personal valour; first of all at the battle of Fornovo [6 July 1495]: in the reign of Louis XII he, with his single arm, defended the bridge at Naples against two hundred knights: in the reign of Francis I he fought so valiantly at the battle of Marignano [15 September 1515], under the eye of his sovereign, that, after the action, Francis insisted upon being knighted by his hand, after the manner of chivalry.

Having given his king the slap on the shoulder, and dubbed him knight, he addressed himself to his sword in these terms: 'How happy art thou, in having this day conferred the order of knighthood on such a virtuous and powerful monarch. Certes, my good sword, thou shalt henceforth be kept as a relic, and honoured above all others, and never will I wear thee except against the infidels.' So saying, he cut a caper twice, and then sheathed his sword. He behaved with such extraordinary courage and conduct on a great number of delicate occasions, that he was promoted to the rank of lieutenant-general, and held in universal esteem. It was at the retreat of Rebec that his back was broken with a musket shot. Perceiving himself mortally wounded, he exclaimed: 'Jesus, my God, I am a dead man.' Then he kissed the cross of his sword, repeated some prayers aloud, caused himself to be laid under a tree, with a stone supporting his head, and his face toward the enemy, observing that he would not, in the last scene of his life, begin to turn his back on the enemy.

VOLTAIRE

41.

The supreme weapon of the Spanish conquistadores against the Indians of Mexico in their first campaigns of 1519–21 was the horse, which the Indians had never seen before and at first believed to be invested with supernatural powers. This illusion was shattered, however, when after a series of striking Spanish victories a Tlascalan force on the battlefield successfully broke through to Hernan Cortes' horsemen.

A body of the Tlascalans, acting in concert, assaulted a soldier named Moran, one of the best riders in the troop. They succeeded in dragging him from his horse, which they despatched with a thousand blows. The Spaniards, on foot, made a desperate effort to rescue their comrade from the hands of the enemy – and from the horrible doom of the captive. A fierce struggle now began over the body of the prostrate horse. Ten of the Spaniards were wounded, when they succeeded in retrieving the unfortunate cavalier from his assailants, but in so disastrous a plight that he died on the following day. The horse was borne off in triumph by the Indians, and his mangled remains were sent, a strange trophy, to the different towns of Tlascala. The circumstance troubled the Spanish commander, as it divested the animal of the supernatural terrors with which the superstition of the natives had usually surrounded it. To prevent such a consequence, he had caused the two horses killed on the preceding day, to be secretly buried on the spot.

WILLIAM PRESCOTT

42.

A tale from the military memoir of Blaise de Monluc (c. 1501–77), of 'a young virgin of Siena' at the 1554–5 siege of the city.

I had made a decree at the time when I was dictator, that no one upon pain of severe punishment should fail to go to The Guard in his turn. This young maid seeing a brother of hers who was concerned to be upon duty not able to go, she took his morian and put it upon her head, his breeches and a collar of buff and put them on, and with his halberd upon her shoulder in this equipage marched the guard, passing when the list was read by her brother's name, and stood sentinel in turn without being discovered till it was fair light day, when she was conducted home with great honour.

BLAISE DE LASSERAN-MASSENCÔME, SEIGNEUR DE MONLUC

43.

The poet-courtier Sir Philip Sidney, author of 108 love sonnets, fell on 22 September 1586, aged thirty-one, during the ill-fated English campaign in the Low Countries. His fate is here described by a contemporary biographer, in terms made more moving by the quaintness of the language.

When that unfortunate stand was to be made before Zutphen, to stop the issuing out of the Spanish Army; with what alacrity soever he went to actions of honour, yet remembring that upon just grounds the ancient Sages describe the worthiest persons to be ever best armed, he had compleatly put on his; but meeting the Marshall of the Camp lightly armed, the unspotted emulation of his heart – to venture without any inequalitie – made him cast off his Cuisses [thigh guards]; and so, by the secret influence of destinie, to disarm that part, where God had resolved to strike him. Thus they go on, every man in the head of his own Troop; and the weather being misty, fell unawares upon the enemie, who had made a strong stand to receive them, near to the very walls of Zutphen; by reason of which accident their Troops fell, not only unexpectedly to be engaged within the levell of the great shot, that played from the Rampiers [ramparts], but more fatally within shot of their Muskets, which were layd in ambush within their own trenches.

Howsoever, an unfortunate hand out of those forespoken Trenches, brake the bone of Sir Philip's thigh with a Musket-shot. The horse he rode upon, was rather furiously cholleric than bravely proud, and so forced him to forsake the field, but not its back, as the noblest and fittest biere to carry a Martiall Commander to his grave. In which sad progress, passing along by the rest of the Army, where his Uncle the General [the Earl of Leicester] was, and being thirstie with excess of bleeding, he called for drink, which was presently brought him; but as he was putting the bottle to his mouth, he saw a poor Souldier carryed along, who had eaten his last at the same Feast, gastly casting up his eyes at the bottle. Which Sir Philip perceiving, took it from his head, before he drank, and delivered it to the poor man, with these words, 'Thy necessity is yet greater than mine.' And when he had pledged this poor souldier, he was presently carried to Arnheim.

SIR FULKE GREVILLE

44.

Henri de La Tour d'Auvergne, Vicomte de Turenne (1611–75) was an outstanding French commander of the age, and was also deemed a pattern of honour.

The deputies of a great metropolis in Germany once offered the great Turenne one hundred thousand crowns not to pass with his army through their city. 'Gentlemen,' said he, 'I can't in conscience accept your money, as I had no intention to pass that way.'

THE PERCY ANECDOTES

45.

The memoirs of mercenary Sednham Poynts, as he usually signed himself, born in Surrey in 1607, are factually unreliable, but illustrate the shifts of fortune that befell his kind – 'free lances' and rogues – during Europe's Thirty Years War. His prosperity rose and fell partly according to the strength of the troop behind his pennant: for a commander, losses in action were a financial disaster, as well as a military one. In 1644, he returned home in time to hold commands in the Parliamentary army during the English Civil War. I have here retained the words of his 1648 memoir, modifying only the punctuation.

When I began first to follow after Mansfield [around 1630], like mad folkes wee knew not whither, I came into Germany with other troopes of souldiers. Wee passed through many brave Princes' Countries in all which wee had supply of Men and Money, and where wee found such plenty of all things for backe and belly that heart could desire and had got pretty store of Crownes. But at the length wee had a Crosse of fortune, for Tilly [Catholic League commander] met with us & stript us naked of all Canon, Amunition and whatsoever wee had, yea with the death of most but those that saved their lives by running away.

Yet at length our Army was encreased againe by those Protestant Princes thorough whose plentifull Countries wee had marched, which came to nothing, and worse then nothing, by the death of Mansfield and Weymar, and most of many brave souldiers fell into miserable captivity where wee were stript of all that wee lightly got in that long journy, but lost in an hower, and made slavish Slaves [of the Turks] & nightly chayned by the feet to a great log after our sharpe dayes' Labour, which was so terrible to fellowes of brave spirits that they did strive to dy & could not, and that which grieved mee as much as for myself was for a brave young gentleman of a Duke of Barlamont's house in Italy, who was beaten to death before our faces, because his Spirit was so great as would not yield to bee a drudge.

But now to myself: I tooke upon mee an humble spirit and fell to my drudgery hoping once for a light night as they say and went merrily

to my Worke and strove to get the language and now & then some
money by hooke or by crooke & hid it in od comers: so after 2 or 3
yeares patience, opportunity fell that I got away and some 40 myles
but was brought backe with a vengeance and had 300 blowes on my
feet [the bastinado] which cooled my running for one yeare. But God
at length did prosper my intentions, for I got a brave horse which at
length brought mee to the skirts of Christendome.

But fortune threw mee againe on my backe, [I] met with theeves
[who] got all my little Mony and horse and all. O how that went to my
heart to part with my horse, which had brought mee out of the Devills
Mouth, and so neare Christendome, I meane Austria, where hee would
have given mee a hundred pound if some other had had hym.

After all these Crosses the Sunne began to shyne clearly upon mee.
I light upon a poore Franciscan, an Englishman by name A. More, and
somewhat allyed by marriage to our name in Sussex. Then I rise by
fortunes from a Lieutenant of a troop of Horse in Saxons Armie. But
beeing taken Prisoner by the Imperialists I lost againe all that I had.
Thus fortune tossed mee up and downe.

But I sped better than I expected, for I was taken Prisoner by Count
Butler with whome after I got in favour hee raysed mee extremely: for
by his favour hee got mee my first Wife, a rich Merchants Daughter,
who though wee lived not two yeares together, shee dying in childbed
to my great grief, yet shee left mee rich, and as she was of an humble
condition and very houswifly, wee should have lived very happily
togeather. And not content with this the good Count Butler got me
another Wife, rich in Land and mony, but of a higher birth & spirit
[Anne Eleanora de Court Stephanus de Cary, of Wurtemberg], and
therefore would live at a higher rate than our meanes would well
afford, for no Lady in this Land wore better close [clothes] than she
did, besides her Coach and 6 Coach-horses wich with Attendants
answearable to it would bee very expensive and had great Kindred that
lay vpon vs.

But I beeing come to this height got to bee by Count Butlers favour
Sergeant Major of a Troop of 200 horse but I was to raise them at my
owne charge, which was no small matter for mee to doe, beeing so well
underlayd and so well aforehand, for I had then £3000 which I carried

into the field with mee besides that I left at home with my Wife. And I made good use of my place for I could and did send home often tymes Mony to my Wife, who it seemes spent at home what I got abroad.

But fortune turned against mee againe for in that cruell bloody Battaille where in the King of Sweveland was killed [Lutzen, 6 November 1632] my horses were all ether killed or ranne away. Great store of prisoners were taken on both sides, I myself was taken prisoner three tymes but twice I was rescued by my fellowes; the third tyme beeing taken hold of by my belt, having my sword in my hand, I threw the belt over my Eares and rescued myself. I lost three horses that day beeing shot under mee, and I hurt under my right side and in my thigh, but I had horses without maisters enough to choose and horse myself; all had pistols at their saddle bowe but shot of and all that I could doe, was with my sword without a scabard, and a daring Pistol but no powder nor shot.

My last horse that was shot had almost killed mee for beeing shot in the guts, as I thinke, hee mounted on a suddaine such a height, yea I thinke on my conscience two yards, and suddaine fell to the ground upon his bum, and with his suddaine fall thrust my bum a foot into the ground and fell upon mee and there lay groveling upon mee, that hee put mee out of my senses. I knew not how I was, but at length coming to myself, with much a doe got up, and found 2 or 3 brave horses stand fighting togeather. I tooke the best, but when I came to mount hym I was so bruised & with the weight of my heavy Armour that I could not get my leg into the saddle that my horse run away with mee in that posture half in my saddle and half out. I followed the Troop having nothing but a daring Pistol and a naked Sword.

After a long march I came nere home, where I heare the true tryall of fortune's mutability, which was that my Wife was killed & my child, my house burned and my goods all pillaged: My Tenants and Neighbours all served in the same sauce, the whole Village beeing burned; nether horse, Cowe, sheep nor Corne left to feed a Mouse. This was donne by a party of french that came out of Italy going homewards. I presently determined to go see my deare friend Count Buttler Governour of that Country, but my hopes were turned upside downe, for it was my good hap to see hym, but he was dying, which strucke more nere to mee, or as much as my owne losse, but there was no remedy, but yet it somewhat revived hym and what show of love a dying man can expresse, hee did grasping my hand with all his strength and calling for his Will gave mee a thousand pound therein and not long after having receaved his Viaticum with a great sylver Crucifix in his hand and in my Armes yielded up the Ghost.

I had thought my heart would have burst with grief, but could get out no teares out of my stony heart: but to my owne heart I cryed *Spes et Fortuna Valete* – my hopes and fortune now farewell – who if hee had lived, I had had fortune almost at my becke; but hee beeing dead about the £1000 hee gave mee, his Wife beeing the Executrix and not so friendly to mee as she might have bene, and her husband's love to me required, kept mee so with delayes, and at last I was forced to goe to Ratisbone and finding there an English Embassadour made as much use of his favour as I could in my passe hither, away I went for England, loosing my frend and his gift.

SEDNHAM POYNTS (OR SYDENHAM POYNTZ)

46.

A soldier writes from Essex's Parliamentary army, telling a tale that illustrates its desperate disarray in the first months of the English Civil War.

Aylesbury, August 1642

On Monday August 8th we marched to Acton; but being the sixth company, we were belated, and many of our soldiers were constrained to lodge in beds whose feathers were above a yard long. Tuesday, early in the morning, several of our soldiers inhabiting the out parts of the town sallied out unto the house of one Penruddock, a papist, and being basely affronted by him and his dog, entered his house and pillaged him to the purpose. This day, also the soldiers got into the church, defaced the ancient and sacred glazed pictures, and burned the holy rails.

Wednesday: Mr Love gave us a famous sermon this day – also the soldiers brought the holy rails from Chiswick, and burned them in our town. At Chiswick they intended to pillage the Lord Portland's house also Dr Duck's, but by our commanders they were prevented. This day our soldiers generally manifested their dislike of our Lieutenant-Colonel, who is a Goddam blade, and will doubtless hatch in hell, and we all desire that either the Parliament would depose him, or the devil fetch him away quick. This day, towards even, our regiment marched to Uxbridge, but I was left behind, to bring up thirty men with ammunition the next morning.

Thursday: I marched towards Uxbridge; and at Hillingdon, one mile from Uxbridge, the rails being gone, we got the surplices, to make us handkerchieves, and one of our soldiers wore it to Uxbridge. This day the rails of Uxbridge, formerly removed was, with the service book, burned. This evening Mr Harding gave us a worthy sermon.

Saturday morning: We came to Wendover, where we refreshed ourselves, burnt the rails and accidentally, one of the Captain Francis's men, forgetting he was charged with a bullet, shot a maid through the head, and she immediately died.

Sabbath day, August 15th: In this town, a pulpit was built in the market place, where we heard two worthy sermons. This evening our ungodly Lieutenant-Colonel, upon an ungrounded whimsy, commanded two of our captains, with their companies, to march out the town, but they went not. I humbly entreat you, as you desire the success of our just and honourable cause, that you would endeavour to root out our Lieutenant-Colonel for if we march further under his command, we fear, upon sufficient grounds, we are all but dead men.

SERGEANT NEHEMIAH WHARTON

47.

On the field of Edgehill, Buckinghamshire, 23 October 1642: as the Royalists marched south from Nottingham, they fought the first significant battle of the conflict, against the Parliamentary army sent from London to intercept them.

The Royalists were in the stronger position, but for the last two days they had come through hostile country where food and shelter were hard to find. The troops were hungry and the tempers of the leaders were at breaking point. Lord Lindsey (1582–1642) who had advised marching through the enclosed land in the valleys among the villages, in the interests of the infantry, had been overruled by Prince Rupert (1619–82), and the advance had been made through the open country on the higher ground where the cavalry could move – and see – best. In the swift seizure of Edgehill and the drawing-up of the battle Rupert's advice had again prevailed.

Rupert, with more military judgment than social tact, protested that the battle could not be planned piecemeal and further insisted that pikemen and musketeers be interspersed with each other in the modern Swedish fashion. Lindsey's sulks flared into rage. In front of the troops, he flung his baton to the ground and declared that if he 'was not fit to be a general he would die a colonel at the head of his regiment'. In the embarrassing circumstances, his place as commander of the foot was taken by Sir Jacob Astley, a mature and competent soldier who had once been Prince Rupert's tutor and understood how to get on with him.

The quarrels concluded, they completed the ordering of the field. Rupert with four cavalry regiments and the King's lifeguards, was on the right wing, Wilmot with five regiments of horse on the left, the infantry in the centre. The King's standard, borne by Sir Edmund Verney, floated at the head of his red-coated foot guard. His lifeguard of cavalry, under his cousin and Richmond's brother, Lord Bernard Stuart, had asked and been given permission to serve with Prince Rupert. The King, in a black velvet coat lined with ermine, now rode along the lines with words of encouragement. He had already briefly addressed the principal officers in his tent: 'Your King is your cause, your quarrel and your captain,' he said, 'come life or death, your King will bear you company, and ever keep this field, this place, and this day's service in his grateful remembrance.'

Rupert also addressed his troops, not on politics, but on tactics. He knew that his cavalry, short of firearms and scantily trained, must achieve the utmost by the impact of their first charge, and consequently instructed them to ride in the closest possible formation, and to hold their fire until they had closed with the enemy. It was afternoon before both armies were in position and Essex, hoping to gain the initiative and cause some preliminary disorder in the King's lines, opened fire with his cannon. At this the King with his own hand ignited the charge and the Royalist guns gave answer. Sir Jacob Astley uttered a brief prayer: 'O Lord, thou knowest how busy I must be this day. If I forget thee, do not thou forget me.' Rupert waited no longer, and suddenly Essex saw the Royalist horse on his left sweep down the slope and hurl themselves upon his wavering lines. Rupert's men came in at an oblique angle, riding down not only the Parliamentary cavalry on that wing, but some of the infantry nearest them.

The opposing forces made no stand but fled 'with the enemy's horse at their heels and amongst them, pell mell'. In their flight they battered their way through their own reserve drawn up in the rear, and although Denzil Holies gallantly 'planted himself just in the way' and tried to rally the fugitives he brought very few of them to a stand. The rest shamefully scattered with Rupert's men hallooing after them. Some stragglers made a wide cross-country circuit and carried the news of Parliament's defeat down the London road as far as Uxbridge. John Hampden and his regiment, marching up with the rest of the delayed artillery, met the dismayed rout, and by expeditiously planting a battery across the road checked or at least deflected the pursuers.

All this time on the slope of Edgehill the King's forces were faring badly. Contrary to instructions the very small reserve of cavalry on the right wing, which should have stayed to give cover to the infantry, had followed Prince Rupert's charge and joined in the pursuit, leaving the centre, with the infantry, the guns, and the King's standard, bare of defence on one side. Before the Prince and his few experienced officers could extricate their men from the enjoyable chase and bring them back to the field, the infantry had been very roughly handled. The resolution of Essex and the skill of the old Scottish veteran, Sir William Balfour, had prevented the total defeat of Parliament. Wilmot, on the Royalist left, had charged when Rupert did, and had driven the greater part of the opposing cavalry from the field, but the wily Balfour, with a party of Parliamentary horse, drew out of the range of Wilmot's onslaught, and while the Royalists pursued the fliers, he made his way up the hill under cover of the hedges until, with a sudden charge, he fell upon the King's guns and the infantry in the centre.

At the silencing of the Royalist guns, the Parliamentary infantry took heart and closed with the now defenceless and disordered Royalist centre, who manfully stood their ground. Lindsey, badly wounded, was taken prisoner; the King's standard-bearer Sir Edmund Verney was killed and the standard taken. The Prince of Wales, to his joy, found himself almost at grips with the rebels. 'I fear them not,' he shouted, and cocked his pistol, but his startled attendants hustled him to the rear.

Some of Rupert's cavalry were now returning, by scattered parties. Captain John Smith rounded up a couple of hundred men and fell in on the Parliamentary flank, diverting them from their prey. Sometime in this hot action he retrieved the King's standard in a hand-to-hand struggle. Exhaustion, and the harassing onslaught of the returning Royalists, forced the Parliamentary infantry to give ground and fall back as the early darkness fell. Both armies camped in the field, neither being willing to allow the other to claim sole possession. Through a night of bitter frost they strove, vainly, to keep warm, and on the next day Essex, while formally announcing his victory, drew off towards Warwick. His cavalry was in total disorder; he had lost about fifty colours and much baggage and equipment.

On the retreat to Warwick, Rupert's cavalry harassed him all the way, forced him to abandon some of his cannon, and blew up four of his ammunition waggons. The Royalists were now between Essex and London with an almost clear road, and he wrote urgently to Westminster to call out all available troops to defend the capital. But his claim of a 'victory' had this much truth: he had, with Balfour's invaluable help and by his own calm and tenacity, retrieved his army from what might easily have been irremediable disaster. The Royalist prisoners in the Parliamentary camp were among the few who at first truly believed in the Parliamentary claim of victory. Lord Lindsey, angry and in pain, declared that he would never fight in a field with boys again; he never did, for he died that day. Those who had watched Prince Rupert's charge from the other side cursed this boy of twenty-three for a different reason. He was a soldier to be reckoned with, and his men had a spirit which needed only a little more discipline to make them irresistible.

Oliver Cromwell in later years recorded a conversation that he had with his cousin Hampden about this time: 'Your troopers, said I, are most of them old decaying serving-men and tapsters, and such kind of fellows. Their troopers are gentlemen's sons, younger sons and persons of quality … You must get men of a spirit that is likely to go on as far as gentlemen will go, or else I am sure you will be beaten still.' Cromwell was not altogether fair to the quality of the Parliamentary cavalry or the sources whence it came, but he saw that the Cavaliers had established a superiority that it would be very hard to challenge.

C. V. WEDGWOOD

48.

One of the most moving communications in the history of warfare: in June 1643 Sir Ralph Hopton (1596–1652), commanding for the King in the West, wrote to his old friend Sir William Waller (1598– 1668), commanding for Parliament, to suggest a meeting at which the Royalist plainly hoped to secure a change of heart in the rebel. This was Waller's reply.

To my Noble friend Sr Ralphe Hopton at Wells.

Sr

The experience I have had of your Worth, and the happinesse I have enjoyed in your friendship are woundinge considerations when I look upon this present distance betweene us. Certainly my affections to you are so unchangeable, that hostility itselfe cannot violate my friendship to your person, but I must be true to the cause wherein I serve: The ould limitation usque ad aras holds still, and where my conscience is interested, all other obligations are swallowed up. I should most gladly waite on you according to your desire, but that I looke upon you as you are ingaged in that partie, beyond a possibility of retraite and consequentlie uncapable of being wrought upon by any persuasion.

And I know the conference could never be so close betweene us, but that it would take wind and receive a construction to my dishonour; That great God, which is the searcher of my heart, knows with what a sad sence I goe upon this service, and with what a perfect hatred I detest this warr without an Enemie, but I looke upon it as *Opus Domini*, which is enough to silence all passion in mee. The God of peace in his good time send us peace, and in the meane time fitt us to receive it: Wee are both upon the stage and must act those parts assigned us in this Tragedy: Lett us do it in a way of honour, and without personall animosities, whatever the issue be, I shall never willingly relinquish the dear title of

Your most affectionate friend and faithful servant,

Wm. Waller.

Bath, 16 June 1643

49.

Lord Falkland (c. 1610–43), who died on the battlefield of Newbury on 20 September 1643, was believed to have deliberately sacrificed himself in despair at the misery of civil war. John Aubrey took a somewhat more cynical view in his Brief Lives.

Falkland adhered to King Charles I, who after Edgehill fight made him Principal Secretary of State, which he discharged with a great deal of wit and prudence, only his advice was very unlucky to his majesty, in persuading him to sit down before Gloucester, which was so bravely defended by that incomparably vigilant [Parliamentary] governor Colonel Massey, and the diligent and careful soldiers and citizens (men and women) that it so broke and weakened the King's army that 'twas the primary cause of his ruin. Anno domini 1643 at the fight at Newbury, my Lord Falkland being there, and having nothing to do, decided to charge; as the two armies were engaging, rode in like a madman (as he was) between them, and was (as he needs must be) shot.

Some that were your superfine discoursing politicians and fine gentlemen, would needs have the reason of this mad action of throwing away his life so, to be his discontent for the unfortunate advice given to his master as aforesaid; but, I have been well informed, by those that best knew him, and knew the intrigues behind the curtain (as they say) that it was the grief of the death of Mrs Moray, a handsome lady at Court, who was his mistress, and whom he loved above all creatures, was the true cause of his being so madly guilty of his own death, as aforementioned: there is no great wit without an admixture of madness.

The next day, when they went to bury the dead, they could not find his lordship's body, it was stripped, trod upon and mangled; so there was one that had waited on him in his chamber would undertake to know it from all other bodies, by a certain mole his lordship had on his neck, and by that mark did find it. He lies interred at Great Tew, but, I think, yet without any monument.

JOHN AUBREY

50.

From the order book of the Staffordshire County Committee of the Parliamentary army.

December 27th 1643:
That whosoever shall committ Fornication, or frequent companie of light women, whether officer or souldier, shall be forthwith casheered and the woman carted.

March 18th 1644:
Ordered, that the Gunner which did committ fornication, shall bee set upon the greate gun, with a marke uppon his backe through the Garrison and then disgracefully expulsed.

MARK BENCE-JONES

51.

Charles I's nephew Prince Rupert of the Rhine (1619–82), ablest of his commanders, ranked foremost in the demonology of his enemies.

His very name struck terror in Puritan hearts. All kinds of legends grew up about him, and were broadcast by the Roundhead pamphleteers. He was said to have observed the strength of Essex's army by disguising himself as an apple vendor and peddling his fruit among the soldiers. Two Roundhead merchants who were brought before him at Henley and found him, for some reason, in bed with his clothes on, were quick to report that he had vowed never 'to undress or shift himself until he had reseated King Charles at Whitehall'.

He was invested with diabolical powers, and his white dog, Boy, was seen as his familiar spirit. The tales told about Boy surpassed those told about his master. He could prophesy, he could make himself invisible, he was endowed with the gift of languages. 'He is weapon-proof himself, and probably hath made his master so too … they lie perpetually in one bed, sometimes the Prince upon the Dog, and sometimes the Dog upon the Prince; and what this may in time produce, none but the close committee can tell.' …

[Then, at Marston Moor on 2 July 1644] in an hour's fighting on that damp July evening he lost both his army and his reputation. And as though that were not enough, he also lost his beloved dog, Boy, who had been with him when he was eating supper and had followed him as he galloped into the fray. His carcase was found next morning among the dead. The Roundheads rejoiced, convinced that without his familiar spirit the Bloody Prince could no longer harm them. And indeed, after Marston Moor, Rupert's luck did not return.

MARK BENCE-JONES

52.

Of all the sieges of history, that of the Royalist Marquis of Winchester's Basing House in Hampshire, from the early spring of 1644 until October 1645, ranks among the more exotic. Here described by a Victorian local parson and amateur historian, the story has a special place in my own affections, because as a teenager I lived within a few yards of the house's ruins.

Dr Thomas Fuller, once the house's chaplain, wrote that Basing 'was the largest of any subject's house in England, yea, larger than most (eagles have not the biggest nests of all birds) of the King's palaces. The motto "Love Loyaltie" was often written in every window thereof, and was well practised in it.' Their enemies styled the Basing garrison 'foxes and wolves', but they showed in many a daring foray that they could bite as well as bark. A plot was formed by some disheartened malcontents within the walls to surrender the house to Sir William Waller, with whom a correspondence was carried on by the Lord Edward Pawlet, brother to the Marquis of Winchester. The plot was discovered, and the conspirators were all executed, with the exception of Lord Edward, who was forced to act as hangman at all future garrison executions.

On June 11th, 1644, the siege of Basing House began in earnest. [Among the Parliamentarians] Colonel Norton, aided by Colonel Onslow and a Surrey contingent, showed himself a daring and resolute foe, and was reinforced by Colonel Herbert Morley with five hundred foot from Farnham. He blockaded the house with his cavalry, occupied Basing village, and cut off supplies. On June 14th there was a smart skirmish near The Vine, and on the same day it was reported in London that the garrison was already suffering severely, Sir William Waller having burnt both their [flour] mills. Salt and other necessaries were also lacking. On June 18th a jet of flame at midnight made the old church tower stand out in bold relief. Half the village of Basing was in a blaze, and it seemed as if a fierce sortie made from the house would raise the siege. But it was not to be.

For eighteen weeks the struggle went on. Sorties, assaults, mines, desertions, famines, and feasting came in quick succession. The story of this period alone would fill a volume. Basing House began to be styled 'Basting House' by rejoicing Cavaliers. The besiegers laboured, like Nehemiah's workmen, with a sword in one hand and a tool in the other, and on June 24th 'three of ours runne to them'. The gallows was always ready for would-be deserters. A heavy fire of shells, some eighty pounds in weight, which the garrison styled 'baubles', and of cannon shot was kept up, and 'they did shoot the Marquisse himselfe through his clothes'. Owing to a lack of salt, on July 24th, 'stinking beef was thrown over Basing walls'. In vain did Colonel Morley summon the Marquis to surrender, in spite of disease making havoc in the ranks of the defenders. Several dashing sorties were made, and once or twice the besiegers were driven off as far as Basingstoke.

In the second week in September, Colonel Sir Henry Gage, a gallant Roman Catholic soldier, led a relieving force from Oxford, and, after a fiercely-contested action on Chineham Down, against desperate odds, with sorely wearied troops, and shrouded in blinding fog, relieved the garrison in masterly fashion. 'That lovers met that day, and blushed and kissed; and old grey-bearded friends embraced each other, and, aye marry, pledged each other, too; that good Catholic comrades exchanged prayers at Basing altar; that brave fathers kissed the wives and children they had left shut up in brave old "Loyalty", needs no telling. But not alone in kissing and quaffing did Gage and his troops spend those two merry days.'

On September 14th the Cavaliers were celebrating their relief, 'drinking in the town, and in no good order'. Parliament's Colonel Norton made an unexpected attack, and 'one hour's very sharp fight followed'. Basing Church was taken and retaken, as, indeed, it was several times during the siege, though, strangely enough, the Virgin and Child on the west front still remained unharmed. The assailants were at length repulsed with heavy loss, but in the struggle the wise and learned Lieutenant-Colonel Johnson, doctor and botanist, was mortally wounded. Ten days later there was another fierce fight. The stern besiegers again closed tenaciously around Basing, and things went on much as before, the gallant little garrison being in vain summoned to surrender.

Famine was now pressing the garrison hard. On November 28th it was said in London that 'Basing garrison had neither shoes nor stockings, drank water, and looked all as if they had been rather the prisoners of the grave than the keepers of a castle.' The diary of the siege closes with these noble words: 'Let no man, therefore, think himself an instrument, only in giving thanks that God had made him so, for here was evidently seen "He chose the weak to confound the strong."' Non nobis Domine. 'Not unto us, not unto us, O Lord, but to Thine own Name be all glory for ever. Amen!'

For some months Basing was left in peace, and many a successful foray and capture of road waggons took place, bold Cavalier riders scouring the country as far as Hindhead. But unfortunately religious dissensions, which have ruined many noble causes, broke out. On May Day, 1645, there was a sorry sight. All the defenders who were not Roman Catholics marched out of Loyalty House some five hundred strong.

At the end of August, the Parliament sent Colonel Dalbier, a Dutchman, from whom it is said that Cromwell learned the mechanical part of soldiering, to reduce Basing at all costs. '*Mercurius Britannicus*' said that the Marquis of Winchester spent his time in bed at the bottom of a cellar, 'out of reach of gunshot, for, you know, generals and governors should not be too venturous'. Dalbier occupied Basing village, and tried in vain to take the house by means of mines. Shells, known as 'granado shells', proved more effective. One mortar, which was sent direct from London – the bridges being strengthened so that it might cross them – fired shells of sixty-three pounds weight and eighteen inches in diameter. Ammunition was sent from Windsor Castle, a Parliamentarian arsenal. 'A compounded stifling smoke', emitted by damp straw, brimstone, arsenic, and other ingredients, made the lives of the besiegers a misery.

On Sunday, September 21st, 1645, the Rev. William Beech, a Wykehamist, gave the besiegers a remarkable sermon, which occupies, in small type, thirty-two small quarto pages. It was entitled 'More Sulphur for Basing', and is a marvellous specimen of the sermon militant. On the following day, Dalbier's guns brought down 'the great tower in the old house'. Deserters and a released prisoner said that 'in the top of this tower was hid a bushel of Scots twopences, which flew about their ears'. Shot and shell now poured in thick and fast, and when on October 8th Lieutenant-General Cromwell, at the head of a brigade detached from General Fairfax's New Model Army, arrived from recently-captured Winchester, the fate of the fortress was sealed.

The besiegers were seven thousand in number, whilst the walls, which needed from eight hundred to a thousand men to hold them, sheltered but three hundred, many of whom were but eighteen and some scarcely twelve years of age, including the priests and the wounded. At five o'clock in the morning of October 14th the attack began, and the invincible Ironsides formed up in column. The garrison was utterly worn out, but it is said that some of them were surprised as they were playing cards. 'Clubs are trumps, as when Basing House was taken', is a well-known Hampshire phrase.

Rush of pike and pistol shot put a speedy end alike to game and players. Four cannon shots boomed out, and, by a breach which is still plainly visible, the storming party entered the New House, and then made their way inch by inch over the huge mounds faced with brick-work into the Old House. In spite of the black flags of defiance which they hung out, and of the heroism of those who 'fought it out at sword's point', superior numbers prevailed. When opposition ceased, plundering began. But in the midst of the pillage, the dread cry of 'Fire' was raised, for a fireball had been left to smoulder unheeded. Ere long, Basing House was but a pile of smoking ruins. Many of the garrison were suffocated or burned to death in the cellars and vaults in which they had taken refuge.

Hugh Peters, 'the ecclesiastical newsmonger', heard them crying in vain for help. 'There were four more Roman Catholic priests beside, who were plundered of their vestments, and themselves reserved for the gallows.' The prisoners were two hundred in number, including the stout old Marquis, who was afterwards allowed to retire to France. Inigo Jones, the celebrated architect, who is said to have designed the west door of Basing Church, 'was carried away in a blanket, having lost his clothes', doubtless borrowed by some trooper. Seventy-four men were killed, but only one woman, the daughter of Dr Griffith, of St Mary Magdalen, Old Fish Street, 'a gallant gentlewoman, whom the enemy shamefully left naked'. We are told by '*Mercurius Veridicus*' of 'the ladies' wardrobe, which furnished many of the soldiers' wives with gowns and petticoats'.

A hundred gentlewomen's rich gowns and petticoats were among the spoil, which was reckoned to be worth £200,000, and was styled by Cromwell 'a good encouragement'. The House of Commons ordered that all and sundry might take brick or stone at will from the ruins. Basing House became the picturesque ruin which it has ever since remained.

G. N. GODWIN

53.

The most romantic campaign of the Civil War was that of James Graham, Marquis of Montrose (1612–50), who rode north with two followers after the Royalist defeat at Marston Moor to raise Scotland for the King. In the autumn of 1644, with a wild little Highland and Irish army, he won a series of extraordinary victories. With the coming of winter, his enemies assumed that campaigning was ended until the

snows melted. Instead, Montrose led his men across the mountains
from Blair Atholl to Inveraray in an epic march that I retraced on
foot through the snows with my labrador in January 1976, to inflict a
devastating surprise upon the Marquis of Argyll and his Campbells,
in arms for the Covenant and against the King. Then Montrose
turned north, and was deep in the mountains when he learned that a
much superior force under Argyll was in pursuit. On 1 February
1645, the Covenanters pitched camp at Inverlochy, just north of the
modern town of Fort William.

It is said that it was Ian Lom Macdonald, the bard of Keppoch, who walked to tell Montrose at Kilcummin the news of the Campbell army at his back. Montrose at first disbelieved him. He had heard nothing to suggest that any Campbell force was anywhere within days. His own army had dwindled to barely 1500 men. All the others had retired home with their plunder. He knew that there were at least 5000 levies under Seaforth at Inverness. He had few doubts of his ability to defeat them. His concern was that whether he did so or not, after fighting one major battle he would be in no condition instantly to turn and grapple the Campbells.

Montrose questioned Macdonald with deadly patience, for the fortunes of the King's cause in Scotland hung upon his answers. He asked about the paths over the hills and the state of the drifts on the higher ground; about the strength of the Campbells and their intentions. If Macdonald lied, said the Lord Marquis coldly, he would hang. Stubbornly the poet insisted that he told the truth. The critical moment had come. Montrose ordered his officers to rouse the army and prepare to march against the Campbells. 'I was willing to let the world see that Argyll was not the man his Highlanders believed him to be,' he wrote cheerfully to the King in his later dispatch, 'and that it was possible to beat him in his own Highlands.'

Montrose reckoned that, left to his own devices, Argyll would dog his footsteps until he was entangled with Seaforth, and then fall upon him from the rear, a strategy that suited the Campbells' temperament. It was essential that Montrose destroy his southern, more dangerous enemy, before moving north to address the vacillating Seaforth. His first move was to place pickets along the main track southwards through the Great Glen, to prevent Campbell scouts or informers reporting the royalist army gone from Kilcummin. Then he led his men from their billets to begin one of the greatest flank marches in the annals of warfare.

The weather was worse than on the hardest passages of the descent upon Inveraray. The Royalists must move fast to have any hope of achieving surprise. Instead of approaching Argyll down the passes of the Great Glen like any lesser mortal, Montrose took to the mountains and marched parallel with the accustomed road south-westwards, hidden by the intervening range of hills. The going, naturally, was ten times more difficult. The snow-drifts lay virgin upon the hillsides. The very deer floundered. The steep faces were untrodden even by cowherds and hunters since winter descended. The cold was appalling. Even if somehow Montrose could complete a passage through the icy wilderness, without benefit of maps or compasses, his exhausted men must then do battle with Argyll against odds of two to one.

Up the valley of the Tarff they laboured, high into the hills towards the summit of Carn Dearg. That first day, Friday, January 31st, they had over two thousand feet to climb before traversing the shoulder of Cam Dearg and making the descent into Glen Roy. Once past the watershed, somewhere high in the pass they may have rested for a few hours. But they had scant time to spare. And to many, it must have seemed worse to lie shivering in their plaids than to stagger on through the snow. Home-crafted deerskin sandals, and ragged clothes worn to shreds by weeks of exposure, offered little protection against the freezing January night. For the thirsty there was only the shocking chill of burn water. When some men brought down a deer, they ate their chunks of bloody flesh unsalted. There was no bread and only a little oatmeal.

Somewhere on the last stages of the march, Montrose's vanguard found themselves face to face with a party of Campbells. There was a startled, brutal little struggle in which most of the Covenanters were cut down. A few escaped, however, and fled to report to Argyll's camp at Inverlochy that an enemy was close at hand. It was unthinkable that Montrose and his army should be nearby: they were assumed to be thirty miles further north, at Kilcummin. This must be some clan raiding party, skirmishers sent to harass the advancing Campbells. The cautious Auchinbreck strengthened his pickets and pushed out patrols to probe the hills, but he left the bulk of his army to its food and sleep.

Montrose's clansmen reached their objective on the hillside above Inverlochy just as night was falling on Saturday, February 1st. As the order was given to halt, men slumped into the shelter of burn beds and the folds of the hill, to rest their exhausted bodies. Montrose's officers reported to him: the army was exhausted; the tail of stragglers stretched behind for miles. There was little to be done to assuage their hunger. They saw Argyll's camp fires on the loch shore below, but had no clues as to his dispositions. 'So be it then,' said the Lord Marquis calmly. 'We will take them at dawn.'

For the Royalists crouched shivering on the hillside, that night seemed unending. There was no fuel for fires. It was almost impossible to sleep amid the cold; a constant confused movement of men and coming of stragglers; the murmur of orders and arguments. A huge moon hung over the mountains, bathing Royalists and Campbells in an eerie grey light, such as seems to promise dragons and fairies whenever it lights Lochaber in winter. Many men lay silent, watching the flicker of the Campbell fires far below, computing how many claymores might lie around each one.

At dawn on the morning of February 2nd, Argyll's general Sir Duncan Campbell of Auchinbreck, still uncertain what enemy he faced, roused his own army. All night his pickets had brushed with mysterious foes on the hills, exchanging shots and cries in the darkness, marking shadows flitting in the moonlight. The Marquis of Argyll announced that he was taking to his galley, which lay offshore. Argyll was rowed out with his companions, including Montrose's brother-in-law Sir James Rollo. As the grey light reluctantly brightened, the Campbells on shore and their chief aboard his galley peered curiously up the hillside.

With the first shaft of sunlight, the sound of a fanfare saluting the raising of the Royal Standard echoed down to the Campbells with the significance of the last trump. The presence of the Standard signified that of the King's Lieutenant, Montrose himself. A shock of astonishment ran through the ranks. Then as the call died away, the pipes took up the tale. They were playing the terrible Cameron pibroch, 'Sons of dogs come and I will give you flesh.' On the hillside as the royalists deployed in their clan regiments, Montrose, Airlie and his closest companions crouched in the snow eating a mess of waterlogged oatmeal off the points of their dirks.

The Irish took up their positions, half under Alasdair Macdonald's command on the right, the rest under Magnus O'Cahan on the left. The order of battle for the centre is a rollcall of the great names of Highland history: men of Glengarry, Maclean, Keppoch, Glencoe, Atholl, Appin, Lochaber. Behind them stood a reserve of Irish musketeers. Thomas Ogilvy's handful of horse gathered around the Royal Standard, their beasts tottering, lame and starved.

Priests walked before the line offering blessings, and the Catholics knelt to pray. Alasdair Macdonald was dismayed to see Ian Lom, the bard of Keppoch who had guided them to this battlefield, walking alone away from the army. 'Ian Lom, wilt thou leave us?' he cried. The poet called back calmly: 'If I go with thee today and fall in battle, who will sing thy praises and thy prowess tomorrow?' He turned again and clambered up the hillside until he reached a knoll from which he could fairly see Loch Eil and Inverlochy below. There, his lonely figure stood unmoving through the morning, watching the battle beneath.

Magnus O'Cahan began the struggle with a headlong charge. Axe, sword and pike clashed and clanged above the snow. Argyll's vanguard recoiled before slashing steel and point-blank musketry. As they fell back on the main Campbell position, Alasdair's men threw themselves on Auchinbreck's left. Then the rest of the Royalist army surged mob-handed down the hillside, hurling themselves at the Campbell battle line. The Lowland musketeers were trained to fight like-minded, ordered regiments. For a minute or two, they rhythmically loaded and fired. Then the Gaelic fury broke upon them. Their ordered ranks disintegrated. Argyll's entire army broke in flight. With a full heart, Montrose wrote to his monarch:

As to the state of affairs in this kingdom, the bearer will fully inform your Majesty in every particular … I doubt not before the end of this summer I shall be able to come to your Majesty's assistance with a brave army which, backed with the justice of your Majesty's cause, will make the Rebels in England, as well as in Scotland, feel the just rewards of Rebellion. Only give me leave, after I have reduced this country to your Majesty's obedience, and conquered from Dan to Beersheba, to say to your Majesty then, as David's General did to his master, 'come thou thyself lest this country be called by my name'.

Your Majesty's most humble, most faithful, and most obedient Subject and Servant,

Montrose.

Inverlochy in Lochaber,

February 3rd, 1645.

Ian Lom Macdonald was true to his pledge. To honour this Candlemas of 1645, he composed one of the greatest of all Gaelic odes. Montrose's army marched north from Inverlochy leaving the shore strewn with the bodies of their enemies, stripped naked by the victors as custom demanded. The Camerons had fulfilled the ghastly promise of their pibroch: the dogs came down to feast upon Campbell flesh. [Montrose fought a series of further campaigns for the Royalist cause until defeated at Carbisdale in April 1650, captured and executed in Edinburgh.]

MAX HASTINGS

54.

General George Monck (1608–70), one of the most experienced if by no means ablest officers at the head of Parliament's forces, laid down a code for the conduct of soldiers at the beginning of his book Observations upon Political and Military Affairs, *written in 1645 but published posthumously, which has been passed down with reverence to successive generations of British warriors.*

The profession of a soldier is allowed to be lawful by the word of God, and so famous and honourable amongst men, that emperors and kings do account it a great honour to be of the profession, and to have experience of it. He that chooseth the profession of a soldier ought to know withal, honour must be his greatest wages, and his enemy his surest paymaster. The two chief parts of a soldier are valour and sufferance, and there is as much honour gained by suffering wants patiently in war, as by fighting valiantly, and as great achievements effected by the one as by the other. It is not virtue but weakness of mind not to be able to endure want a little. The greatest virtue which is required in a soldier is obedience, as a thing wherein the force of all discipline consisteth. Let every soldier arm his mind with hopes and put on courage. Whatsoever disaster falleth, let not his heart sink. For the passage of Providence liest through many crooked ways, and a despairing heart is the true prophet of approaching evil.

GEORGE MONCK

55.

One of the last skirmishes of the First Civil War took place on 21 March 1646 at Stow-on-the-Wold, Gloucestershire, where Sir Jacob Astley (1579–1652) conducted a gallant, vain last stand for the King with a body of Welsh levies, who soon surrendered.

Astley himself, unhorsed and surrounded, gave up his sword to one of Colonel Birch's men. 'You have done your work, boys,' said the old cavalier, 'you may go play, unless you fall out among yourselves.'

C. V. WEDGWOOD

56.

Cromwell's dispatch from Dublin of 17 September 1649, describing his capture of Drogheda five days earlier.

Divers of the enemy retreated into the Millmount, a place very strong and of difficult access; being exceedingly high, having a good graft [ditch], and strongly palisadoed. The Governor, Sir Arthur Aston, and divers considerable officers, being there, our men, getting up to them, were ordered by me to put them all to the sword. And, indeed, being in the heat of action, I forbade them to spare any that were in arms in the town; and, I think, that night they put to the sword about two thousand men – divers of the officers and soldiers being fled over the bridge into the other part of the town, where about one hundred of them possessed St Peter's church steeple, some the west gate, and others a strong round tower next the gate called St Sunday's.

These, being summoned to yield to mercy, refused. Whereupon I ordered the steeple of St Peter's church to be fired, when one of them was heard to say, in the midst of the flames, 'G—d d—n me! G—d confound me! I burn, I burn!' The next day, the other two towers were summoned; in one of which was about six or seven score; but they refused to yield themselves: and we, knowing that hunger must compel them, set only good guards to secure them from running away until their stomachs were come down. From one of the said towers, notwithstanding their condition, they killed and wounded some of our men.

When they submitted, their officers were knocked on the head, and every tenth man of the soldiers killed; and the rest shipped for the Barbadoes. The soldiers in the other tower were all spared, as to their lives only, and shipped likewise for the Barbadoes. I am persuaded that this is a righteous judgment of God upon these barbarous wretches, who have imbrued their hands in so much innocent blood; and that it will tend to prevent the effusion of blood for the future.

OLIVER CROMWELL

57.

An incident during King Louis XIV's (1638–1715) 1667 siege of Lille.

The Compte de Brouay, governor of the province on behalf of Spain, had occasion to send a flag of truce into the besiegers' camp. When the officer who bore it was returning, the Duc de Charrost, captain of the King's Guard, called out, 'Tell Brouay not to follow the example of the governor of Douai, who yielded like a coward.' The King turned round laughing, and said, 'Charrost, are you mad?' 'How, sir?' answered he; 'Brouay is my cousin.'

LORD DE ROS

58.

Childhood pranks of the future Tsar Peter the Great (1672–1725).

His favourite game was war. A small parade ground had been laid out in the Kremlin where he could drill the boys who were his playmates. Unlike most boys who play at war, Peter could draw on a government arsenal to supply his equipment. Its records show that his requests were frequent. In January 1683, he ordered uniforms, banners and two wooden cannon, their barrels lined with iron, mounted on wheels to allow them to be pulled by horses. On his eleventh birthday, in June 1683, Peter abandoned wooden cannon for real ones with which, under the supervision of artillerymen, he was allowed to fire salutes. He enjoyed this so much that messengers came almost daily to the arsenal for more gunpowder. In May 1685, Peter, nearing thirteen, ordered sixteen pairs of pistols, sixteen carbines with slings and brass mountings and, shortly afterward, twenty-three more carbines and sixteen muskets.

By the time Peter was fourteen, his martial games had transformed the summer estate into an adolescent military encampment. Peter's first 'soldiers' were the small group of playmates who had been appointed to his service when he reached the age of five. The ranks were further swelled by other young noblemen presenting themselves for enrolment, either on their own impulse or on the urging of fathers anxious to gain the young Tsar's favour. The sons of clerks, equerries, stable grooms and even serfs in the service of noblemen were set beside the sons of boyars. Eventually 300 of these boys and young men had mustered on the Preobrazhenskoe estate. They lived in barracks, trained like soldiers, used soldiers' talk and received soldiers' pay. From this collection of young noblemen and stableboys he eventually created the proud Preobrazhensky Regiment.

Soon, all the quarters available in the little village were filled, but Peter's boy army kept expanding. New barracks were built in the nearby village of Semyonovskoe; in time, this company developed into the Semyonovsky Regiment, and it became the second regiment of the Russian Imperial Guard. Barracks, staff offices and stables were built, more harnesses and caissons were drawn from the equipment of the regular horse artillery, five fifers and ten drummers were detached from regular regiments to pipe and beat the tempo of Peter's games. Western-style uniforms were designed and issued: black boots, a black three-cornered hat, breeches and a flaring, broad-cuffed coat which came to the knees, dark bottle green for the Preobrazhensky company and a rich blue for the Semyonovsky.

Rather than taking for himself the rank of colonel, Peter enlisted in the Preobrazhensky Regiment at the lowest grade, as a drummer boy, where he could play with gusto the instrument he loved. Eventually, he promoted himself to artilleryman or bombardier, so that he could fire the weapon which made the most noise and did the most damage. He performed the same duties, stood his turn at watch day and night, slept in the same tent and ate the same food. When earthworks were built, Peter dug with a shovel. When the regiment went on parade, Peter stood in the ranks, taller than the others but otherwise undistinguished. Peter's boyhood refusal to accept senior rank in any Russian military or naval organisation became a lifelong characteristic. Later, when he marched with his new Russian army or sailed with his new fleet, it was always as a subordinate commander. If he, the Tsar, did this, no nobleman would be able to claim command on the basis of title.

ROBERT MASSIE

59.

Irishwoman Christian Cavanagh – 'Mother Ross' – was born in 1667, and enlisted in 1693 to pursue her lost husband. Daniel Defoe wrote her story after meeting her in old age as a Chelsea Pensioner.

In the morning I thought of going in search of my dear Richard, and this gave some ease to my tortured mind. I began to flatter myself that I should meet no great difficulty in finding him out, and resolved in one of his suits, for we were both of a size, to conceal my sex, and go directly for Flanders, in search of him whom I preferred to everything else the world could afford me, which, indeed, had nothing alluring, in comparison with my dear Richard, and whom the hopes of seeing had lessened every danger to which I was going to expose myself. The pleasure I found in the thoughts of once more regaining him, recalled my strength, and I was grown much gayer than I had been at any time in my supposed widowhood.

I was not long deliberating, after this thought had possessed me, but immediately set about preparing what was necessary for my ramble; and disposing of my children, my eldest with my mother, and that which was born after my husband's departure, with a nurse (my second son was dead), I told my friends, that I would go to England in search of my husband, and return with all possible expedition after I had found him. My goods I left in the hands of such friends as had spare house room, and my house I let to a cooper. Having thus ordered my affairs, I cut off my hair, and dressed me in a suit of my husband's having had the precaution to quilt the waistcoat, to preserve my breasts from hurt, which were not large enough to betray my sex, and putting on the wig and hat I had prepared, I went out and bought me a silver-hilted sword, and some Holland shirts: but was at a loss how I should carry my money with me, as it was contrary to law to export above £5 out of the kingdom; I thought at last of quilting it in the waistband of my breeches, and by this method I carried with me fifty guineas without suspicion.

I had nothing upon my hands to prevent my setting out; wherefore, that I might get as soon as possible to Holland, I went to the sign of the Golden Last, where Ensign Herbert Laurence, who was beating up for recruits, kept his rendezvous. He was in the house at the time I got there, and I offered him my service to go against the French, being desirous to show my zeal for his majesty King William, and my country. The hopes of soon meeting with my husband, added a sprightliness to my looks, which made the officer say, I was a clever brisk young fellow; and having recommended my zeal, he gave me a guinea enlisting money, and a crown to drink the king's health, and ordered me to be enrolled – having told him my name was Christopher Walsh – in Captain Tichbourn's company of foot.

We stayed but a short time in Dublin after this, but, with the rest of the recruits, were shipped for Holland, weighed anchor, and soon arrived at Williamstadt, where we landed and marched to Gorcum. Here our regimentals and first mountings were given us. The next day we proceeded forward to Landen, where we joined the grand army. Having been accustomed to soldiers, when a girl, and delighted with seeing them exercise, I very soon was perfect, and applauded by my officers for my dexterity.

MOTHER ROSS AND DANIEL DEFOE

60.

On leaving Oxford in 1694, Irish writer Richard Steele (1672–1729) became a cadet in the Life Guards. He represents the story below as one that he heard from a corporal of his regiment who had served in the civil wars, which would have made the man a very great age indeed.

This gentleman was taken by the enemy; and the two parties were upon such terms at that time, that we did not treat each other as prisoners of war, but as traitors and rebels. The poor corporal, being condemned to die, wrote a letter to his wife when under sentence of execution. He writ on the Thursday, and was to be executed on the Friday: but, considering that the letter would not come to his wife's hands until Saturday, the day after execution, and being at that time more scrupulous than ordinary in speaking exact truth, he formed his letter rather according to the posture of his affairs when she should read it, than as they stood when he sent it: though, it must be confessed, there is a certain perplexity in the style of it, which the reader will easily pardon, considering his circumstances.

'Dear Wife,

Hoping you are in good health, as I am at this present writing; this is to let you know, that yesterday, between the hours of eleven and twelve, I was hanged, drawn, and quartered. I died very penitently, and everybody thought my case very hard. Remember me kindly to my poor fatherless children.

Yours until death,

W.B.'

It so happened, that this honest fellow was relieved by a party of his friends, and had the satisfaction to see all the rebels hanged who had been his enemies. I must not omit a circumstance which exposed him to raillery his whole life after. Before the arrival of the next post, that would have set all things clear, his wife was married to a second husband, who lived in the peaceable possession of her; and the corporal, who was a man of plain understanding, did not care to stir in the matter, as knowing that she had the news of his death under his own hand, which she might have produced upon occasion.

RICHARD STEELE

61.

In 1704 Peter the Great stormed the city of Narva, then in Sweden but afterwards incorporated into Russia, and since 1918 part of Estonia.

The soldiers, in defiance of the express orders of the Tsar, carried fire and destruction into every quarter of the town, slaughtering the inhabitants without mercy. Peter threw himself, sword in hand, into the midst of the massacre; and rescued many of the defenceless women and children from his merciless and savage troops. He killed, with his own hand, one of his ferocious soldiers, whom the heat of the carnage rendered deaf to his voice; and at last succeeded in curbing the fury of this unlicensed scene. Covered with dust, sweat, and blood, he then hastened to the townhouse, where the principal inhabitants of the place had taken refuge. His terrible and threatening air greatly alarmed these unhappy people. As soon as he had entered the hall, he laid his sword on a table; and then addressing himself to the affrighted multitude, who waited their doom in anxious silence, 'It is not,' said he, 'with the blood of your fellow-citizens that this sword is stained; but with that of my own soldiers, whom I have been sacrificing for your preservation.'

LORD DE ROS

62.

*The 1st Duke of Marlborough (1650–1722) met for the first time
Prince Eugene of Savoy (1663–1736), who was to become his close
friend and comrade-in-arms, on 10 June 1704 at Gross Heppach.
Each strove to outdo the other in courtesies.*

Next morning, Eugene inspected Marlborough's cavalry escort. 'My
Lord,' said Eugene, 'I never saw better horses, better clothes, finer belts
and accoutrements; but money, which you don't want in England, will
buy clothes and fine horses, but it can't buy that lively air I see in every
one of these troopers' faces.' 'Sir,' said Marlborough, 'that must be
attributed to their heartiness for the public cause and the particular
pleasure and satisfaction they have in seeing your Highness.'

DR HARE

63.

*It often happens that when an unpopular officer falls on the
battlefield, questions are asked as to which side shot him.*

A commander who served under Marlborough, Brigadier-General
Richard Kane, warned that officers who ill-treated their men could
expect to 'meet with their fate in the day of battle from their own
men'. A major of the 15th Foot, on the field of Blenheim, turned to
address his regiment before the assault, and apologized for his past
ill behaviour. He requested that, if he must fall, it should be by the
bullets of the enemy. If spared, he would undertake to mend his
ways. To this abject performance, a grenadier said: 'March on, sir;
the enemy is before you, and we have something else to do than
think of you now.' After several attacks, the regiment carried its position and the major, gratified, no doubt, to be still alive, turned to his
troops and removed his hat to call for a cheer. No sooner had he said
'Gentlemen, the day is ours' than he was struck in the forehead by a
bullet and killed. There was a decided suspicion that the bullet was
no accident.

E. S. TURNER

64.

After his great victory over the French at Blenheim on 13 August 1704, Marlborough invited his defeated opponent, Marshal Camille Tallard (1652–1728), to accompany him as he inspected the Allied army. Their conversation is said to have gone something like this:

Marlborough: 'I am very sorry that such a cruel misfortune should have fallen upon a soldier for whom I have had the highest regard.'

Tallard: 'And I congratulate you on defeating the best soldiers in the world.'

Marlborough: 'Your Lordship, I presume, excepts those who had the honour to beat them.'

LIVES OF THE TWO ILLUSTRIOUS GENERALS

65.

The end of the story of 'Mother Ross'.

Almost the last shot fired by the French at the battle of Ramillies, in 1706, wounded a trooper in Lord Hay's Regiment of Dragoons (now the Royal Scots Dragoon Guards). With a fractured skull, the soldier underwent the operation of trepanning, when it was discovered that the supposed man was really a woman. It turned out that she had followed her husband to the war, and after discovering him had continued to serve, making her partner promise not to disclose her sex. Until her discovery thirteen years later, she had served in different regiments through several campaigns. Naturally, the news of the exploit of Mrs Richard Walsh, which was her married name, spread rapidly through the army, and the plucky woman received many kindnesses from officers and men.

The great Duke of Marlborough himself took an interest in her, and persuaded her to be remarried to her husband. The ceremony was attended by a large number of officers, who all kissed the bride before leaving. 'Mother Ross', as she afterwards was called, was appointed cook in her husband's regiment; but at the siege of Ath she could not resist the sound of battle, so, seizing a musket, she killed one of the enemy. Unfortunately, at the same moment, a ball from the enemy struck her in the mouth, splitting her underlip and knocking one of her teeth into her mouth. Mrs Walsh's husband was killed at the battle of Malplaquet (1709), but at the end of eleven weeks she married Hugh Jones, a grenadier in the same regiment. Her second husband being killed, she married a soldier of the Welsh Fusiliers, named Davies, who survived her. Eventually 'Mother Ross' retired on a pension of a shilling a day, given by Queen Anne, and on her death in 1739 was buried with military honours in the cemetery belonging to Chelsea Hospital.

JAMES SETTLE

66.

Private Matthew Bishop (1688–?) wrote just three letters home in six years' campaigning under Marlborough. Yet he professed astonishment and indignation, on eventual discharge at Dover, to discover that his wife had meanwhile married another man.

I will now make bold to trouble the Reader with a line or two that I sent to my Wife before I went to Camp [in 1707], then I will proceed forward to the rest of my Actions. 'My Dear, be so candid as to excuse my annual Letters, as my Thoughts hitherto have been taken up with Business of great Importance that required great Attention, and rendered me unfit to think of my private Affairs. But as we have accomplished all we undertook the last Campaign, I am in Hopes we shall not meet with any Obstruction the next ensuing; and I don't doubt in the least but that our Desires will be fulfilled, if we trust to the Providence of God: For he is our Rock, our Shield, and strong Tower: For my part, I trust in him to aid and assist me, knowing that he is our only Support. So I conclude with Fervency of Heart to what is abovementioned; likewise give me Leave to subscribe myself yours for ever, not forgetting my Mother.'

MATTHEW BISHOP

67.

A masterclass in scavenging.

There was [a] … man that was remarkable for a great eater, his name was John Jones, who belonged to Captain Cutler's company: He said he was prodigious hungry. With that the men asked him how many cannonballs he had eaten for his breakfast. Then I said to him, Thou deservest preferment, if thou canst digest cannonballs. Then Sergeant Smith came up to me, and told me, he had eaten four or six twenty-four pounders, and as many as six twelve-pounders in a morning. Now this sergeant was not addicted to tell fabulous stories, though it seemed incredible to any one's thinking. But he explained it in this manner, that the man often frequented the fields in search of those cannonballs; that he used to dig them out of the banks, and had brought a great number in a morning to the artillery, in order to dispose of them for money; and the money he bought his provision with. Had there been no cannonballs flying, he certainly could not have subsisted; for he both eat and drank more than ten moderate men: So that his daily study was to provide for his belly.

MATTHEW BISHOP

68.

From the diary of a British soldier who served in the garrison of Gibraltar under siege, 1727.

March 9th. Came a deserter who reports that while our guns were firing at them a [Spanish] officer pulled off his hat, huzzaed and called God to damn us all, when one of our balls with unerring justice took off the miserable man's head and left him a wretched example of the Divine justice.

April 12th. A recruit who refused to work, carry arms, eat or drink was whipped for the fifth time, after which being asked by the officer he said he was now ready to do his duty.

May 7th. This morning Ensign Stubbs of Colonel Egerton's regiment retired a little out of the camp and shot himself.

June 17th. Today two corporals of the Guards boxed over a rail until both expired, but nobody can tell for what reason.

October 11th. One of Pearce's regiment went into the belfry of a very high steeple, threw himself into the street, and broke his skull to pieces.

October 16th. Will Garen, who broke his back, was hanged.

December 9th. Last night a deserter clambered up within a little of Willis's battery and was assisted by a ladder of ropes by our men. When the officers came to examine his face, they found him to have deserted out of the Royal Irish two months ago. Asking the reason of his return, he said he chose rather to be hanged than continue in the Spanish service, so is to have his choice.

January 2nd 1728. Here is nothing to do nor any news, all things being dormant and in suspense, with the harmless diversions of drinking, dancing, revelling, whoring, gaming and other innocent debaucheries to pass the time – and really, to speak my own opinion I think and believe that Sodom and Gomorrah were not half so wicked and profane as this worthy city and garrison of Gibraltar.

AN UNKNOWN BRITISH SOLDIER WHO SIGNED HIMSELF 'S.H.'

69.

The Duc de Richelieu (1696–1788) found himself serving at the 1734 French siege of Phillipsburg with his wife's cousin, the Prince de Lixin.

The Prince de Conti, who was commanding a regiment, gave a party to celebrate his own seventeenth birthday. Richelieu was an old friend of the Prince's father; he felt he could go to the party straight from a day in the trenches, without changing his clothes. When the Prince de

Lixin saw him he remarked in a loud voice that M. de Richelieu, in spite of his marriage, still seemed to have a good deal of dirt clinging to him.

The Duke called him out; they decided to fight at once, because fighting among officers was forbidden and they were afraid of being stopped. So they proceeded, with their friends, to a deserted place behind the trenches and told the servants to light flares. These attracted the enemy's fire, and the duel took place amid falling shells; the Germans soon found the range and one of the servants was killed. Lixin almost immediately wounded Richelieu in the thigh. The Duke's seconds, who were liking the situation less and less, urged him to give up. He refused and the fight went on a good long time. In the end Richelieu ran Lixin through the heart. The officers present, thankful to be alive themselves, carried the two principals off the field, one to his grave and the other to the hospital.

NANCY MITFORD

70.

The caption to a portrait of Hannah Snell.

Born at Worcester 1723 – Inlisted herself by the name of James Gray in Colonel Guise's regiment then at Carlisle 1745, where she received 500 lashes. Deserted from thence and went to Portsmouth, where she Inlisted in Colonel Fraser's Regiment of Marines, went in Admiral Boscawen's Squadron to the East Indies, at the siege of Pondicherry where she received 12 shot, one in her Groin, Eleven in her legs; 1750 came to England without the least discovery of her Sex, and on her petitioning His Royal Highness the Duke of Cumberland he was pleased to order her a Pension of £30 a year.

ANON.

71.

A celebrated incident at Fontenoy, on 11 May 1745, as the opposing armies closed in silence to within fifty paces of each other, recounted by a French pen.

A regiment of the English guards, that of Campbell, and the Royal Scots, formed the front rank: M. de Campbell was their lieutenant-general; the Count of Albemarle their major-general; and M. de Churchill, natural grandson of the great Duke of Marlborough, their brigadier. The English officers saluted the French by raising their hats. The Count de Chabannes, the Duke de Biron, who had advanced along with all the officers of the French guards, returned the salute. Lord Charles Hay, captain of the English guards, cried: 'Gentlemen of the French guards, open fire.' The Count d'Auteroche, then a lieutenant of grenadiers, called back in a loud voice: 'Gentlemen, we shall not be the first to fire; fire yourselves!' The English then loosed a rolling volley [though some spoilsport historians assert that the French fired first].

VOLTAIRE

72.

Following the 1748 peace of Aix-La-Chapelle, many unemployed British army officers faced penury.

One of these gentlemen accidentally introducing himself into a subscription billiard room, at a coffee house near St James's, found the Duke of Cumberland, his late Majesty's uncle, at play with a colonel of the Guards; it was a match for a considerable sum, and the termination of it was looked for with apparent eagerness by the numerous spectators. His Royal Highness lost the game and immediately putting his hand into his pocket, discovered that he had lost a gold snuff box on the top of which was a fine portrait of Frederick of Prussia, set round with brilliants. A general confusion ensued, the door was immediately locked, and a search called for, which was readily assented to by all present except the stranger, who declared that he would lose his life before he would submit to the proposal; little doubt was then entertained but he was the pickpocket; and resistance appeared useless. The indignant soldier then requested that His Royal Highness would honour him with a private interview; to this the Duke instantly assented, and the company remained in the greater suspense. On entering the room, the officer thus addressed the Duke: 'May it please your Royal Highness, I am a soldier; but my sword is no longer of service to me or my country, and the only means I have to support the character of a gentleman (which no distress shall induce me to forfeit) is the half pay which I receive from the bounty of my sovereign.

'My name is C—, my rank a lieutenant in the Old Buffs. I dined this day at a chop house, where I paid for a rump steak; but eating only half of it I have the remainder wrapped up in paper in my pocket, for another scanty meal at my humble lodgings'; and immediately producing it, added, 'I am now, Sir, ready to undergo the strictest search.' 'I'll be d—d if you shall,' replied the Duke, and on their returning to the billiard room the flap of His Highness's coat struck against the entrance; when it was discovered that the seam of his pocket was unsewed and the lost valuable was safe in the silk lining. A few days after, the gallant officer received a Captain's Commission, with a flattering letter from the Royal Duke offering a prospect of future promotion.

THE SOLDIER'S COMPANION

73.

A nice exercise of military diplomacy.

King Frederick William of Prussia [1688–1740] asked Sir Robert Sutton, at a review of his tall grenadiers, if he thought an equal number of Englishmen could beat them? 'Sir,' replied Sir Robert, 'I do not venture to assert that; but I know that half the number would try.'

NAVAL AND MILITARY ANECDOTES

74.

King Frederick William achieved a reputation as a disciplinarian,
here displayed during his 1741–2 first war in Silesia.

The King, being desirous of making, in the night time, some alterations in his camp, ordered that under pain of death, neither fire nor candle should be burning in the tents after a certain hour. He went round the camp himself, to see that his orders were obeyed; and, as he passed by Captain Zietern's tent, he perceived a light. He entered, and found the captain sealing a letter, which he had just written to his wife, whom he tenderly loved. 'What are you doing there?' said the king: 'Do you not know the orders?' Zietern threw himself at his feet, and begged mercy, but he neither could nor attempted to deny his fault. 'Sit down,' said the king to him, 'and add a few words I shall dictate.' The officer obeyed, and the king dictated: 'Tomorrow I shall perish on the scaffold.' Zietern wrote it, and he was executed the next day.

THE PERCY ANECDOTES

75.

On the evening of 24 August 1758, King Frederick the Great of Prussia (1712–86) lay in a mill on the river Mutzel, preparing to fight a desperate battle at Zorndorf the next morning against an invading Russian army.

On this occasion as on so many in his past, Frederick preferred the company of his young Swiss *lecteur*, de Catt, to that of his generals. Attracted by his tact, modesty and good sense, Frederick used him during these war years as a kind of sounding-board for his opinions, an audience for his recitations and a critic and editor of his verses. Whenever he had the time, in the afternoons or evenings, he would summon him for a chat and de Catt kept a record of their conversations. 'When you look at it in later years,' the King told him, 'you will say: this is what that garrulous old warrior told me who was always bemoaning his lot, always in a fever-heat, wondering anxiously how matters would end and shouting at me, his life was no better than a dog's ...'

That evening, on 24th August, de Catt was summoned to the King: 'I found him writing busily in a very small room in the mill. I thought he was making his plans for the battle. Not at all. He was writing verses. "Verses, Sire? And tomorrow Your Majesty intends to fight a battle!" "What's so unusual about that? My thoughts have been on the main business all day, my plan is ready, I have made my decisions. So now I think I might be allowed to scribble verses like any other man." The King then remarked that Racine's odes were not nearly so good as his tragedies.'

The record continues: 'I think, Sire, it would be very difficult to write verse in the style of Racine, starting, for instance, with: "*Celui qui des flots assouvit la fureur* …" "You are right, that would be difficult. But, *mon cher*, suppose I try?" He had seized his pen when the generals were announced. "Wait a moment in my room. I want to give them their instructions. Everyone must know exactly what he has to do …" The King went out and returned after talking to his generals for half an hour. "Well, all's said. Now, what can I say about '*Celui qui des flots* …'?" In a quarter of an hour he had finished the verses, which he allowed me to keep. "Now, my friend," said the King, "let us eat these grapes, for who knows who may eat them tomorrow. We move at dawn.'" Next morning, Frederick's servant told de Catt that the King had slept so soundly he had had difficulty in waking him. [Frederick's army held its ground on the battlefield, at terrible cost in casualties.]

LUDWIG REINERS

76.

The story of Frederick that every schoolboy once learned.

It was customary with Frederick, whenever a new soldier appeared in his Guards, to ask him three questions, viz., 'How old are you? How long have you been in my service? Are you satisfied with your pay and treatment?' It happened that a young soldier, a native of France, who had served in his own country, desired to enlist into the Prussian service, and his figure was such as to cause him to be immediately accepted. He was, however, totally ignorant of the German language, but his captain gave him notice that the King would ask him certain questions in that language the first time he saw him, and therefore instructed him to learn by heart the three answers which he was to make the King. The soldier learned them by the next day; and as soon as he appeared in the ranks, Frederick came up to interrogate him. His Majesty, however, happened to begin with the second question first; and asked him 'How long have you been in my service?' 'Twenty-one years,' answered the soldier.

The King, struck with his youth, which contradicted his answer, said to him much astonished, 'How old are you?' 'One year, an't please your

Majesty.' Frederick, still more astonished, cried, 'You or I must certainly be bereft of our senses.' The soldier, who took this for the third question about 'pay and treatment', replied firmly, 'Both, an't please your Majesty.' 'This is the first time I ever was treated as a madman at the head of my army,' rejoined Frederick. The soldier, who had exhausted his stock of German, stood silent; and when the King again addressed him, in order to penetrate the mystery, the soldier told him in French, that he did not understand a word of German. The King laughed heartily, and, after exhorting him to perform his duty, left him.

NAVAL AND MILITARY ANECDOTES

77.

Edward Gibbon (1737–94), historian of The Decline and Fall of the Roman Empire, *recalls his own military experience.*

In the act of offering our names and receiving our commissions, as major and captain in the Hampshire regiment (June 1759), we had not supposed that we should be dragged away, my father from his farm, myself from my books, and condemned, during two years and a half (May 1760–December 1762), to a wandering life of military servitude. But a weekly or monthly exercise of thirty thousand provincials would have left them useless and ridiculous; and after the pretence of [a French] invasion had vanished, the popularity of Mr Pitt gave a sanction to the illegal step of keeping them till the end of the war under arms, in constant pay and duty, and at a distance from their respective homes.

When the King's order for our embodying came down, it was too late to retreat, and too soon to repent. The south battalion of the Hampshire militia was a small independent corps of four hundred and seventy-six, officers and men. My proper station, as first captain, was at the head of my own, and afterwards of the grenadier company; but in the absence, or even in the presence, of the two field officers, I was entrusted by my friend and my father with the effective labour of dictating the orders, and exercising the battalion.

From Winchester, the first place of assembly (June 4, 1760), we were removed, at our own request, for the benefit of a foreign education. By the arbitrary, and often capricious orders of the War Office, the battalion successively marched to the pleasant and hospitable Blandford (June 17); to Hilsea barracks, a seat of disease and discord (September 1); to Cranbrook in the Weald of Kent (December 11); to the seacoast of Dover (December 27); to Winchester camp (June 25, 1761); to the populous and disorderly town of Devizes (October 23); to Salisbury (February 28, 1762); to our beloved Blandford a second time (March 9): and finally, to the fashionable resort of Southampton (June 2); where the colours were fixed till our final dissolution (December 23).

On the beach at Dover we had exercised in sight of the Gallic shores. But the most splendid and useful scene of our life was a four months' encampment on Winchester Down. We improved our time and opportunities in morning and evening field-days; and in the general reviews the South Hampshire were rather a credit than a disgrace to the line. The loss of so many busy and idle hours was not compensated by any elegant pleasure; and my temper was insensibly soured by the society of our rustic officers. In every state there exists, however, a balance of good and evil. The habits of a sedentary life were usefully broken by the duties of an active profession: in the healthful exercise of the field I hunted with a battalion, instead of a pack.

But my principal obligation to the militia was the making me an Englishman, and a soldier. After my foreign education, with my reserved temper, I should long have continued a stranger to my native country, had I not been shaken in this various scene of new faces and new friends: had not experience forced me to feel the characters of our leading men, the state of parties, the forms of office, and the operation of our civil and military system. In this peaceful service I imbibed the rudiments of the language, and science of tactics. The discipline and evolutions of a modern battalion gave me a clearer notion of the phalanx and the legion; and the captain of the Hampshire grenadiers (the reader may smile) has not been useless to the historian of the Roman empire.

EDWARD GIBBON

78.

One of Maj.-Gen. James Wolfe's biographers describes the meeting at
10 Downing Street following his appointment in 1759, aged just
thirty-two, to command in Canada. Scholars question some details,
but romantics should allow them to stand, along with Wolfe's
reputation as victor at Quebec.

On the day preceding his embarkation for America, Pitt, desirous of
giving his last verbal instructions, invited him to dinner, Lord
Temple being the only other guest. As the evening advanced, Wolfe,
heated perhaps by his own aspiring thoughts and the unwonted society of statesmen, broke forth into a strain of gasconade and bravado.
He drew his sword, he rapped the table with it, he flourished it
round the room, he talked of the mighty things which that sword
was to achieve. The two ministers sat aghast at an exhibition so
unusual from any man of real sense and real spirit. And when at last
Wolfe had taken his leave, and his carriage was heard to roll from the
door, Pitt seemed for the moment shaken in the high opinion which
his deliberate judgment had formed of Wolfe; he lifted up his eyes
and arms, and exclaimed to Lord Temple, 'Good God! that I should
have entrusted the fate of the country and of the Administration to
such hands.'

The night before the battle [for Quebec, 12 September 1759]
Wolfe told John Jervis, an old school-fellow and later Earl St Vincent,
that he did not expect to survive the action. Taking from his neck a
miniature of Miss Lowther, he asked Jervis to deliver it to her in
person. She was the sister of the first Earl of Lonsdale, and Wolfe had
become engaged to her. For full two hours the procession of boats,
borne on the current, steered silently down the St Lawrence. The stars
were visible, but the night was moonless and sufficiently dark. The
General was in one of the foremost boats, and near him was a young
midshipman, John Robison, afterwards professor of natural philosophy in the University of Edinburgh. He used to tell in his later life
how Wolfe, with a low voice, repeated Gray's 'Elegy in a Country
Churchyard' to the officers about him. Probably it was to relieve the

intense strain of his thoughts. Among the rest was the verse which his own fate was soon to illustrate –

The paths of glory lead but to the grave.

'Gentlemen,' he said, as his recital ended, 'I would rather have written those lines than take Quebec.'

As they neared their destination, the tide bore them in towards the shore, and the mighty wall of rock and forest towered in darkness on their left. The dead stillness was suddenly broken by the sharp *Qui vive!* of a French sentry, invisible in the thick gloom. *'France!'* answered a Highland officer of Fraser's regiment from one of the boats of the light infantry. He had served in Holland, and spoke French fluently.

'A quel regiment?'

'De la Reine,' replied the Highlander. He knew that a part of that corps was with Bougainville. The sentry, expecting the convoy of provisions, was satisfied, and did not ask for the password. Soon after, the foremost boats were passing the heights of Samos, when another sentry challenged them, and they could see him through the darkness running down to the edge of the water, within range of a pistol-shot. In answer to his questions, the same officer replied, in French: 'Provision-boats. Don't make a noise; the English will hear us.' In fact, the sloop-of-war *Hunter* was anchored in the stream not far off. This time, again, the sentry let them pass.

In a few moments they rounded the headland above the Anse du Foulon. There was no sentry there. The strong current swept the boats of the light infantry a little below the intended landing-place. They disembarked on a narrow strand at the foot of heights as steep as a hill covered with trees can be. The twenty-four volunteers led the way, climbing with what silence they might, closely followed by a much larger body. When they reached the top they saw in the dim light a cluster of tents at a short distance, and immediately made a dash at them. Vergor leaped from bed and tried to run off, but was shot in the heel and captured. His men, taken by surprise, made little resistance. One or two were caught, and the rest fled.

The main body of troops waited in their boats by the edge of the strand. The heights nearby were cleft by a great ravine choked with forest trees; and in its depths ran a little brook called Ruisseau St-Denis, which, swollen by the late rains, fell plashing in the stillness over a rock. Other than this no sound could reach the strained ear of Wolfe but the gurgle of the tide and the cautious climbing of his advance-parties as they mounted the steeps at some little distance from where he sat listening. At length from the top came a sound of musket-shots, followed by loud huzzas, and he knew that his men were masters of the position. The word was given; the troops leaped from the boats and scaled the heights, some here, some there, clutching at trees and bushes, their muskets slung at their backs. Tradition still points out the place, near the mouth of the ravine, where the foremost reached the top. Wolfe said to an officer near him: 'You can try it, but I don't think you'll get up.' He himself, however, found strength to drag himself up with the rest.

The narrow slanting path on the face of the heights had been made impassable by trenches and abattis; but all obstructions were soon cleared away, and then the ascent was easy. In the grey of the morning the long file of red-coated soldiers moved quickly upward, and formed in order on the plateau above.

The battle on the Heights of Abraham began about 9 a.m. on 13 September 1759 with a charge by the French which was repulsed by volley fire from the British ranks, Wolfe being wounded in the wrist. He was hit twice more, this time mortally, as he advanced with his men against the reeling French regiments, but he ordered an officer to support him as his van swept past, lest they should see that he was wounded.

Captain John Knox wrote later: 'After our late worthy General, of renowned memory, was carried off wounded to the rear of the front line, he desired those who were about him to lay him down; being asked if he would have a Surgeon he replied, "it is needless; it is all over with me." One of them cried out, "They run, see how they run." "Who runs?" demanded our hero with great earnestness, like a person aroused from sleep. The Officer answered, "The Enemy, Sir; Egad, they give way everywhere." Thereupon the General rejoined, "Go one of you, my lads, to Colonel Burton; tell him to march Webb's regiment with all speed down to Charles's river, to cut off the retreat of the fugitives from the bridge." Then, turning on his side, he added, "Now, God be praised, I will die in peace"; and thus expired.'

FRANCIS PARKMAN

79.

The field of Landeshut, 23 June 1760.

Field-Marshal Ernst von Laudon, a German commanding the Austrian army, attacked the Prussian general, Heinrich Fouqué, 'one of those tigerish lieutenant-generals whom the Prussian service bred in such profusion'. Fouqué was badly cut up by the Austrian cavalry whilst making a last stand with his few remaining men. Colonel Voit of the Austrian Lowenstein Dragoons brought up his parade horse and courteously begged Fouqué to mount. The Prussian declined, saying 'the blood would spoil your fine saddlery.' Voit replied 'it will become far more precious, when it is stained with the blood of a hero.' There was one officer who was vulgar enough to taunt the captured commander for his misfortune to his very face. But all the officers who were present condemned his bad manners. Fouqué interrupted, and merely said, 'let him speak, gentlemen! You know how it goes in war; it's my turn today, and tomorrow it's yours.'

CHRISTOPHER DUFFY

80.

The beginnings of George Washington's (1732–99) American army.

Officers of the militia were, up to the end of 1776, appointed by popular vote or ballot. A New Jersey man or a Maryland man looked upon the whole proceedings as being of the nature of a vestry meeting. Wherever he went he carried his voting ticket in his pocket, and whether the post to be filled was that of a mayor, sheriff, magistrate, collector, or captain, the principle for him remained the same. The militiamen enrolled themselves according to a form of contract drawn on democratic lines; as thus: 'We the subscribers do hereby severally enlist ourselves into the service of the United American Colonies until the first day of January next ... and we severally consent to be formed by such person or persons as the general court shall appoint into a company of ninety men, including one captain, two lieutenants, one ensign, four sergeants, four corporals, one drum, and one fife to be elected by the company.

The practice of election extended even to field officers. A Maryland regiment having been called together to poll for a field officer, the colonel fixed a day for the poll and appointed himself returning officer for the declaration of the poll. The men of the regiment disregarded the date fixed by the colonel, refused his services as returning officer, and threatened him with personal violence should he dare to interfere, so fiercely was the democratic principle asserted even in a state of the aristocratic origin which Maryland was proud to claim.

HENRY BELCHER

81.

Washington's surprise attack of 26 December 1776 on Trenton, New Jersey, across the Delaware River, became one of his most important early successes for the revolutionary cause, and the inspiration for a famous heroic painting by Emanuel Gottlieb Leutz.

Back and forth the boats shuttled, several dozen troops in each Durham, and one gun or a few wild-eyed horses on each flatboat, the hours slipping past as the storm howled and the temperature eased to just above freezing. 'It was as severe a night as ever I saw,' one captain wrote. The men, miserable but stoic, 'nearly half dead with cold,' huddled together above the steep-banked New Jersey riverfront and waited. 'It was all the same,' a Massachusetts fifer later recalled, 'owing to the impossibility of being in a worse situation than their present one.' Washington waited too, eating and drinking in the saddle atop his sorrel mount once he reached Jersey, his broad, ruddy face stung with sleet. 'He appeared at that time pensive and solemn in the extreme,' one officer noted. When Sullivan, struggling to impose order on the ranks nearby, sent a messenger to report that his wet muskets would likely be useless, Washington replied, 'Tell General Sullivan to use the bayonet.'

Several hours behind schedule, the last lifts disembarked. At four a.m. a command sounded through the ranks – 'Shoulder your firelocks!' – and the regiments surged forward. As Knox would write Lucy, 'Perseverance accomplished what at first seemed impossible.' The column stretched for more than a mile, first rising east from the river, then turning sharply south onto Bear Tavern Road. As the road angled through dripping copses of hickory and black oak, soldiers draped handkerchiefs, greased rags, coat skirts, and blankets over their musket priming pans, to small effect. Whenever the march stalled momentarily, as night marches inevitably did, men fell asleep on their feet and had to be forcibly roused. Sergeants prodded the sick and lame who lingered by the roadside, but at least two soldiers fell behind and froze to death on the tableland that night.

'Press on. Press on, boys!' Washington urged, trotting along the line and at one point grabbing his horse's mane to avoid being thrown when his mount slipped. 'For God's sake, keep by your officers.' Knox's artillerymen used drag ropes to ease their guns down a steep defile and across Jacob's Creek, then hauled them up again. Officers tendered advice. 'Fire at their legs,' a colonel urged. 'One man wounded in the leg is better [than] a dead one, for it takes two more to carry him off … Leg them, damn 'em, I say leg them.'

RICK ATKINSON

82.

The American army straggled into winter quarters at Valley Forge, Pennsylvania, on 21 December 1777, to begin an experience that passed into national legend as an epic of endurance, and ended only with the coming of spring.

In all, Washington's army numbered 11,000 officers and men, of whom 8,200 were fit for duty. They made a camp in a fine strategic site, but there was much about it that added to their misery – and they were miserable when they arrived. They lacked almost everything an army needs for survival. They had been hungry for several weeks, and their new quarters were in a part of Pennsylvania barren of provisions. They had lived for weeks in the open and required barracks or housing that would give them protection from the winter. Valley Forge had virtually no buildings; the troops would have to put up their own.

The recent campaign had worn out shoes and clothing as well as men. The hills offered no more in the way of clothing than of food. Almost everything else was in short supply as well. A few days after their arrival Washington remarked that there was no soap in the army but, he concluded, there was not much use for it since few men had more than one shirt, and some none at all. And he might have noted that, though Valley Creek and the Schuylkill bordered the camp, water for all uses had to be carried for considerable distances, in some places a mile or more.

The woods afforded the materials for housing, and the soldiers fell to building huts almost immediately. Washington ordered that the camp be carefully laid out. Huts, fourteen by sixteen feet, were to be constructed of logs, roofed with 'split slabs'. Clay sealed the sides and was used to make fireplaces. Nails were not to be had of course, and the logs had to be notched. Each hut housed a squad of twelve men. Washington promised to share his soldiers' hardships until the first huts were completed and lived in a tent before finally moving into one of the few houses nearby.

Comfort did not abound inside the huts' walls. Many had only the ground for floors, and straw for beds was not readily available. Worst of all, the troops frequently had nothing to eat. At the time of their arrival the commissary seems to have contained only twenty-five barrels of flour – nothing else, neither meat nor fish. During the days that followed the soldiers chopped down trees and put up huts with empty stomachs. At night, according to Albigence Waldo, a surgeon of the Connecticut line, there was a general cry that echoed through the hills – 'No meat! No meat!' The troops added to this 'melancholy sound' their versions of the cawing of crows and the hooting of owls.

Imitating bird calls suggests that the troops' sense of humour saw them through the worst of their sufferings. They had their hatreds, too, and these also may have helped sustain them. One was firecake, a thin bread made of flour and water and baked over the campfire. Another was the commissaries who were supposed to provide food for the army. Waldo reconstructed a number of conversations along the following lines: 'What have you for your dinners, boys?' 'Nothing but firecake and water, Sir.' At night: 'Gentlemen, the supper is ready.

What is your supper, lads?' 'Firecake and water, Sir.' In the morning: 'What have you got for breakfast, lads?' 'Firecake and water, Sir.' And from Waldo, the snarl: 'The Lord send that our Commissary of Purchases may live [on] firecake and water till their glutted guts are turned to pasteboard.'

ROBERT MIDDLEKAUFF

83.

A matter of honour.

In 1777 British army officer Captain John Pennington was posted to a unit serving against the American revolutionaries. He begged a berth to New York on a frigate commanded by a friend, Captain Hon. John Tollemache RN. During the Atlantic passage Pennington made remarks about Tollemache's wife, deemed so disobliging that the sailor demanded mortal satisfaction. When the warship reached New York, an immediate meeting took place at which pistol shots were exchanged without effect. Swords were then drawn. Captain Pennington received seven wounds so severe that his life was despaired of, but he finally ran Tollemache through. The naval officer was buried in Trinity Church, Broadway and Wall Street, while Pennington recovered to resume his army career. Tollemache was survived by an only son, Lionel, who on reaching adolescence joined the army, with which in July 1793 he was serving under the Duke of York. On the final day of the siege of Valenciennes, a French grenade struck the eighteen-year-old, who became the action's only British officer fatality. A colonel, seeing a body being carried to the rear, asked its identity. On being told 'Ensign Tollemache', the shocked John Pennington – for it was he – said: 'Good God, sixteen years ago I killed his father in a duel.'

THE TOLLEMACHE FAMILY ARCHIVE

84.

One of the Franco-American force besieging the British garrison of Savannah in 1779 was Samuel Warren, formerly an officer in the King's service.

An English aunt sent him word that if the report was true she hoped he would have an arm or a leg shot off in his first battle. She had her wish, as he lost one on 9th October. After the war, he placed the leg bone in an elegant mahogany case to which he affixed a plate bearing the date of its loss. This Warren sent to his aunt with a note to the effect that, while her wish had been fulfilled, he would rather be a rebel with one leg than a royalist with two.

WILLIAM STEVENS

85.

A volunteer in the South Carolina infantry describes an incident at Eutaw Springs on 8 September 1781.

One of his [Colonel William Washington's] dragoons, Billy Lunsford, requested of his captain leave to steal upon and shoot a British sentinel. The captain told him it could not do the cause any good, and, as the sentinel was doing his duty, it was a pity to shoot him. Billy swore his time was out, and, as he was going home to Virginia, he would have it to tell that he had killed 'one damned British son of a bitch'. Accordingly, Billy commenced passing backwards and forwards with a pistol, creeping on his all fours and grunting like a hog. The sentinel was heard to slap his cartouche box and fired, and Billy changed his grunting to groaning, being shot through the body, entering his right and coming out his left side. It was as pretty a shot as could have been made in daylight. The British sentinel, being reinforced, carried Billy a prisoner into their camp, where, by the kind attention of a British surgeon who had him nursed all night, he recovered.

JOHN CHANEY

86.

Washington's army suffered terribly from the inadequacy of medical care.

There were some 3,500 medical practitioners of various sorts in America when the war began. This figure probably includes quacks as well as reputable physicians and a great number of indifferently trained men who treated the sick and worked at other occupations as well. Probably fewer than four hundred had a medical degree. Although generalizations about such a motley group cannot be reliable, it is unlikely that any theory of disease or therapy found wide acceptance among them. The physicians among them probably believed that sickness generally represented some variation from the normal pattern of the human system, an old idea which persisted through the eighteenth century. There were diseases identified as diseases – smallpox, syphilis, and tuberculosis, for example – but both theory and practice usually dealt with body conditions, such symptoms as fevers, fluxes, and dropsies. The assumption behind this practice was that a fever indicated that the state of the system was off, not that the body was afflicted by a disease. To be sure, some physicians had come to recognize that diseases were objectively real. While treating their patients, they had observed that a medicine might be effective against one set of symptoms but not another. From this experience they inferred that they faced two different diseases.

These physicians easily reconciled this inference with the ancient assumption that there was one basic cause of all disease. The most common theory held that the body's humours were somehow awry, perhaps impure or out of balance, with one or more present in excessive or insufficient amounts. The treatment followed from the diagnosis, with bleeding, purging, and sweating all calculated to reduce excessive amounts, and diets and drugs intended to build up volume. Another basic cause of sickness, it was widely thought, might be a chemical imbalance, with body fluids showing an improper blend of acidity or alkalinity. The treatments in such cases often resembled those prescribed to restore humoral balance.

The ordinary soldier, of course, lived largely oblivious to theory, though he, his officers, and the regimental surgeons may have shared a good deal of common lore about health and medicine. Judging from the orders that came down from on high in every American camp, one belief they did not share was that cleanliness was next to godliness. Away from home, the American soldier did not mind the filth that piled up in crowded camps – or if he minded, refused nevertheless to follow rudimentary practices which would have kept them cleaner.

Soldiers throughout the war apparently disdained use of the vaults, as latrine pits were called, preferring to void whenever taken by the urge. They also scattered food scraps, carrion and garbage throughout camps. They had to be forced to change the straw that served as bedding. And some had to be ordered to bathe. The British, professionals in this sort of thing as in all things pertaining to military life, kept clean camps and probably suffered less from disease.

ROBERT MIDDLEKAUFF

87.

At the port of Leith late one April evening in 1779, Highland soldiers of the 42nd and 71st regiments, abruptly drafted to serve with a detested Lowland regiment, refused the order and fought a brief, bloody little battle with men of the South Fencibles sent to compel them. Fourteen of the combatants died and many more were wounded. A court martial sentenced three of the captured Highlanders to death.

At Holyrood, early on Friday, May 28, they were brought up to the chamber of the Great Hall and out into the sunlight of Palace Yard, their heads bare and their hands tied behind them. Preceded by muffled drums and escorted by a platoon of the Argyll Fencibles, now the garrison regiment and soon to be in revolt itself, they were slow-marched from the citadel under the narrow arch of Foog's Gate, northward past the foot of the Lang Stairs to the Portcullis and Inner Barrier until they crossed the drawbridge to the mound of Castle Hill. Here five companies of the Argyll Fencibles were drawn up in hollow square facing the fortress, and three ranks of red dragoons held back the press of the crowd from the Royal Mile. The sun was strong, but a brisk wind moved across the high ground, pulling at the belted plaids of the Argylls and of the other Highland prisoners who had been brought out to watch the death of their friends.

By the wall of the dry ditch and below the Half Moon Battery were three open coffins to which the condemned men were led. Each man knelt beside the plain box allotted to him, MacGregor with them, but as the minister began to pray he was stopped by Major Hugh Montgomerie of the Argylls. Before the condemned men could submit their souls to God they must listen to the General Orders issued two days before.

Although these orders were brief the reading took some minutes, for the Major frequently paused, awaiting MacGregor's whispered translation [from English into Gaelic]. The court having found the accused guilty of mutiny and a breach of the second, third, fourth and fifth Articles of War, and having duly considered the evil tendency to mutiny and sedition, especially when carried to such enormous lengths in the present case, did adjudge Charles Williamson, Archibald MacIver and Robert Budge to be shot to death. Which sentence had been transmitted to the King, and His Majesty had been pleased to signify his pleasure. Having regard to the former commendable behaviour of the 42nd Regiment, and that Robert Budge, only now recovering from the wounds he had received in the affray, did not appear to have any forward part in the mutiny, His Majesty was most graciously pleased to grant the condemned a free pardon 'in full consideration that they will endeavour upon every future occasion, by a prompt obedience and orderly demeanour, to atone for this unpremeditated but atrocious offence'.

Quickly stripped of their muffling crepe, the Argyll drums now beat a spirited march, the companies turning and wheeling toward the Royal Mile in a river of green tartan. The bewildered prisoners rose from their knees. Their hands were untied and they stumbled from the coffins to the waiting, weeping embrace of their comrades.

JOHN PREBBLE

88.

After three days and nights of continuous gunfire in the streets of Gibraltar, which was besieged by the Spanish from September 1779 to February 1783, on 15 April 1781 Sergeant Samuel Ancell (1760– 1802) encountered a comrade of his own 58th Regiment.

I met Jack Careless in the street, singing with uncommon glee (notwithstanding that the enemy were firing with prodigious warmth) part of the old song:

A soldier's life's a merry life,
From care and trouble free.

He ran to his comrade with eagerness and, presenting his bottle, cried: 'Damn me if I don't like fighting! I'd like to be ever tanning the Dons. Plenty of good liquor for carrying away – never was the price so cheap – fine stuff – enough to make a miser quit his gold.'

'Why, Jack,' said he, 'what have you been about?'

With an arch grin he replied: 'That would puzzle a heathen philosopher, or yearly almanack-maker to unriddle – I scarce know myself, I have been constantly on foot and watch, half-starved and without money, facing a parcel of pitiful Spaniards. I have been fighting, wheeling, marching and counter-marching – sometimes with a firelock, then with a handspike, and now my bottle,' brandishing it in the air, 'I am so pleased with the melody of great guns that I consider myself a Roman general, gloriously fighting for my country's honour and liberty.'

A shell that instant burst, a piece of which knocked the bottle out of his hand. With the greatest composure he replied (having first graced it with an oath): 'This is not any loss. I have found a whole cask by good luck.' And he brought his comrade to view his treasure.

'But Jack,' says he, 'are you not thankful to God for your preservation?'

'How do you mean?' answered Jack, 'Fine talking of God with a soldier, whose trade and occupation is cutting throats. Divinity and slaughter sound very well together. They jingle like a cracked bell in the hand of a noisy crier. Our King is answerable to God for us! I fight for him. My religion consists in a firelock, open touch-hole, good flint, well-rammed charge and seventy rounds of powder and ball. This is the military creed. Come, comrade, drink success to the British arms!' On his asking him for a glass, he seemed surprised.

'Why,' says he, 'you may well know there is not one to be had. But there is something that will do as well,' and he took up a piece of shell. 'Here is a cup fit for a monarch. This was not purchased with gold or friendship, but with the streams of our countrymen's blood.'

Having filled the piece of shell, he gave it to his comrade to drink. 'Come, Jack,' said he. 'Here is to King George and victory!'

'And he that would not drink the same,' replied Jack, 'I'd give him an ounce of lead to pay Charon to ferry him over the river Styx!'

SAMUEL ANCELL

89.

The later traveller and pamphleteer William Cobbett (1763–1835), a footloose Surrey farmer's son, on a youthful whim 'went for a soldier'.

I happened to cast my eye on an advertisement, inviting all loyal young men, who had a mind to gain riches and glory, to repair to a certain rendezvous, where they might enter into His Majesty's Marine Service, and have the peculiar happiness and honour of being enrolled in the Chatham division. I was not ignorant enough to be the dupe of this morsel of military bombast; but a change was what I wanted; besides, I knew that marines went to sea, and my desire to be on that element had rather increased than diminished by my being penned up in London. In short, I resolved to join this glorious corps; and, to avoid all possibility of being discovered by my friends, I went down to Chatham, and enlisted into the marines as I thought, but the next morning I found myself before a Captain of a marching regiment. There was no retreating; I had taken a shilling to drink His Majesty's health, and his further bounty was ready for my reception.

When I told the Captain that I thought myself engaged in the marines, 'By Jasus, my lad,' said he, 'and you have had a narrow escape.' He told me, that the regiment into which I had been so happy as to enlist was one of the oldest and boldest in the whole army, and that it was at that time serving in that fine, flourishing and plentiful country, Nova Scotia. He dwelt long on the beauties and riches of this terrestrial paradise, and dismissed me, perfectly enchanted with the prospect of a voyage thither.

I enlisted in 1784, and, as peace had then taken place, no great haste was made to send recruits off to their regiments. I remember well what sixpence a day was, recollecting the pangs of hunger felt by me, during the thirteen months that I was a private soldier at Chatham. Of my sixpence, nothing like fivepence was left to purchase food for the day.

Indeed, not fourpence. For there was washing, mending, soap, flour for hair-powder, shoes, stockings, shirts, stocks and gaiters, pipe-clay and several other things to come out of the miserable sixpence! Judge then of the quantity of food to sustain life in a lad of sixteen, and to enable him to exercise with a musket (weighing fourteen pounds) six to eight hours every day. The best battalion I ever saw in my life was composed of men, the far greater part of whom were enlisted before they were sixteen, and who, when they were first brought up to the regiment, were clothed in coats made much too long and too large, in order to leave room for growing.

We had several recruits from Norfolk (our regiment was the West Norfolk); and many of them deserted from sheer hunger. They were lads from the plough-tail, all of them tall, for no short men were then taken. I remember two that went into a decline and died during the year, though when they joined us, they were fine hearty young men. I have seen them lay in their berths, many and many a time, actually crying on account of hunger. The whole week's food was not a bit too much for one day.

My leisure time was spent, not in the dissipations common to such a way of life, but in reading and study. In the course of this year I learnt more than I had ever done before. I subscribed to a circulating library at Brompton, the greatest part of the books in which I read more than once over. The library was not very considerable, it is true, nor in my reading was I directed by any degree of taste or choice. Novels, plays, history, poetry, all were read, and nearly with equal avidity. Such a course of reading could be attended with but little profit: it was skimming over the surface of everything.

One branch of learning, however, I went to the bottom with, and that the most essential too: the grammar of my mother tongue. I had experienced the want of knowledge of grammar; but it is very probable that I never should have thought of encountering the study of it, had not accident placed me under a man whose friendship extended beyond his interest. Writing a fair hand procured me the honour of being copyist to Colonel Debieg, the commandant of the garrison. The Colonel saw my deficiency, and strongly recommended study. He enforced his advice with a sort of injunction, and with a promise of reward in case of success. I procured me a Lowth's grammar, and applied myself to the study of it with unceasing assiduity.

The edge of my berth, or that of the guard-bed, was my seat to study in; my knapsack was my bookcase; a bit of board lying on my lap was my writing desk; and the task did not demand anything like a year of my life. I had no money to purchase candle or oil; in winter time it was rarely that I could get any evening light but that of the fire, and only my turn even of that. To buy a pen or a sheet of paper I was compelled to forgo some portion of food, though in a state of half-starvation; I had no moment of time that I could call my own; and I had to read and to write amidst the talking, laughing, singing, whis-tling and brawling of at least half a score of the most thoughtless of men, and that, too, in the hours of their freedom from all control.

Think not lightly of the farthing that I had to give, now and then, for ink, pen, or paper. That farthing was, alas! a great sum to me! I was as tall as I am now, I had great health and great exercise. I remember, and well I may! that, upon one occasion, I, after all absolutely neces-sary expenses, had, on a Friday, made shift to have a halfpenny in reserve, which I had destined for the purchase of a red herring in the morning; but, when I pulled off my clothes at night, so hungry then as to be hardly able to endure life, I found that I had lost my half-penny! I buried my head under the miserable sheet and rag, and cried like a child.

Though it was a considerable time before I fully comprehended all that I read, still I studied with such unremitted attention, that, at last, I could write without falling into any very gross errors. The pains I took cannot be described: I wrote the whole grammar out two or three times; I got it by heart. I repeated it every morning and every evening, and, when on guard, I imposed on myself the task of saying it all over once every time I was posted sentinel. To this exercise of my memory I ascribe the retentiveness of which I have since found it capable, and to the success with which it was attended, I ascribe the perseverance that has led to the acquirement of the little learning of which I am master.

I was soon raised to the rank of Corporal, a rank, which, however contemptible it may appear in some people's eyes, brought me in a clear twopence per diem, and put a very clever worsted knot upon my shoulder, too. As promotion began to dawn, I grew impatient to get to my regiment, where I expected soon to bask under the rays of royal favour. The happy days of departure at last came: we set sail from Gravesend, and, after a short and pleasant passage, arrived at Halifax in Nova Scotia. [He later married an American wife, and was discharged as a sergeant-major in 1791.]

WILLIAM COBBETT

90.

The poet Samuel Taylor Coleridge (1772–1834) enlisted in 1794 for the same reason as Cobbett – poverty – but found soldiering far less rewarding.

Towards morning he came upon a recruiting poster for the 15th (Elliott's) Light Dragoons. It seemed suddenly that the pacifist and the pantisocrat must face his antithesis; the country lad who wore so uneasily the brilliant trappings of a man of the world must immerse his misery in the thing he feared most, to experience the physical discomfort, to face the fact of being a soldier. He went at once to the recruiting office. An old and benevolent sergeant listened to his agitated request to be recruited into the army, and asked him if he had been in bed. The answer was apparent before the pale young man with the black dishevelled hair had time to shake his head. The sergeant gave him breakfast and persuaded him to rest for a few hours. In the evening he offered a guinea, and told him to go to the play and cheer himself up. He did so, but came back afterwards. To withdraw now would seem to be weakness. The sergeant shook his head sadly, almost burst into tears. 'Then it must be so,' he said. Next morning he mustered his recruits. The General made an inspection. He stopped before the drooping and somewhat ungainly figure. 'Your name, sir?'

'Cumberbatch, sir,' he extemporized quickly.

'What do you come here for, sir?'

'To be a soldier, sir.'

'Do you think you can run a Frenchman through the body?'

'I don't know, sir, as I never tried. But I'll let a Frenchman run me through the body before I'll run away.'

'Good enough,' said the General, and passed on.

So he became a cavalryman. 'Never make a proper soldier out of he,' said the drill sergeant. His horse was liable to bite. He couldn't even rub her down himself. With charm and volubility he bribed a lad to do it for him, paying him by writing letters and love stanzas. As soon as he mounted his horse on one side, he was off on the other. 'Silas is off again!' they cried. They thought him a 'talking natural'. He won them over with his stories. He sat at the foot of his bed with a group of dragoons round him. He told them of the Peloponnesian war, which lasted 27 years.

'There must have been some fine fine promotion there,' said one soldier. 'Aye,' said another. 'How many rose from the ranks, I wonder.' He related the feats of Archimedes, but they couldn't swallow them. 'Silas, that's a lie,' they shouted. 'Do you think so?' he said mildly, and went on. 'That Silas – his fancy's always on the stretch.' He switched to Alexander the Great and recaptured their interest. 'Ah, he were a great general. Who were his father? Was he a Cornishman? I did know an Alexander in Truro.' They protested when he told them of the retreat of the ten thousand. 'Don't like to hear about retreat. I'm for marching on.'

'What rations did they have on that campaign?'

'Sure every time the sun rose they had two pounds of good ox beef and plenty of whisky,' said an Irishman, and that clinched the matter.

That was probably later, after he had proved hopeless as a trooper, and had been removed to the medical department, where his stories were better than physic. He had told the story of his army experiences so many times, such a maze of fantasy and legend had grown from it, that even he could not remember the true story of his release. He had told Cottle he was acting as sentry at the door of the officers' ballroom. Two officers passed, talking of the Greek drama, quoting what they thought was Euripides. 'Excuse me, sir,' he interrupted. 'The lines are not quite accurately quoted. Besides, they're not from Euripides; they're from the second antistrophe of the Oedipus of Sophocles.'

'Who the devil are you?' said the officers in astonishment, 'Old Faustus grown young again?'

'I am your honour's humble sentinel.'

'Damme, sir,' said the second officer. 'The fellow must be a gentleman.'

'An odd fish.'

'Not an odd fish, but a stray bird from the Oxford or Cambridge aviary.'

Or was the truth that Captain Ogle found pencilled on the door of his stable: '*Eheu! quam infortunii miserrimum est fuisse felicem*'? ['Alas, how much the worst part of misfortune is once to have known happiness'.]

A confidential letter to one of his Cambridge friends must have been passed on to his brother George. The family had got to work to procure his release. Investigations were made, and about this time George Cornish, a school friend, arrived and lent him a guinea. Colonel James, the soldier of the family, proved his interest at the War Office. On April 10th 1794 S. T. Cumberbatch was discharged as insane.

MAURICE CARPENTER

91.

The Russian army in the last years of the eighteenth century found the Tsar Paul (1754–1801) a difficult master; mad, dangerous and subject to extraordinary military obsessions.

He had a passion for parades and for useless ceremonials. Soldiers must be made beautiful, he declared, regardless of military efficiency. New uniforms were designed which were so tight that the wearers were scarcely able to breathe, let alone fight. Soldiers staggered under the weight of wigs which had thick, stinking grease plastered over them, and which had iron rods inserted into the queues to make them fall straight. Men were obliged to wear a type of straitjacket in order to train them to stand erect, and steel plates were strapped round their knees to prevent legs bending when marching on parades. Hours had to be spent polishing weapons, buttons and buckles, pipe-claying belts and powdering the greasy wigs. Musket-butts were hollowed out and filled with loose shot to make them rattle nicely as the men went through the various exercises. Discipline was intensified so that the troops would behave in puppet fashion.

'The Guard-parade became for him the most important institution and focal point of government,' wrote Frederic Masson, tutor to [Paul's son] Alexander. 'Every day, no matter how cold it might be, he dedicated the same time to it, spending each morning in plain deep green uniform, great boots and a large hat exercising his Guards. Surrounded by his sons and aides-de-camp he would stamp his heels on the stones to keep himself warm, his bald head bare, his nose in the air, one hand behind his back, the other raising, and falling, a baton as he beat time, crying out 'One, two – one, two.'

ROGER PARKINSON

92.

One day in 1805 fourteen-year-old schoolboy Joseph Anderson (1790–1877) found himself in serious trouble at Banff Academy.

He had been shooting sparrows with a gun which he had no right to possess, and when taxed with the offence he denied it. His form master, a Mr Simpson, exclaimed: 'You have told a lie, sir, and I must punish you; so down with your breeches.'

Indignantly the culprit refused, saying: 'I am an officer and won't submit.'

Mr Simpson was no doubt aware that his pupil had recently been commissioned an ensign in the 78th Foot – all the more reason, he may have thought, why prevarication should not be overlooked. But young Anderson kicked, thumped and created such a din that the rector of the Academy came hurrying up. After hearing explanations from both sides, he said to the pupil: 'I will not disgrace you, sir. You are an officer.' Joseph Anderson went on to soldier with honour in the Peninsula and to rule a convict island off the coast of Australia, ending his service as a lieutenant-colonel, a Commander of the Bath and a Knight of Hanover.

E. S. TURNER

93.

For those who imagine that Sir Arthur Conan Doyle's enchanting, vainglorious, absurd fictional creation Brigadier Etienne Gerard caricatured France's cavalry officers, the memoirs of such imperial grandees as Thiébault, Marbot and Caulincourt suggest that reality outdid Gerard. This is one of Thiébault's stories of Bonaparte's Army of Italy in 1796.

A review took place at the gates of Verona. Complete full-dress had been ordered, and the care taken to execute the order caused all the more surprise at the appearance of La Salle, who, usually the most brilliant as he was the handsomest officer in the army, turned up in an old pelisse, pantaloons, and dirty boots, and riding an Austrian hussar's horse, on which he had been careful to leave its saddle, bridle, and even rope-halter. The surprise caused by this get-up was universal, and the commander-in-chief's first question was: 'What horse have you got there?' The answer was ready: 'A horse I have just taken from the enemy!' 'Where?' 'At Vicenza, general.' 'Are you mad?' 'I have just come thence; indeed I bring news from thence, which you will, perhaps, deem not unimportant.' Bonaparte at once took him aside, talked with him for a quarter of an hour, and came back to the group formed by Generals Berthier, Masséna and Augereau, and by the staff-officers present, announcing that he had just promoted La Salle major. Here is the rest of the story.

La Salle, who was a man of many accomplishments and a highly susceptible temperament, found, amid all his enthusiasm for military duties, some time at his disposal for love-affairs. He was carrying on one of these with a Marchesa di Sale, one of the cleverest and most charming women of Upper Italy, who afterwards poisoned herself in despair at the loss of him. She lived at Vicenza, and the withdrawal of our army across the Adige had interrupted the liaison. The lovers had found means to correspond across the Austrian army, but correspondence was not enough for La Salle, and he resolved on one of those enterprises which success alone will justify. Selecting twenty-five men from the 1st Regiment of Cavalry – one of the best that we then had

– he assembled them after nightfall and set out at once, without orders, without letting anyone know, without even a show of authority.

He passed the enemy's vedettes unperceived, escaped his pickets, got through the hills to the rear of the Austrian army, and, marching without cockades and with cloaks unfolded, by mountain roads which he knew, reached Vicenza, where he knew there was no garrison, toward midnight, concealed his little troop, and hastened to the Marchesa. About half-past two in the morning, as he was preparing to be off some pistol-shots were heard. He mounted at once and rejoined his escort, learning then that he had been discovered and surrounded.

The most direct roads were strongly guarded, but he recollected one point which was likely still to be open, and hastened thither. Thirty-six hussars were occupying it; he charged them without knowing their numbers, overturned them, captured and brought away nine horses; then he returned by a different road which involved a long way round, avoided cantonments, spoke German, and passed himself off for an Austrian to the men of a picket through which he had to pass. Lastly, marching as fast as possible, he fell upon the rear of the last Austrian advanced post, sabred all that he could get at, and returned by daylight to San Martino d'Albaro, whence he had started, without having lost a single man.

But the fleeting moments which La Salle had passed at Vicenza were not devoted solely to making love. The Marchesa, prepared for the interview, had procured some valuable information, which she had passed on to him. Moreover, he had chosen for his prank the night preceding the commander-in-chief's review. On his return he had avoided showing himself, so as not to have to report to anyone, and then had waited for the moment when, by appearing before Bonaparte in the get-up and on the horse which I have mentioned, he might make the most he could of an attempt which would have either to be punished or rewarded.

GENERAL PAUL THIÉBAULT

94.

1 December 1805, the eve of Austerlitz.

Towards evening an order of the day announced the forthcoming battle to the army. One phrase especially roused the troops; that, namely, in which the Emperor proclaimed that, if they justified his expectations, he should confine himself to directing the movements; but, in the contrary event, he should expose himself where danger was greatest. No sooner had the order been read to all the corps than the Emperor passed incognito and without escort along the front of several regiments. He was at once recognized, and was the object of the greatest enthusiasm. Just as he was in front of the 28th a soldier cried, 'We promise that you will only have to fight with your eyes tomorrow!' Halting in front of Ferny's brigade, composed of the 46th and 57th, he asked the men if their supply of cartridges were complete. 'No,' answered one; 'but the Russians taught us in the Grisons that only bayonets were needed for them. We will show you tomorrow!'

Between one army excited to this point and another composed of fanatics, the battle of the next day was bound to be decisive and merciless. But the next day was December 2, the anniversary of the coronation. This coincidence, joined with the conviction of success, put the troops into such a state of enthusiasm that they wished to give some sign of general rejoicing as an announcement to the Emperor of the celebration they were preparing for him; in which they also wanted to let the enemy share, as he had just allowed us to witness those last manoeuvres executed under our very noses, as if to terrify us. Hardly had night fallen when, by a spontaneous impulse hardly credible, nearly 80,000 men, distributed among a dozen bivouacs, suddenly armed themselves with long poles bearing bundles of lighted straw. They kept renewing them for half an hour, carrying them about and waving them, as they danced a farandole and shouted 'Long live the Emperor!' Out of the 80,000, 25,000 to 30,000 were to transform this festival field into a field of slaughter; but it is the way of our soldiers to mingle the gayest with the most terrible images.

GENERAL PAUL THIÉBAULT

95.

Engineer officer Major John Blakiston (1788–1867), a baronet's younger son and author of pleasing memoirs of many hazards and adventures, enjoyed his service in India in the first decade of the nineteenth century.

One thing, however, struck me as disagreeable, that was the parade and nonsense kept up in the army, and which, without adding to discipline, only served to create disgust to the service. However hot the weather might be, an officer could scarcely stir out of his tent without being buckled up in sword and sash, for fear of meeting some jack-in-office of a staff-officer, who, if he found him straying out of his lines not altogether *en militaire*, would send him back to his tent with a flea in his ear. In truth, the airs which these favourites of fortune gave themselves, towards those who continued to trudge on in the beaten path of their profession, were insufferable.

I recollect about this time a brother officer of mine asking one of these upstarts, with whom he had formerly been on most intimate terms, whether we should halt the next day? 'I really do not know the intentions of the general,' was the reply. Returning to his tent somewhat disgusted with the airs of his former companion, and soliloquizing on the nature of man, and the fantastic tricks which he plays, when 'dressed in a little brief authority', he was met by his maty-boy with the information that the army was to halt the next day. 'Where did you learn that?' said my friend. 'Major M—'s washerman tell.'

This circumstance reminds me also of Captain Grose of the Madras army. He was remarkable for his wit and humour, and his memory is still cherished by all the lovers of fun who knew him. Having occasion to make some communication to headquarters, he was received much in the usual manner by one of the understrappers, who told him that no verbal communication could be received, but that what he had to say must be sent through the medium of an official letter. He happened, some days afterwards, to have a party dining with him, and among others were a few members of the staff. In the midst of dinner a jackass came running among the tent-ropes, exerting his vocal organs in a manner by no means pleasing to the company. Grose immediately rose, and thus addressed the intruder: 'I presume, sir, you come from headquarters. I receive no verbal communications whatever, sir. If you have anything to say to me, sir, I beg you will commit it to paper.' The will which Grose made the night before the storming of Seringapatam (May 1799), under a presentiment of his fate, was quite in character. Among other bequests, it contained the present of a wooden sword to an officer of rank to whom he bore no good will, and who was supposed not to be endowed with any superfluous quantity of personal valour.

JOHN BLAKISTON

96.

Marshal Francois Joseph Lefèbvre (1755–1820) was made Duke of Danzig for his success in seizing that city for Napoleon.

On March 11th, 1807, the town was invested. The first shot was fired by the French batteries on April 24th – Lefèbvre was no hustler – and by the end of May the town had fallen, and not so long afterwards Madame la Maréchale-Duchesse de Danzig, the buxom, kind-hearted, jolly old washerwoman, was visiting the Tuileries to thank the Empress on behalf of herself and her husband the Duke. They were simple souls, the new Duke and Duchess. She used to begin half her sentences with the words 'When I used to do the washing', and he was naively proud of his new grandeur. But at the same time he was very conscious of the years of hard work that had raised him so high. On an occasion when an old friend of his youth was admiring enviously the splendours of his house in Paris: 'So you're jealous of me,' exclaimed the veteran, 'very well; come out into the courtyard and I'll have twenty shots at you at thirty paces. If I don't hit you, the whole house and everything in it is yours.' The friend hastily declined to take the chance, where-upon Lefèbvre remarked drily, 'I had a thousand bullets fired at me from much closer range before I got all this.'

A. G. MACDONELL

97.

Marshal Louis-Alexandre Berthier (1753–1815), the great manager
of the marches of Napoleon's armies, tried his hand at organizing a
rabbit shoot in woods outside Paris, to divert his emperor.

Every detail for the day's sport was worked out with the same meticulous accuracy with which the Grande Armée had been swept from Boulogne to Austerlitz. The carriages arrived on the stroke at the Tuileries, the beaters were ready, the keepers in their best clothes, a beautiful lunch waiting to be eaten, and a thousand rabbits, brought the night before and dumped in the park, waiting to be shot. But poor, ugly little Berthier made one trivial mistake. Instead of buying wild rabbits, he bought tame ones and did not know that they were accustomed to be fed twice a day.

When the Emperor took his gun in hand and advanced into the park, the rabbits, all thousand of them, mistook him for the man who provided their daily lettuce, and leapt to their feet and charged towards him. Berthier and his staff beat them off with horse-whips, but the rabbits, who were more expert in the Napoleonic warfare than some of the Marshals, wheeled round on both flanks and reached the Emperor's carriage before the Emperor could mount and drive off back to Paris.

[At another such shooting party] Napoleon, who was a better hand with a field-gun than he was with a fowling-piece, accidentally shot Masséna in the eye. With characteristic readiness, the Emperor put the blame of the accident on Berthier, who with characteristic subservience accepted the blame, while Masséna, who lost his eye, with characteristic tact accepted the transference of blame.

A. G. MACDONELL

98.

Aged twenty-three, Nadezhda Durova (1783–1866) disguised herself
as a man, sneaked out of her parents' home, and joined the Russian
cavalry. She never again referred to the husband and infant son whom

she abandoned, so nothing is known of their fate, but she became a
minor contemporary celebrity. She was presented to the Tsar, who
granted her request to be allowed to remain in the army in the guise of
a man. She served until 1813, being wounded at Borodino. This is
her own account of her first action.

May 22, 1807. Guttstadt
For the first time I have seen a battle and been in it. What a lot of
absurd things they told me about the first battle, about the fear, timidity, and the last, desperate courage! What rubbish! Our regiment went
on the attack several times, not all at once but taking turns by squadron. I was berated for joining the attack with each new squadron.
However, this was honestly not from any excess of bravery, but simply
from ignorance; I thought that was how it was done, and was amazed
when the sergeant-major of another squadron, alongside which I was
racing like a whirlwind, shouted at me, 'Get the hell out of here! What
are you galloping here for?'

I returned to my squadron, but instead of taking my place in formation, I went on riding around nearby. The novelty of the scene absorbed
all my attention: the menacing and majestic boom of cannon fire, the
roar or kind of rumble of the flying balls, the mounted troops galloping by, the glittering bayonets of the infantry, the roll of drums, and
the firm pace and calm look with which our infantry regiments
advanced on the enemy – all this filled my soul with sensations that I
have no words to express.

I came close to losing my priceless Alcides. While I was riding
around, as I said before, near my squadron and looking over the curious scene of battle, I caught sight of several enemy dragoons
surrounding a Russian officer and knocking him off his horse with a
pistol shot. He fell, and they prepared to hack at him as he lay. Instantly
I rushed toward them with my lance tilted. I can only suppose that this
scatterbrained audacity frightened them, because in a flash they abandoned the officer and scattered. I galloped over to the wounded man
and stopped above him; for a couple of minutes I watched him in
silence. He lay with his eyes closed and gave no sign of life; he obviously thought that it was the enemy who stood over him. At last he

risked a glance, and I at once asked him if he wanted to get on my horse.

'Oh, be so kind, my friend!' he said in a barely audible voice. At once I dismounted from my horse and with great effort managed to raise the wounded man, but here the aid I could render him came to an end: he fell chest down across my arm and I, barely able to keep my feet, had no idea what to do and how to get him onto Alcides, whom I was also holding by the reins with my other hand. This situation would have ended very disadvantageously for us both – that is, for the officer and for me – except that fortunately a soldier from his regiment rode over to us and helped me to seat the wounded man on my horse. I told the soldier to send the horse to Recruit Durov in the Polish Horse regiment, and the dragoon told me that the officer I had saved was Lieutenant Panin of the Finnish Dragoons and that they would return my horse immediately.

The officer was carried off to his regiment, and I set out for mine. I felt at a complete disadvantage, left on foot among charges, gunfire, and swordfights. Seeing everywhere men either flying by like lightning or quietly galloping in various directions with complete confidence in their good steeds, I exclaimed, 'Alas, my Alcides! Where is he now?' I deeply repented having so rashly given up my horse – and even more when my captain, after first asking me with concern, 'Did they kill your horse, Durov? Are you wounded?' shouted at me in vexation, 'Get away from the front, you scamp!' when he heard how I happened to be wandering about on foot. Quickly, albeit sadly, I headed for the spot where I saw lances with the pennons of the Polish Horse. The men I passed said with compassion, 'Oh, my God! Look, what a young boy has been wounded!' Nobody who saw an Uhlan on foot in a uniform covered with blood could think anything else. As I mentioned, the wounded officer had lain chest down across my arm, and I have to assume that his was a chest wound, because my sleeve was all bloody.

To my inexpressible joy, Alcides [was] returned to me – not quite the way I hoped, but at least returned. I was walking pensively through the fields to my regiment, when suddenly I saw our Lieutenant Podwyszacki riding away from the enemy position on my horse. I was beside myself with joy and, without stopping to wonder how my horse had turned up under Podwyszacki, ran over to stroke and caress Alcides, who also expressed his joy by frisking and neighing loudly. 'Is this really your horse?' asked the astonished Podwyszacki. I recounted my adventure to him. He too had no praise for my rashness. He said that he had bought my horse from Cossacks for two gold pieces. I begged him to return Alcides and take from me the money he had paid for him. 'Very well, but let me keep him today. My horse was killed, and I have nothing to ride in action!' And with this he spurred Alcides and galloped off on him. I was close to weeping as I saw my comrade-in-arms in strange hands, and I swore with all my heart never again to give up my horse as long as I live! At last this agonizing day came to an end. Podwyszacki returned Alcides to me, and our army is now pursuing the retreating enemy.

NADEZHDA DUROVA

99.

At Corunna on 16 January 1809, the enlightened and much-beloved Gen. Sir John Moore (1761–1809), son of a Glasgow doctor, was fighting a rearguard action against a superior French force under Marshal Soult, to cover the evacuation of 14,000 British troops under his command. His death inspired Charles Wolfe's famous 1816 poem, which began 'Not a drum was heard, not a funeral note/As his corse to the rampart we hurried'.

The battle raged fiercely: fire flashing amidst the smoke, and shot flying from the adverse guns; when Hardinge rode up and reported that the Guards were coming quickly. As he spoke, Sir John Moore was struck to the ground by a cannon-ball, which lacerated his left shoulder and chest. He had half-raised himself, when Hardinge having dismounted, caught his hand: and the General grasped his strongly, and gazed with anxiety at the Highlanders, who were fighting courageously: and when Hardinge said, 'They are advancing,' his countenance lightened.

Colonel Graham now came up, and imagined, from the composure of the General's features, that he had only fallen accidentally, until he saw blood welling from his wound. Shocked at the sight, he rode off for surgeons. Hardinge tried in vain to stop the effusion of blood with his sash: then, by the help of some Highlanders and Guardsmen, he placed the General upon a blanket. In lifting him, his sword became entangled, and Hardinge endeavoured to unbuckle the belt; when he said with soldierly feelings, 'It is as well as it is; I had rather it should go out of the field with me.' His serenity was so striking, that Hardinge began to hope the wound was not mortal; he expressed his opinion, and said, that he trusted the surgeons would confirm it, and that he would still be spared to them.

Sir John turned his head, and cast his eyes steadily on the wounded part, and then replied, 'No, Hardinge, I feel that to be impossible – You need not go with me; report to General Hope, that I am wounded and carried to the rear.' He was then raised from the ground by a Highland serjeant and three soldiers, and slowly conveyed towards Corunna. The soldiers had not carried Sir John Moore far, when two surgeons came running to his aid. They had been employed in dressing the shattered arm of Sir David Baird; who, hearing of the disaster which had occurred to the commander, generously ordered them to desist, and hasten to give him help. But Moore, who was bleeding fast, said to them, 'You can be of no service to me: go to the wounded soldiers, to whom you may be useful'; and he ordered the bearers to move on. But as they proceeded, he repeatedly made them turn round to view the battle, and to listen to the firing; the sound of which, becoming gradually fainter, indicated that the French were retreating.

Before he reached Corunna it was almost dark, and Colonel Anderson met him; who, seeing his general borne from the field of battle, and steeped in blood, became speechless with anguish. Moore pressed his hand, and said in a low tone: 'Anderson, don't leave me.' As he was carried into the house, his faithful servant François came out, and stood aghast with horror: but his master, to console him, said smiling, 'My friend, this is nothing.' He was then placed on a mattress on the floor, and supported by Anderson, who had saved his life at St Lucia; and some of the gentlemen of his staff came into the room by turns. He asked each if the French were beaten, and was answered affirmatively. They stood around; the pain of his wound became excessive, and deadly paleness overspread his fine features; yet, with unsubdued fortitude, he said, at intervals, 'Anderson, you know that I have always wished to die this way. I hope the people of England will be satisfied! I hope my country will do me justice!' 'Anderson, you will see my friends as soon as you can. Tell them – everything – Say to my mother –.' Here his voice faltered, he became excessively agitated, and not being able to proceed, changed the subject.

'Hope! – Hope! I have much to say to him – but cannot get it out. Are Colonel Graham, and all my aides-de-camp, safe?' (At this question, Anderson, who knew the warm regard of the General towards the officers of his staff, made a private sign not to mention that Captain Burrard was mortally wounded.) He then continued, 'I have made my will, and have remembered my servants. Colborne has my will, and all my papers.' As he spoke these words, Major Colborne, his military secretary, entered the room. He addressed him with his wonted kindness; then, turning to Anderson, said, 'Remember you go to Willoughby Gordon, and tell him it is my request, and that I expect he will give a Lieutenant-Colonelcy to Major Colborne; – he has been long with me – and I know him to be most worthy of it.'

He then asked the Major, who had come last from the field, 'Have the French been beaten?' He assured him they had on every point. 'It's a great satisfaction', he said, 'for me to know that we have beat the French. Is Paget in the room?' On being told he was not, he resumed, 'Remember me to him; he is a fine fellow.' Though visibly sinking, he then said, 'I feel myself so strong – I fear I shall be long dying – It's great uneasiness – it's great pain!' – 'Everything François says is right – I have great confidence in him.' He thanked the surgeons for their attendance. Then seeing Captains Percy and Stanhope, two of his aides-de-camp, enter, he spoke to them kindly, and repeated to them the question 'if all his aides-de-camp were safe'; and was pleased on being told they were. After a pause, Stanhope caught his eye and he said to him, 'Stanhope! remember me to your sister.' He then became silent.

JAMES CARRICK MOORE

100.

Arthur Wellesley (1769–1852), later 1st Duke of Wellington, was probably the greatest commander in the history of the British Army, a whisker ahead of Cromwell and Marlborough. He disapproved of his soldiers cheering, as 'too nearly an expression of opinion', but was obliged to endure frightful emotional outbursts from some of his subordinates in the Spanish Peninsular campaign.

Few of the new officers sent out to him by the Horse Guards [in 1810] possessed the solid ability and agreeable temperament of an Edward Pakenham or a Lowry Cole. There was General Sir William Erskine, drunken, 'blind as a beetle', according to a fellow officer, and probably mad, whom he had sent home 'indisposed' the year before. Back he came in 1810, along with other known disasters such as Generals Lumley and Lightburne and Colonel Landers. Landers had also been sent home once already, by Sir John Moore from Sicily. Wellington would at least keep Landers off the battlefield by appointing him 'perpetual President of General Courts-Martial', with Lightburne, if Wellington had his way, as the perpetual President's first customer – Lightburne's conduct having been 'scandalous'. Wellington gave Colonel Torrens, Military Secretary at the Horse Guards, the full blast of his indignation: 'Really when I reflect upon the characters and attainments of some of the General officers of this army on whom I am to rely against the French generals, I tremble: and, as Lord Chesterfield said of the generals of his day, "I only hope that when the enemy reads the list of their names he trembles as I do."' Wellington always liked to give his favourite quotations a good run. The Chesterfield epigram would have often sparkled at his dinner table, enhanced with characteristic 'By God!' and divested of references to authorship, until in due course his enchanted aides-de-camp handed it down to posterity as the great man's own work. He was eventually delivered from Erskine when the unfortunate general committed suicide at Lisbon in 1813.

ELIZABETH LONGFORD

101.

This story, from one of Masséna's officers, gave birth to a host of literary legends.

A French sergeant, wearied of the misery in which the army was living, resolved to decamp and live in comfort. To this end he persuaded about a hundred of the worst characters in the army, and going with them to the rear, took up his quarters in a vast convent deserted by the monks, but still full of furniture and provisions. He increased his store largely by carrying off everything in the neighbourhood that suited him; well-furnished spits and stewpans were always at the fire, and each man helped himself as he would; and the leader received the expressive if contemptuous name of 'Marshal Stockpot'. The scoundrel had also carried off numbers of women; and being joined before long by the scum of the three armies attracted by the prospect of unrestrained debauchery, he formed a band of some three hundred English, French, and Portuguese deserters, who lived as a happy family in one unbroken orgy.

This brigandage had been going on for some months, when one day, a foraging detachment having gone off in pursuit of a flock as far as the convent which sheltered the so-called 'Marshal Stockpot', our soldiers were much surprised to see him coming to meet them at the head of his bandits, with orders to respect his grounds and restore the flock which they had just taken there. On the refusal of our officers to comply with this demand, he ordered his men to fire on the detachment. The greater part of the French deserters did not venture to fire on their compatriots and former comrades, but the English and Portuguese obeyed, and our people had several men killed or wounded. Not being in sufficient numbers to resist, they were compelled to retreat, accompanied by all the French deserters, who came back with them to offer their submission.

Masséna pardoned them on condition that they should march at the head of the three battalions who were told off to attack the convent. That den having been carried after a brief resistance, Masséna had 'Marshal Stockpot' shot, as well as the few French who had remained with him. A good many English and Portuguese shared their fate, the rest were sent off to Wellington, who did prompt justice on them.

BARON DE MARBOT

102.

A Gerardian affray described by an English pen, outside Salamanca in 1812.

I had an encounter in single combat this day with a very young French officer between the two lines of skirmishers, French and English, who stood still, by mutual consent, to witness it. The French officer showed great cunning and skill, seeing the superiority of my horse, for he remained stationary to receive me, and allowed me to ride round and round him while he remained on the defensive. He made several cuts at the head of my horse and succeeded in cutting one of my reins and the forefinger of my bridle hand which was, however, saved by the thick glove I wore, though the finger was cut very deeply to the joint. As my antagonist was making the last cut at me I had the opportunity of making a thrust at his body which staggered him, and he made off. I thought I had but slightly wounded him, but I found on enquiry the next day, when sent on a flag of truce, that the thrust had proved mortal, having entered the pit of his stomach.

I felt deeply on this occasion and was much annoyed, as I had admired the chivalrous and noble bearing of this young officer. He was a mere youth, who, I suppose, thought it necessary to make this display as a first essay, as French officers usually do on their first appearance on the field and, indeed I believe, it is expected of them by their comrades. I shall never forget his good-humoured, fine countenance during the whole time we were engaged in this single combat, talking cheerfully and politely to me as if we were exchanging civilities instead of sabre cuts. The cut I received on the forefinger of my bridle hand proved a great grievance for some time, as it prevented me from playing the violin for weeks – a great deprivation, as I always played in the bivouac at night.

CAPTAIN T. W. BROTHERTON

103.

Not all Wellington's officers were heroes.

We had in my old corps, amongst other 'characters', one that, at the period I am writing about, was well-known in the army to be as jovial a fellow as ever put his foot under a mess-table. His name was Fairfield; and though there were few who could sing as good a song, there was not in the whole British army a worse duty officer. Indeed, it was next to impossible to catch hold of him for any duty whatever; and so well-known was his dislike to all military etiquette, that the officer next to him on the roster, the moment Fairfield's name appeared for guard-mounting or court-martial, considered himself as the person meant. The frequent absence of Fairfield from drill, at a time too when the regiment was in expectation of being inspected by the general of division, obliged the officer commanding to send the surgeon to ascertain the nature of his malady, which from its long continuance (on occasions of duty!) strongly savoured of a chronic complaint. The doctor found the invalid traversing his chamber rather lightly clad for an indisposed person; he was singing one of Moore's melodies, and accompanying himself with his violin, which instrument he touched with great taste.

The doctor told him the nature of his visit, and offered to feel his pulse, but Fairfield turned from him, repeating the lines of Shakespeare, 'Canst thou minister,' etc. etc. 'Well,' replied the surgeon, 'I am sorry for it, but I cannot avoid reporting you fit for duty.' – 'I'm sorry you cannot,' rejoined Fairfield; 'but my complaint is best known to myself! and I feel that were I to rise as early as is necessary, I should be lost to the service in a month.' 'Why,' said the doctor, 'Major Thompson says you have been lost to it ever since he first knew you, and that is now something about six years.'

The Major's orderly was soon at Fairfield's quarters with a message to say that his presence was required. 'Mr Fairfield,' said the Major, 'your constant habit of being absent from early drill has obliged me to send the surgeon to ascertain the state of your health, and he reports that you are perfectly well, and I must say that your appearance is anything but that of an invalid – how is this?' 'Don't mind him, sir,' replied Fairfield; 'I am, thank God! very well now, but when the bugle sounded this morning at four o'clock a cold shivering came over me – I think it was a touch of ague! – and besides, Dr Gregg is too short a time in the Connaught Rangers to know my habit.' – 'Is he?' rejoined the old Major, 'he must be d—d stupid then. But that is a charge you surely can't make against me. I have been now about nineteen years in the regiment, during six of which I have had the pleasure of knowing you, and you will allow me to tell you, that I am not only well acquainted with "your habit", but to request you will, from this moment, change it' – and with this gentle rebuke he good-humouredly dismissed him. Fairfield was an excellent duty officer ever after.

WILLIAM GRATTAN

104.

Contrasting approaches to morale-boosting.

Both Wellington and Napoleon were eminently practical men. They knew there was a time for dressing-up and a time for undressing too. I am reminded of the anecdote told by Wellington to his confidante, Mrs Arbuthnot, on the methods Napoleon wished to employ in order to distract the Parisian public's attention from the appalling losses in the Russian campaign then in progress. He ordered that the ballet dancers at the opera were to appear *sans culotte*. The order was given, but the dancers flatly refused to comply. 'Wellington added', says Mrs Arbuthnot in her journal, 'that if the women had consented he did not doubt but that it would have obliterated all recollection of the Russian losses.' The Iron Duke had a sound understanding of human nature: and he was being realistic, rather than cynical, when, during the Peninsular campaigns he set a limit of forty-eight hours for his officers' leaves in Lisbon, or behind the lines. This, he said, was as long as any reasonable man could wish to spend in bed with any woman.

LESLEY BLANCH

105.

An Irish officer, wounded at Badajoz, is brought to his tent on the morning of 7 April 1812.

Bray and Macgowan, the two faithful soldiers that conducted me there found my truss of straw occupied by Mrs Nelly Carsons, the wife of my batman, who, by the way of banishing care, had taken to drinking divers potations of rum to such an excess that she lay down in my bed, thinking, perhaps, that I was not likely again to be its occupant; or, more probably, not giving it a thought. Macgowan attempted to wake her, but in vain – a battery of a dozen guns might have been fired without danger of disturbing her repose! 'Why then, sir,' said he, 'sure the bed's big enough for yees both, and she'll keep you nate and warm, for, be the powers, you're kilt with the cold and the loss ov blood.'

I was in no mood to stand on ceremony, or, indeed, to stand at all. I allowed myself to be placed beside my partner, without any further persuasion; and the two soldiers left us to ourselves and returned to the town. Weakness from loss of blood soon caused me to fall asleep, but it was a sleep of short duration. I awoke, unable to move, and, in fact, lay like an infant. The fire of small arms, the screams of the soldiers' wives, and the universal buzz throughout the camp, acted powerfully upon my nervous and worn-out frame; but Somnus conquered Mars, for I soon fell into another doze, in which I might have remained very comfortable had not my companion awoke sooner than I wished; discharging a huge grunt, and putting her hand upon my leg, she exclaimed, 'Arrah! Dan, jewel, what makes you so stiff this morning?'

WILLIAM GRATTAN

106.

Wellington's Judge-Advocate dispenses cavalier justice.

We arrived at Guinaldo in two hours, finished a case and tried a man for shooting a Portuguese, acquitted him of murder, but found him guilty of very disorderly conduct, and sentenced him to receive eight hundred lashes. I then walked round the town, looked into the church, and came back; wrote the whole out fair on six sides of folio paper; dined with the president at six, had a hospitable reception; and in the evening went to a sort of frolicsome masked ball, given extra on account of the courts-martial.

FRANCIS LARPENT

107.

As Napoleon's Grand Army approached Moscow in October 1812, the Russian rearguard marched out of it.

The final Russian regiments moved through Moscow unmolested by the French. A military band began to beat out a defiant tune in an attempt to lift the spirits of the soldiers, but General Mikhail Miloradovich rode in a fury to the commander, thrusting forward his prominent jaw and shouting: 'What idiot told your band to play?' The band officer replied that a garrison must play suitable music when leaving a fortress, under a regulation laid down by Peter the Great. 'Where do the regulations of Peter the Great provide for the surrender of Moscow?' bellowed back Miloradovich. 'Order that damned music to be stopped!'

ROGER PARKINSON

108.

Yet the occupation of Moscow proved an illusory French triumph. At the Niemen on 14 December, the ruins of the Grand Army fought their last action of the Russian campaign. It is bewildering that modern France continues to revere Bonaparte, whose legacy of a mountain of corpses attained its summit with more than 200,000 dead among his own 1812 army. Out of an invading force of 615,000 French and Allied troops, barely 100,000 emaciated survivors returned to France. The even more catastrophic loss of horses crippled his 1813 campaign.

Awakened by the sound of cannon-fire, Marshal [Michel] Ney (1769– 1815) ran to the Vilna gate [of Kovno – Kaunas in modern Lithuania].

He found his own cannon had been spiked, and that the artillerymen had fled! Enraged, he darted forward, and elevating his sword, would have killed the officer who commanded them had it not been for his aide-de-camp, who warded off the blow and allowed the miserable fellow to escape. Ney then summoned his infantry, but only one of the two feeble battalions of which it was composed had taken up arms; these were the three hundred Germans. He drew them up, encouraged them, and as the enemy was approaching, was just about to give them the order to fire when a Russian cannonball, grazing the palisade, came in and broke the thigh of their commanding officer. He fell, and without the least hesitation, finding that his wound was mortal, he coolly drew his pistol and blew out his brains before his troops. Terrified at this act of despair, his soldiers were completely scared. All of them at once threw down their arms and fled in disorder.

Ney, abandoned by all, neither deserted himself nor his post. After vain efforts to detain these fugitives, he collected their muskets, which were still loaded, became once more a common soldier, and with only a few others kept facing thousands of the Russians. His very audacity stopped them; it made some of his artillerymen ashamed, who then returned to join their Marshal; and it gave time to another aide-de-camp, Heymès, and to General Marchand, to assemble thirty soldiers and bring forward two or three light pieces. Meanwhile Marchand went to collect the only battalion which remained intact.

At about 2 p.m. the second Russian attack began from the other side of the Niemen, although still directed against the bridge. Obviously the last desperate action of the 1812 campaign was now approaching its climax. Ney sent Marchand and his four hundred men forward to secure the bridge. As to himself, without giving way, or disquieting himself further as to what was happening at the rear, he kept on fighting at the head of his thirty men and maintained himself until night at the Vilna gate. He then traversed the town and crossed the Niemen, constantly fighting, retreating but never flying, marching after all the others, supporting to the last moment the honour of Napoleon's arms, and for the hundredth time during the last forty days and forty nights, putting his life and liberty in jeopardy just to save a few more Frenchmen.

At Gumbinnen in East Prussia on 15 December General Matthieu Dumas was just sitting down to his first decent breakfast in months when someone kicked the door open. 'There stood before him a man in a ragged brown coat, with a long beard, dishevelled and with his face darkened as if it had been burned, his eyes red-rimmed and glaring. Underneath his coat he wore the rags and shreds of a discoloured and filthy uniform.'

'Here I am then,' the newcomer exclaimed.

'But who are you?' the general cried, alarmed.

'What! Don't you recognize me? I am Marshal Ney: the rearguard of the Grande Armée! I have fired the last shot on the bridge at Kovno. I have thrown the last of our muskets into the Niemen. I have made my way here across a hundred fields of snow. Also I'm damnably hungry. Get someone to bring me a plate of soup.'

RAYMOND HORRICKS

109

Among witnesses to the catastrophe of the Grand Army, in which many of Napoleon's Prussian temporary allies were among the victims, was Prussian officer Karl von Clausewitz (1780–1831), author of On War, *the greatest tract on strategy ever penned. In 1812, he was wearing Russian uniform.*

Clausewitz was always something of an introvert; solitary, bookish, shy, intellectually arrogant. But beneath the scholarly, withdrawn exterior there burnt an ambition for military glory worthy of Stendhal's Julian Sorel, an ambition deeply repressed, given vent only in his letters to his wife; never to be fulfilled in the series of staff appointments for which his superiors considered, probably rightly, that his intellectual talents best fitted him. All Clausewitz's writings bear the stamp of a passionate temperament, as often at war with as in the service of a powerful analytic mind. In 1803 he met and fell in love with Marie (1779–1836) daughter of the Count von Bruhl, a lively and well-educated girl high in the favour of Queen Louise of Prussia. The family resistance to this unsuitable match and the demands of military service delayed the marriage for seven years, which made possible the long, passionate correspondence in which Clausewitz developed many of his ideas. Once married, Marie was to identify herself wholeheartedly with her husband's work, act as his amanuensis and after his death as his editor and preside over what still remains the most complete edition of his works, which she published.

[In 1806 Clausewitz was captured following the Prussian defeat at Auerstadt, and spent the ensuing two years as a prisoner in France.] It was a humiliating experience that stoked the fires of his patriotic zeal and gave him a lifelong dislike for all things French. Then in the spring of 1812 the king whose uniform he wore, and whose claims on his loyalty he had never questioned, concluded an alliance with the French enemy Clausewitz so detested. It was too much. In company with some thirty other officers he resigned his commission, parted again from his wife, and took service with Emperor Alexander I of Russia. Although Clausewitz spoke no Russian, employment was found for him in various advisory positions. He took part in his second great battle at Borodino. He witnessed the disastrous crossing of the Beresina by the retreating French army and wrote a horrifying account of it. Finally he acted as an intermediary when in December 1812 the commander of the Prussian corps serving under Napoleon's command, Yorck von Wartenberg, took his historic decision to capitulate and go over with his forces to the Russians. It was wearing the uniform of a Russian officer that Clausewitz served as adviser to the Prussian army commander, Marshal Blucher, during the Leipzig campaign. The last task assigned to Clausewitz was to organise a cordon sanitaire to check the 1831 advance of the cholera epidemic into Germany, but it was a problem his strategic insights could not solve. He himself caught the disease and died within twenty-four hours at Breslau, on 16 November 1831, at the age of 51.

MICHAEL HOWARD

110.

A soldier of the 51st Foot writes home from Spain.

Moimento, 28 April 1813
I am sorry to say we have had a Serjeant, Corporal and private punished

by a division court martial for what the Earl of Dalhousie considers a great crime. A mail had just arrived from England and the General and his staff were looking over the newspapers. One of our men, sentry at the tent, having got some rum had drank so much that he was quite intoxicated. Seeing the papers he began to be anxious to know how things were going on, so without any ceremony he walks up to the tent door, but unfortunately the tent peg or cords caught his toe and in he bolted head foremost and lay prostrate at the Earl's feet, and not having the benefit of a polite education without making an apology for his abrupt intrusion with much sangfroid asked the General 'What news from England?' A Division Court Martial followed. The Serjeant was tried for passing the man, the Corporal for planting him on sentry, and the man for being drunk on duty. As a matter of course they were all found guilty. The Serjeant and Corporal to be reduced to the ranks and receive 300 lashes and the private 500. These sentences were carried into execution in presence of as many troops of the Division as could be conveniently assembled. You have often hinted in your letters that I ought to endeavour to get promoted. I have always declined every offer made to me. I am young and I have so far kept out of trouble. I endeavour to perform my duty in the best way I am able. I have therefore nothing to fear. But should I be entrusted with a guard perhaps I might get into a scrape through their neglect. There is time enough after the razor has passed round my chin some hundred times more before I think about accepting of any responsible situation. [He finally retired as a sergeant in 1828.]

WILLIAM WHEELER

111.

After Lutzen, 2 May 1813.

Napoleon expressed great alarm at the sight of Ney. 'My dear Cousin! But you are covered in blood!' The Marshal looked down at his gory uniform. 'It isn't mine, Sire,' he replied calmly; '... except where that damned bullet passed through my leg!'

RAYMOND HORRICKS

112.

Posterity owes a debt to Earl Stanhope, an unabashed hero-worshipper, for minutely recording his many post-war conversations with Wellington, including this one in 1831, about the respective systems of discipline in the British and French armies.

The Duke spoke strongly in favour of having a strong military punishment in reserve, were it only to give efficacy to the milder ones. I think he must have alluded to flogging. I asked him: Do they beat them in the French Army? 'Oh, they bang them about very much with ramrods and that sort of thing, and then they shoot them. Besides, a French army is composed very differently from ours. The conscription calls out a share of every class – no matter whether your son or my son – all must march; but our friends – I may say it in this room – are the very scum of the earth. People talk of their enlisting from their fine military feeling – all stuff – no such thing. Some of our men enlist from having got bastard children – some for minor offences – many more for drink; but you can hardly conceive such a set brought together, and it really is wonderful that we should have made them the fine fellows they are.'

EARL STANHOPE

113.

Aftermath of a skirmish, November 1813.

I was resting myself on a bank, thinking to what trouble and misery many lovely young women of respectable connections had brought themselves into by marrying soldiers who but a few years since I had seen in old England in the full enjoyment of health etc. when I was roused from my reverie by a well-known voice saying 'Then you have caught it at last, Corporal.' I looked up and there stood Marshall, his arm was in a sling. I was about to spake to him when I was interrupted by a female voice in a trembling accent, 'Oh W— have you seen my husband or can you give me any account of him.' I shook my head. 'Oh it is too true, your silence confirms what I have just been told, he is dead.' She then ran away towards the hill where the severe conflict had taken place.

'Ah Marshall,' said I, 'her fears are but too true, he is dead indeed, he fell not many yards from where I lay.' 'Who the D— is she' said M— 'I did not see her to notice her.' 'Not know her?' said I, 'it was Mrs Foster.' 'Oh Damn it' said he, 'was it? I am sorry for her, but you know there is so many of these damned women running and blubbering about, enquiring after their husbands. Why the D—l don't they stop at home where they ought to be. This is no place for them. Come let us go in and get something to drink, not stop here to be pestered to death by a parcel of women. Come make haste, here comes Cousins' wife, snivelling as if she was a big girl going to school without her breakfast.' 'She has reason to snivel as you calls it' said I 'she is the most unfortunate creature in the army.' 'Unfortunate indeed' said M— 'why I think she is devilish lucky in getting husbands, she has had a dozen this campaign.' M— was drunk and had rather stretched the number. This unfortunate woman was now a widow for the third time since the battle of Vittoria.

WILLIAM WHEELER

114.

His Majesty's Foot Guards not seen to advantage in the rain.

During the action of the 10th of December 1813, in the neighbourhood of Bayonne, the Grenadier Guards, under the command of Colonel Tynling, occupied an unfinished redoubt. The Duke of Wellington happened to pass with Freemantle and Lord Hill, having satisfied himself that the fighting was merely a feint on the part of Soult. His Grace on looking around saw, to his surprise, a great many umbrellas, with which the officers protected themselves from the rain that was then falling. Arthur Hill came galloping up to us saying 'Lord Wellington does not approve of the use of umbrellas during the enemy's firing, and will not allow "the gentlemen's sons" to make themselves ridiculous in the eyes of the army.' Colonel Tynling, a few days afterwards, received a wigging from Lord Wellington for suffering his officers to carry umbrellas in the face of the enemy; his Lordship observing, 'The Guards may in uniform, when on duty at St James's, carry them if they please; but in the field it is not only ridiculous but unmilitary.'

REES HOWELL GRONOW

115.

Rejoicing in the ranks of the British 51st Foot in Belgium, at the return of their great commander in April 1815.

Grammont, 29th May, 1815

We were delighted by a General Order issued by H.R.H. the Prince of Orange, in which he 'Surrenders the Command of the Army into the more able hands of His Grace the Duke of Wellington'. I never remember anything that caused such joy, our men were almost frantic. I was accosted by every one, thus 'Sergt. W— have you seen the order? Glorious news, Nosey has got the command, wont we give them a drubbing now?' I had a bottle of gin thrust up to my mouth, and twenty voices shouting 'drink hearty to the health of our old Commander, we don't care a d—n for all France, supposing everyone was a Napoleon etc.' Let it suffice to say that it caused a general fuddle, the evening was spent by reminding each other of the glorious deeds done in the Peninsular, mingled with song and dance, good Hollands [gin] and tobacco.

WILLIAM WHEELER

116.

Wellington received news of Napoleon's advance towards Quatre Bras on 15 June 1815, but declined to hasten precipitately from Brussels to the battlefield. 'The numerous friends of Napoleon who are here', he declared, 'will be on tiptoe; the well-intentioned must be pacified; let us therefore go all the same to the Duchess of Richmond's ball, and start for Quatre Bras at 5 a.m.'

The most famous ball in history was the climax of Wellington's psychological warfare which always involved 'pleasure as usual'. The question of holding it or not had first come up in May. 'Duke,' said the Duchess of Richmond one day, 'I do not wish to pry into your secrets. I wish to give a ball, and all I ask is, may I give my ball? If you say, "Duchess, don't give your ball", it is quite sufficient, I ask no reason.'

'Duchess, you may give your ball with the greatest safety, without fear of interruption.' At that date, indeed, the Duke had intended to give a ball himself on 21 June, the second anniversary of the battle of Vitoria. Operations were not expected to begin before 1 July.

Since those dignified ducal exchanges, circumstances had altered with a vengeance, more radically than Wellington even now supposed. That very afternoon there had been a close-run thing, though a small one, at Quatre Bras. It has often been asked why Wellington did not cancel the ball at 3 p.m. instead of going to hear the fiddlers while Rome burned, or at any rate did not ride out to Quatre Bras at midnight to see for himself what was on the other side of the hill. Apart from Wellington's extreme sensitivity to the chances of a Belgian stab in the back, his place was in Brussels.

Having at last redirected his whole army towards Quatre Bras, nothing more remained for him to do there that night. He was personally to lead out the reserve in the morning. Orders had still to be distributed among officers in Brussels and personal interviews held. Why not under the convenient camouflage and at the ready-made rendezvous of a ball? This was to be Wellington's explanation to his friends during later post-mortems of Waterloo, and it is confirmed by Lord Fitzroy Somerset's own brief statement: 'As it [the ball] was the place where every British officer of rank was likely to be found, perhaps for that reason the Duke dressed & went there.'

Morale-building, duty, convenience – they all played their part in getting Wellington to the ball. Why not admit also that the Irish devil in him wanted to go? He would go; and see 'those fellows' damned. Wellington's decision gave Byron his chance to include Brussels in *Childe Harold's Pilgrimage* and Thackeray to make Becky Sharp roll her green eyes and flaunt her pink ball dress in a perfect setting.

> *There was a sound of revelry by night,*
> *And Belgium's Capital had gather'd then*
> *Her Beauty and her Chivalry – and bright*
> *The lamps shone o'er fair women and brave men …*

The ballroom, situated on the ground floor of the Richmonds' rented house in the Rue de la Blanchisserie, had been transformed into a glittering palace with rose-trellised wallpaper, rich tent-like draperies and hangings in the royal colours of crimson, gold and black, and pillars wreathed in ribbons, leaves and flowers. Byron's 'lamps' were the most magnificent chandeliers and the list of chivalry, if not beauty, was headed by H.R.H. the Prince of Orange, G.C.B. All the ambassadors, generals and aristocrats were present as well as dashing young officers like Arthur Shakespear of the Light Dragoons, and Captain Pakenham of the Royal Artillery – Sir Charles Stuart, General Alava, the Mountnorrises and Wedderburn-Websters, the Capels, Grevilles and Mrs Pole.

Creevey's stepdaughters, the Misses Ord, got their tickets, for the amusing Radical moved in the best circles, treasuring the probably correct conviction that he was the illegitimate son of a former Lord Sefton. The rear was brought up by the diplomat Mr Chad, Wellington's surgeon Dr John Hume and his chaplain the Reverend Samuel Briscall whose name as usual was spelt wrong.

Wellington arrived 'rather late' at the entrance, where streams of light poured through the open windows into the warm streets and over the thronged carriages. In the ballroom those officers whose regiments were at any distance were already beginning to slip quietly away. The seventeen-year-old Lady Georgiana Lennox, whose sisters used to enjoy riding across the Phoenix Park with 'the great Sir Arthur' in 1807, was dancing. She immediately broke off and went up to Wellington to ask whether the rumours were true (Arthur Shakespear wrote in his diary, 'about twelve o' clock it was rumoured that we were to march in the morn!'). Wellington replied, as she thought, very gravely, 'Yes, they are true, we are off tomorrow.'

As this terrible news (Georgiana's words) rapidly circulated, the ballroom was like a hive that someone had kicked: an excited buzz arose from all the tables and elegantly draped embrasures. The Duke of Brunswick felt a premonition of death and gave such a shudder that he dropped the little Prince de Ligne off his lap. Some officers flew to and fro saying their goodbyes and departed, others clung so desperately to the loved one's hand or to the champagne bottle that when the hour struck there was no time to change and they had to march in their dancing pumps. The Duke meanwhile appeared to the two youthful Miss Ords as composed as ever, while the even younger William Lennox, covered with plaster after a riding accident, particularly noticed the serenity that 'beamed' all over his face.

'On with the dance! Let joy be unconfined.' A rather more perspicacious guest, however, Lady Hamilton-Dalrymple, who sat for some time beside Wellington on a sofa, was struck by his preoccupied and anxious expression beneath the assumed gaiety: 'Frequently in the middle of a sentence he stopped abruptly and called to some officer, giving him directions, in particular to the Duke of Brunswick and Prince of Orange, who both left the ball before supper.' But even the lady on the sofa did not suspect the degree of drama with which the Prince of Orange's departure was attended.

Shortly before supper, as Wellington stood with Lady Charlotte Greville on his arm, a despatch was brought in by Lieutenant Henry Webster from Quatre Bras for the Prince of Orange. Slender Billy, merry as a marriage bell, handed it unopened to Wellington who quietly slipped it into his pocket for the moment. The message, dated about 10 p.m. that night, announced the repulse of Prussian forces from Fleurus on the road north-east of Charleroi and less than eight miles as the crow flies from Quatre Bras. As soon as Wellington had read this enlightening but grim piece of news he recommended the Prince to miss supper and return straight to his headquarters in the field.

'Webster!' he called to the Prince's aide-de-camp, 'four horses instantly to the Prince of Orange's carriage ...' After other instructions now made necessary had been delivered in whispers or scribbles, Wellington proceeded to the supper-room. Hardly had he sat down before the Prince of Orange reappeared and whispered something to him for several minutes. Wellington looked incredulous but said nothing except to repeat that the Prince should go back to his quarters at Braine-le-Comte and to bed. Wellington kept up an animated and smiling conversation for twenty minutes more, when a lesser man would have fled the moment he heard Slender Billy's news. A notable Belgian aristocrat, indeed, the Marquise d'Assche, who sat next to the

Duke of Richmond and opposite the Duke of Wellington at supper, did not relish the English nonchalance. Painfully conscious that her brother was somewhere out there where the cannon had been booming at dusk she looked across at the Duke with a jaundiced eye.

His own placement was agreeable: Georgy Lennox on one side, who received from him a miniature of himself painted by a Belgian artist, and Frances Webster on the other. 'I would willingly have throttled him,' recalled the Marquise d'Assche, 'from the impatience which his phlegm caused me, and the ease of his conversation with Lady Withesburne [sic] to whom he paid ardent court.' At last the necessary interval was up and Wellington turned casually to the Duke of Richmond. 'I think it is time for me to go to bed likewise ...' The party rose and moved into the hall. As Wellington was saying goodnight to his host he whispered something in Richmond's ear – the last recorded and most celebrated whisper of an evening remarkable for its undertones. 'Have you a good map in the house?' He needed to discover the exact implications of the almost incredible message verbally passed on to him at the supper table by the Prince of Orange. The written message which the Prince had received from his headquarters at Braine-le-Comte was dated 15 June 1815 '10½ p.m.' and signed by Baron Jean de Constant Rebecque, the Prince's chief-of-staff.

The enemy, de Constant Rebecque reported, were said to have pushed up the *chaussée* towards Brussels as far as Quatre Bras. The Duke of Richmond took him into his study next to the ballroom and spread out a map. Wellington looked at it wryly: 'Napoleon has humbugged me, by God! He has gained twenty-four hours' march on me.'

'What do you intend doing?'

'I have ordered the army to concentrate at Quatre Bras; but we shall not stop him there, and if so, I must fight him here.' Wellington passed his thumbnail over the map just south of the Waterloo position. Then he left the scene of his acute discomfort, avoiding for once the hall-door.

ELIZABETH LONGFORD

117.

*Wellington's second-in-command, the cavalryman and dedicated rake
Lord Uxbridge, later Marquess of Anglesey (1768–1854), faced a
dilemma on 17 June, the eve of battle.*

That night Uxbridge slept at Waterloo, but before he went to bed he
consulted Sir Hussey Vivian upon a subject that was weighing heavily
on his mind: 'I find myself in a very difficult position. A great battle
will take place tomorrow. The Duke, as you know, will not economize
his safety. If any accident happens to him, I shall suddenly find myself
Commander-in-Chief. Now, I have not the slightest idea what are the
projects of the Duke. I would give anything in the world to know the
dispositions which, I have no doubt, have been profoundly calculated.
It will be impossible for me to frame them in a critical moment. I dare
not ask the Duke what I ought to do.'

Vivian advised him to consult Count Alava, the Spanish general,
whose friendship with Wellington led to his being attached to his staff.
Uxbridge at once went to him and found that Alava agreed that the
question was a serious one, but suggested that it was for Uxbridge
himself to tackle it. Uxbridge then explained to the Duke 'the motive
of his visit with all the delicacy imaginable'. The Duke listened to him
without saying a single word, and then asked: 'Who will attack the
first tomorrow, I or Bonaparte?' 'Bonaparte,' was the reply. 'Well,'
continued the Duke, 'Bonaparte has not given me any idea of his
projects: and as my plans will depend upon his, how can you expect
me to tell you what mine are?' The Duke then rose, and putting his
hand on Uxbridge's shoulder, added, 'There is one thing certain,
Uxbridge, that is, that whatever happens, you and I will do our duty.'
The two men then shook hands, and Uxbridge retired to sleep.

LORD ANGLESEY

118.

*The repulse of the French assault upon the Chateau d'Hougoumont
was one of the decisive actions of the next day, on the field of Waterloo.*

A gigantic subaltern named Legros and reinforced with the nickname of *l'Enfonceur*, the Smasher, stove in a panel of the great north door and followed by a handful of wildly cheering men, dashed into the courtyard. Pandemonium broke out. The defenders slashed and hewed at the invaders in desperate hand-to-hand duels. But the real thing was to prevent any more of the enemy from entering the yard. Five powerful Coldstreamers – Macdonnell, three other officers and a sergeant – threw themselves bodily against the huge door and slowly, slowly, by main force pushed it back against the pressure outside. This done they turned their attention to the invaders. 'The success of the battle of Waterloo depended on the closing of the gates of Hougoumont.' So said Wellington afterwards. Among the heroic five was Henry Wyndham. It was said afterwards by his niece when she found herself sitting in a draught, that no Wyndham had ever closed a door since Hougoumont.

ELIZABETH LONGFORD

119.

The Prince of Orange's (1792–1856) towering Lion mound, erected with vast displacement of soil in 1820 to commemorate his role, disfigured in perpetuity Waterloo's ridge of Mont St-Jean where many of the Prince's soldiers, in truth, played inglorious parts.

Trip's Dutch-Belgian cavalry was now at hand. Lord Uxbridge, pleased with their fine appearance, and desirous of exciting in them a courageous enthusiasm, placed himself conspicuously in their front, and ordering the 'charge', led them towards the enemy. He had proceeded but a very short distance, when his aide-de-camp, Captain Horace Seymour, galloped close up to him, and made him aware that not a single man of them was following him. Turning round his horse, he instantly rode up to Trip, and addressed himself to this officer with great warmth. Then, appealing to the brigade in terms the most exhorting and encouraging, and inciting them by gestures the most animated and significant, he repeated the order to charge, and again led the way in person. But this attempt was equally abortive; and Uxbridge, exasperated and indignant, rode away from the brigade, leaving it to adopt any course which its commander might think proper.

WILLIAM SIBORNE

120.

John Keegan, in his classic 1974 work The Face of Battle, *compared the experiences of combat at Agincourt, Waterloo and the Somme. His book had more influence than any other modern work on the way that historians of war who wish to be read now focus upon human experience, rather than mark arrows on maps.*

If smoke oppressed the senses, the noise of Waterloo assaulted the whole being. At Agincourt noise would have been chiefly human and animal and would have overlaid the clatter of weapon-strokes. There was still a good deal of perceptible human noise at Waterloo: an officer of Picton's division had remembered the noise of the army preparing for battle as similar to that of the 'distant murmur of the waves of the sea, beating against some ironbound coast'. Once the battle got under way there was cheering – Leeke, like several others, mentions hearing 'continued shouts of "*Vive L'Empereur*" and at the time of the Imperial Guard's attack, shrieking – the 32nd set up a death howl' when the French reached within forty yards of their lines, and confused shouting – an officer of the 73rd describes a French advance as 'very noisy and evidently reluctant'. There were cries of pain and protest from the wounded – though here the testimony is contradictory, Mercer being pierced 'to the very soul' by the scream of a gunner whose arm had just been shattered, Leeke insisting that the wounded kept unnaturally silent. And there were of course shouts of command.

There was also music: Gronow, Leek and Standen recall hearing the [drum]beating of the *pas de charge* (which, one of Picton's officers says, was called by his men, who remembered it from Spain, 'Old Trousers'); and there was piping in the squares of the Scottish regiments. The 71st's pipers played and replayed 'Hey, Johnnie Cope' and Piper McCay of the 79th stepped outside the square under French fire to play '*Cogdah na sith*'.

But it was weapons which made by far the loudest and most insistent noise. Some of the sounds were incidental and unexpected. Lt. Wyndham of the Scots Greys remarked on the 'extraordinary manner in which the bullets struck our swords', a phenomenon which set up a weird harmonic vibration; something of the same sort was produced by shot hitting bayonets, quite a frequent occurrence, though that could also sound like a stick being drawn along park railings. Leeke noted the 'rattle' which grape made when striking arms and accoutrements and Gronow likened the impact of his Guardsmen's musket-balls on the breastplates of Kellermann's and Millhaud's cuirassiers to 'the noise of a violent hailstorm beating upon panes of glass'. These were sounds which could only be caught at close range, however, for at any distance they would be drowned by the much louder and pervasive crash and rumble of firearms and artillery. Mercer's medical officer, for the first time 'hearing this infernal carillon about his ears, began staring round in the wildest and most comic manner imaginable, twisting himself from side to side, exclaiming 'My God … what is that? What is all this noise? How curious! How very curious!' and then, when a cannon-shot rushed hissing past, 'There! – There! What *is* it all?' Mercer, at the end of the day, was 'almost deaf' – and we may take him quite literally.

JOHN KEEGAN

121.

Wellington was running short of aides-de-camp – the 'gallopers' who bore orders to his regiments, at mortal risk.

It was said that once or twice he was reduced to using stray civilians, with whom the battlefield was still supplied, to carry his messages – a young Swiss, perhaps a traveller in buttons from Birmingham, as well as a small Londoner on a pony who turned out to be a commercial traveller for a City firm. 'Please, Sir, any orders for Todd and Morrison?'

'No; but will you do me a service? Go to that officer and tell him to refuse a flank.'

ELIZABETH LONGFORD

122.

It became plain that a French battery had identified Wellington and his staff, who became the targets of its fire in a fashion the Duke plainly considered unsporting, and declined to reciprocate.

The shot fell fast about them, generally striking and turning up the ground on which they stood. Their horses became restive, and Copenhagen himself [Wellington's mount] so fidgety that the Duke, getting impatient, and having reasons for remaining on the spot, said to those about him, 'Gentlemen, we are rather too close together – better to divide a little.' Subsequently, at another point of the line, an officer of the artillery came up to the Duke, and stated that he had a distinct view of Napoleon, attended by his staff; that he had the guns of his battery well pointed in that direction, and was prepared to fire. His Grace instantly and emphatically exclaimed, 'No! no! I'll not allow it. It is not the business of commanders to be firing upon each other.'

WILLIAM SIBORNE

123.

It was one of the miracles of the day that the British commander-in-chief survived, when officers were repeatedly killed and wounded at his side.

All along the battered ridge Wellington pursued his charmed course, reining in Copenhagen wherever the tension was greatest to speak a word of caution or encouragement. 'Are we to be massacred here? Let us go at them, let us give them Brummegum!' the men shouted at him, brandishing their bayonets [forged in Birmingham]. 'Wait a little longer, my lads, you shall have at them presently.' To their officers he said, 'Hard pounding, this, gentlemen; try who can pound the longest.' Once there was an echo of Henry V: 'Standfast … we must not be beat – what will they say in England?'

ELIZABETH LONGFORD

124.

*Among the decisive French failures of the day was that of Grouchy
and three corps, detached to engage the Prussians, who failed to
receive the emperor's order to rejoin the main army. Napoleon had
good cause to lament the death of Berthier, his former chief of staff.*

Napoleon having asked Soult whether he had sent for Grouchy, Soult
answered that he had despatched an officer at a particular hour. '*Un
officier!*' exclaimed Napoleon, turning round to his suite; '*un officier!
Ah, mon pauvre Berthier! S'il avait été ici il en aurait envoyé vingt!*'

EARL STANHOPE

125.

*Wellington's second-in-command was hit at the commander-in-chief's
side.*

'Just as Sir Hussey Vivian's brigade were going down to the charge,'
wrote Wildman the day after the battle, 'Lord Uxbridge was struck by
a grape shot on the right knee which shattered the joint all to pieces.'
The Duke told his brother William that Uxbridge was wounded when
talking to him 'during the last attack, almost by the last shot'. To
Stanhope Wellington explained that he was on the side from which the
shot proceeded, and that it passed over the neck of his horse till it
reached Uxbridge. The Duke supported him and prevented his falling
from the saddle.

Uxbridge himself remembered that he was hit 'in the low ground
beyond La Haye Sainte, and perhaps ¼ of an hour before Dusk, at the
moment when I was quitting the Duke to join Vivian's Brigade of
Hussars which I had sent for, being the only fresh Corps I had'. In the
popular version, Uxbridge exclaims 'By God, sir, I've lost my leg!'
Wellington momentarily removes the telescope from his eye, considers
the mangled limb, says 'By God, sir, so you have!' and resumes his
scrutiny of the victorious field. Men of a Hanoverian infantry battalion, advancing rapidly behind the cavalry, helped to remove the
wounded hero from his horse, and six of them, with the faithful aide-

de-camp Seymour walking at their head, bore him from the field. A number of old soldiers claimed in years to come that they had assisted in this 'melancholy duty', among them one with the name of Esau Senior of the Inniskilling Dragoons.

Back at his headquarters in Waterloo, the surgeons who examined the wound all agreed that it would be at the imminent danger of his life to attempt to save the limb. His comment was typical: 'Well, gentlemen,' he said, 'I thought so myself. I have put myself in your hands and, if it is to be taken off, the sooner it is done the better.' He at once wrote a letter to Char [his wife], saying that had he been a young single man he might have run the risk of keeping his leg, but that as it was he would, if possible, preserve his life for her and his children. Then, while the surgeons prepared for their task, he put the coming agony quite out of his mind, and conversed at length with his staff about the action, forgetting his wound 'in the exultation for the Victory'.

Wildman, who was present at the amputation, tells how he never moved or complained: no one even held his hand. He said once perfectly calmly that he thought the instrument was not very sharp. When it was over, his nerves did not appear the least shaken and the surgeons said his pulse was not altered. He said, smiling, 'I have had a pretty long run, I have been a beau these forty-seven years and it would not be fair to cut the young men out any longer' and then asked if we did not admire his vanity. I have seen many operations [continues Wildman], but neither Lord Greenock nor myself could bear this, we were obliged to go to the other end of the room. Thank God he is doing as well as possible. He had no fever and the surgeons say nothing could be more favourable.

Later that night Vivian looked in, fresh from the pursuit. He was greeted with: 'Vivian, take a look at that leg, and tell me what you think of it. Some time hence, perhaps, I may be inclined to imagine it might have been saved, and I should like your opinion upon it.' Confronted with the gruesome object, Vivian readily confirmed that it was best off, and left his chief to compose himself for sleep.

The owner of the house where the operation was performed, a M. Paris, placed the leg in a wooden coffin and asked its owner's permission '*de placer le membre du noble et intéressant Milord dans notre petit jardin*'. Permission was granted, and in due course a weeping willow was planted over the site. M. Paris erected a commemorative plaque, which may still be seen today; upon it these words appear:

Ci est enterré la Jambe
 de l'illustre et vaillant Comte Uxbridge,
 Lieutenant-General de S.M. Britannique,
 Commandant en chef la cavalerie anglaise, belge et hollandaise,
blessé le 18 juin, 1815, à la mémorable bataille de Waterloo; qui, par son héroisme, a concouru au triomphe de la cause du genre humain; glorieuseinent décidée par l'éclatante victoire du dit jour.

Some wag is said to have scribbled beneath the inscription:

Here lies the Marquis of Anglesey's limb;
The Devil will have the remainder of him.

The wife of the Bishop of Norwich, visiting the house on the first anniversary of the battle, was shown 'as a relic almost as precious as a Catholic bit of bone, the blood upon a chair in the room where the leg was cut off, which M. Paris had promised my lord "*de ne jamais effacer*"'. When in later years Uxbridge visited the place, it was said that he found the very table on which he had lain for the amputation of the limb, and that by his direction dinner was spread upon it for himself and two of his sons.

LORD ANGLESEY

126.

Having played a critical role in repulsing French cavalry charges throughout the day, in the last minutes of the battle Captain Cavalié Mercer's (1783–1868) G Troop of the Royal Horse Artillery found itself under devastating 'friendly fire'.

One shell I saw explode under the two finest wheel-horses in the troop. In some instances the horses of a gun or ammunition waggon remained, and all their drivers were killed. The whole livelong day had cost us nothing like this. Our gunners too – the few left fit for duty – were so exhausted that they were unable to run the guns up after firing, consequently at every round they retreated nearer to the limbers; and as we had pointed our two left guns towards the people who were annoying us so terribly, they soon came altogether in a confused heap, the trails crossing each other, and the whole dangerously near the limbers and ammunition waggons, some of which were totally unhorsed, and others in sad confusion, many of them lying dead in their harness attached to their carriages. I had dismounted, and was assisting at one of the guns to encourage my poor exhausted men, when through the smoke a black speck caught my eye, and I instantly knew what it was. The conviction that one never sees a shot coming towards you unless directly in its line flashed across my mind, together with the certainty that my doom was sealed. I had barely time to exclaim 'Here it is then!' – much in that gasping sort of way one does when going into very cold water takes away the breath – 'whush' it went past my face, striking the point of my pelisse collar, which was lying open, and smash into a horse close behind me. I breathed freely again.

Under such a fire, one may be said to have had a thousand narrow escapes; and, in good truth, I frequently experienced that displacement of air against my face caused by the passing of shot; but the two above recorded were remarkable, and made me feel in full force the goodness of Him who protected me. Whilst in position on the right of the second line, I had reproved some of my men for lying down when shells fell near them until they burst. Now my turn came. A shell, with a long fuse, came slop into the mud at my feet, and there lay fizzing and flaring. After what I had said on the subject, I felt that I must act up to my own words, and, accordingly, there I stood, endeavouring to look quite composed until the cursed thing burst – and, strange to say, without injuring me, though so near. The effect on my men was good. We had scarcely fired many rounds at the enfilading battery when a tall man in the black Brunswick uniform came galloping up to me from the rear, exclaiming, 'Ah! mine Gott! – mine Gott! vat is it you doos, sare? Dat is your friends de Proosiens; an you kills dem! Ah mine Gott! – mine Gott! vill you no stop, sare? – vill you no stop? Ah! mine Gott! – mine Gott! vat for is dis? De Inglish kills dere friends de

Proosiens! Vere is de Dook von Vellington? – vere is de Dook von Vellington? Oh, mine Gott! – mine Gott!' etc. etc., and so he went on raving like one demented.

I observed that if these were our friends the Prussians they were treating us very uncivilly; and that it was not without sufficient provocation we had turned our guns on them, pointing out to him at the same time the bloody proofs of my assertion. Apparently not noticing what I said, he continued his lamentations, and, 'Vill you no stop, sare, I say?' Wherefore, thinking he might be right, to pacify him I ordered the whole to cease firing, desiring him to remark the consequences. Psieu, psieu, psieu, came our friends' shot, one after another; and our friend himself had a narrow escape from one of them. 'Now, sir,' I said, 'you will be convinced; and we will continue our firing, whilst you can ride round the way you came, and tell them they kill their friends the English; the moment their fire ceases, so shall mine.' Still he lingered, exclaiming, 'Oh, dis is terreeble to see de Proosien and de Inglish kill vonanoder!' At last darting off, I saw no more of him.

CAVALIÉ MERCER

127.

It was 9 p.m. on 18 June 1815 and nearly dark when Blücher and Wellington rode forward to greet one another on the Brussels road between La Belle-Alliance and Rossomme. The former afterwards described the moment.

'We were both on horseback, but he embraced and kissed me, exclaiming "*Mein lieber Kamerad*" and then "*Quelle affaire!*", which was pretty much all he knew of French. At my supper that night, I had forty or fifty people. There came in several of the French officers prisoners – there was Cambronne – you know his mot: *La garde meurt et ne se rend pas?* – there was Lobau – but I told them that I could not allow them to sup with me until after they had made their peace with the King of France.' [Wellington slept that night on a pallet on the floor of his quarters, because an officer was dying in his bed.]

EARL STANHOPE

128.

Looting the enemy's dead was perceived as a victorious soldier's right,
and that night fortunes lay on the battlefield for the taking.

Dead officers, in particular, had purses, watches, pistols, swords, lockets and sentimental charms. Their epaulettes and gold braid were worth money. When all those were gone, there were clothes and equipment, and when even the clothes were gone there were teeth. False teeth were either carved out of ivory or made up of human teeth, and dentists would pay well for the raw materials. Such a haul was made from the field of Waterloo that dentures for years afterwards were often called Waterloo teeth.

DAVID HOWARTH

129.

The morning after the battle, Cavalié Mercer and his surviving
artillerymen ate for the first time in twenty-four hours, watched by a
ghoulish party of civilian 'war tourists'.

Having made ourselves seats of cuirasses, we soon had that most agreeable of animal gratifications – the filling of empty stomachs. Never was a meal more perfectly military, nor more perfectly enjoyed. We had not yet finished, when a carriage drove onto the ground from Brussels, the inmates of which, alighting, proceeded to examine the field. It was amusing to see the horror with which they eyed our frightful figures; they all, however, pulled off their hats and made us low bows. One, a smartly dressed middle-aged man, in a high cocked-hat, came to our circle, and entered into conversation on the events of yesterday. He approached holding a delicately white perfumed handkerchief to his nose; stepping carefully to avoid the bodies (at which he cast fearful glances *en passant*), to avoid polluting the glossy silken hose that clothed his nether limbs. May I be pardoned for the comparison: Hotspur's description of a fop came forcibly to my mind as we conversed; clean and spruce, as if from a bandbox, redolent of perfume, he stood ever and anon applying the 'kerchief to his nose.

I was not leaning on my sword, but I arose to receive him from my seat of armour, my hands and face begrimed and blackened with blood and smoke – clothes too. 'I do remember when the fight was done,' etc. etc. It came, as I said, forcibly to my mind as I eyed my friend's costume and sniffed the sweet-scented atmosphere that hovered round him. The perfumed handkerchief, in this instance, held the place of Shakespeare's 'pouncet-box'! With a world of bows my man took leave, and proceeded, picking his steps with the same care as he followed the route of his companions in the direction of Hougoumont.

[Mercer then heard a Frenchman upbraid a wounded comrade for screaming in his agony with 'unsoldierlike want of fortitude'.] The speaker was sitting on the ground, with his lance stuck upright beside him – an old veteran, with a thick bushy, grizzly beard, countenance like a lion – a lancer of the Old Guard, and no doubt had fought in many a field. One hand was flourished in the air as he spoke, the other, severed at the wrist, lay on the earth beside him; one ball (case-shot, probably) had entered his body, another had broken his leg. His suffering, after a night of exposure so mangled, must have been great; yet he betrayed it not. His bearing was that of a Roman, or perhaps of an Indian warrior, and I could fancy him concluding appropriately his speech in the words of the Mexican king, 'And I too; am I on a bed of roses?'

I could not but feel the highest veneration for this brave man, and told him so, at the same time offering him the only consolation in my power – a drink of cold water, and assurances that the waggons would soon be sent round to collect the wounded. He thanked me with a grace peculiar to Frenchmen, and eagerly inquired the fate of their army. On this head I could tell him nothing consolatory, so merely answered that it had retired last night, and turned the conversation to the events of yesterday. This truly brave man spoke in most flattering terms of our troops, but said they had no idea we should have fought so obstinately, since it was generally understood that the English Government had, for some inexplicable reason, connived at Napoleon's escape from Elba, and therefore had ordered their army only to make a show of resistance. After a very interesting conversation, I begged his lance as a keepsake. The old man's eyes kindled as I spoke, and he assured me that it would delight him to see it in the hands of a brave soldier, instead of being torn from him, as he had feared, by those vile peasants.

CAVALIÉ MERCER

130.

Earl Stanhope recorded a timeless example of civilian scepticism about tidings received at home from the battlefield, related to him by the Secretary to the Admiralty.

John Wilson Croker gave me an interesting account of his receiving the news of Waterloo – I believe from naval officers in the Channel – and his communication to the Duchess of Angoulême [niece by marriage of the Bourbon King Louis XVIII]. It was in the middle of the night, and Her Royal Highness was residing at a house near Fulham. With some difficulty he obtained admittance, had her awakened, and saw her come down to him in her nightcap and *robe de chambre*. With a profound bow he began to announce the gain of a great battle in Flanders where – 'Stop,' she cried, 'stop!' and to his great astonishment immediately left the room.

But in a few moments she returned, bringing a map. 'Now,' she said, 'show me where was the allied army before the battle?' 'Here, madam.' 'And where after the battle?' – 'Here.' Then on hearing the account – she sunk on her knees and remained a few minutes in silent prayer. On rising, she turned to Croker, and thanked him for the trouble he had taken, expressing also her joy at the event, but added that from her experience of military news, she never trusted the account of any success unless she found that the army said to have gained it was in a more forward position after than before the engagement.

EARL STANHOPE

131.

After Waterloo the wounded horses of the Household Brigade were sold by auction, prompting this rare contemporary example of concern for animal welfare.

Sir Astley Cooper attended the sale, and bought twelve, which he considered so severely hurt as to require the greatest care and attention. Having had them conveyed, under the care of six grooms, to his park in the country, the great surgeon followed, and with the assistance of his servants, commenced extracting bullets and grapeshot from the bodies and limbs of the suffering animals. In a very short time after the operations had been performed, Sir Astley let them loose in the park; and one morning, to his great delight, he saw the noble animals form in line, charge and then retreat, and afterwards gallop about, appearing greatly contented with the lot that had befallen them. These manoeuvres were repeated generally every morning, to his great satisfaction and amusement.

REES HOWELL GRONOW

132.

The 51st Foot on the road from Waterloo to Paris, June 1815.

On the morning of the 24th inst. we marched on Cambray. We soon came in sight of the town, saw the tricolour flag flying on the citadel. A great many stragglers were collected here, and these with the national guard seemed to threaten us with some resistance. A flag of truce was sent to the town but they were fired at, which caused them to return, and a ball had passed through the trumpeter's cap. We now pushed on to the works, near the gate, got into the trenches, fixed our ladders and was soon in possession of the top of the wall. The opposition was trifling, the regular soldiers fled to the citadel, and the shopkeepers to their shops. We soon got possession of the gate and let in the remainder of the brigade, formed and advanced to the great square. We were received by the people with *vivas*, many of whom had forgot to wash the powder off their lips caused by biting off the cartridges when they were firing at us from the wall. Piquets were established at the citadel, and about dusk the remainder of the division were marched out of the town and encamped. We had picked up some money, or more properly speaking we had made the people hand it over to save us the trouble of taking it from them, so we were enabled to provide ourselves with what made us comfortable. About an hour after we had left the town we heard an explosion and soon learned that a Serjt. Corporal and four men fell in with a barrel of gunpowder. They being drunk took it for brandy, and Corporal C— fired into it as he said to make a bunghole, while the others were waiting with tin canteens to catch the supposed liquor, but it blew up. The loss of the regiment was 2 rank and file killed, and ten wounded, exclusive of the brandy merchants, who are so dreadfully scorched it is feared that four cannot recover, and other two will not be fit for service again.

The 25th His pottle belly Majesty, Louis 18th, marched into the loyal town of Cambray. His Majesty was met by a deputation of his beloved subjects who received their father and their King with tears of joy. Louis blubbered over them like a big girl for her bread and butter, called them his children, told them a long rigmarole of nonsense about France, and his family, about his heart, and about their hearts, and I don't know what. The presence of their good old fat King had a wonderful effect on their tender consciences, the air rent with acclamations. The Loyal and faithful soldiers of the Great Napoleon followed their example and surrendered the citadel to their beloved master Old Bungy Louis. No doubt the papers will inform you how his loyal subjects welcomed their beloved King, how the best of monarchs wept over the sufferings of his beloved people, how the Citadel surrendered with acclamations of the joy to the best of kings, and how his most Christian Majesty effected all this without being accompanied by a single soldier. But the papers will not inform you that 4th Division and a brigade of Hanoverian Huzzars were in readiness within half a mile, and if the loyal citizens had insulted their kind, it was very probable we should have bayoneted every Frenchman in the place. The people well knew this, and this will account for the sudden change in their allegiance from their Idol Napoleon (properly named) the Great, to an old bloated poltroon, the Sir John Falstaff of France.

WILLIAM WHEELER

133.

Cavalié Mercer, in France with the victorious Allied army, witnesses the sort of wanton destruction which accompanies every army in every war, in this case the work of Prussian hussars.

A handsome lodge and grille gave a view of a long avenue terminated by a chateau. In this place about twenty or thirty hussar horses were standing, in charge of one man. We learned that his comrades were at the chateau, and thither we went. We were not quite so much shocked at the scene of ruin and havoc which presented itself as we should have been a week ago; they are becoming familiar now. The fragments of sofas, chairs, tables, etc., lying about the grass, bespoke a richly furnished house, and the nearer we drew the thicker became these signs of vengeance. Large pieces of painted paper torn from the walls, remnants of superb silk window-curtains, with their deep rich fringe, hung amongst the bushes; broken mirrors and costly lustres covered the ground in such a manner as to render it difficult to avoid hurting our horses' feet – the brilliant drops of these last, scattered amongst the grass, might, with a little stretch of imagination, have induced us to believe ourselves traversing Sinbad's valley of diamonds; slabs of the rarest marble, torn from the chimney-pieces, lay shattered to atoms; even the beds had been ripped open, and the contents given to the winds, and conveyed by them to all parts of the park, covering in some places the ground like newly fallen snow.

The trees of the avenue were cut and hacked, and large patches of bark torn off – many were blackened and scorched by fires made at the foot of them, with the mahogany furniture for fuel; the shrubs cut down or torn up by the roots; the very turf itself turned up or trampled into mud by the feet of men and horses. Hitchins and I dismounted at the grand entrance into the house; and, by way of securing our horses, shut them up in a little room to which a door was still left. Shouts and laughter resounded through the building. As we passed the magnificent stairs leading up from the hall, we narrowly escaped being crushed under a large mirror which these gentlemen at that very moment launched over the banisters above with loud cheers.

The ground-floor consisted of a suite of magnificent rooms, lofty, finely proportioned, and lighted by a profusion of windows down to the floor. These had been most luxuriously and richly furnished; now they were empty, the papering hanging in rags from the walls, and even the cornices destroyed. Every kind of abuse of France and the French was written on the walls. In one room was the remnant of a grand piano. The sad reflections awakened by this sight may be more easily conceived than described, and I turned from it with a sickening and overwhelming sensation of disgust. The next room seemed to have been chosen as the place of execution of all the porcelain, which had there been collected for a grand smash. The handsomest Sèvres and Dresden vases, tea and dinner services, formed heaps of fragments, and a large porcelain stove had shared the same fate. Another room had been lined with mirrors; these had been made targets of, for many were the marks of pistol-balls on the walls they had covered; little remained except some parts of their rich gilt frames. The last room of the suite was painted to represent the interior of a forest, and on one side was a pool of water, in which several naked nymphs were amusing themselves. The plaster was torn down in large patches, and the nymphs stabbed all over with bayonets. The upper floor consisted of bedrooms, dressing-rooms, and baths, and exhibited the same melancholy destruction; even the leaden lining of the baths, the leaden water-pipes, etc., were cut to pieces. On inquiring of one hussar why they so particularly wreaked their vengeance on this house, he said because it belonged to Jerome Buonaparte, whom every German detested.

CAVALIÉ MERCER

134.

A dinner at Hythe, Kent, in honour of the Peninsular veteran Sir George Murray, around 1822.

There were present Officers of the 95th Rifles, the Tipperary Militia and the South Devon Militia, making some fifty persons. In many regiments it was normal to allow a bottle of port a head and sometimes two or three would be drunk. No one was allowed to escape a bumper toast without a certificate from a doctor. In order to avoid the inevitable drunkenness our Militia Captain, with a softer head than his companions, used to quietly draw his bumper glass off the table and pour his wine into his jackboots, which were then worn high up to the knee with knee-breeches before Wellington boots and trousers were introduced.

The dinner to Sir George Murray commenced at six o'clock in the evening and did not terminate until seven o'clock the following morning. Drunkenness prevailed, many dropped off their seats, while others fell when trying to get out of the room without assistance.

The Tipperary boys were noisy and full of obscene and indecent toasts, while the General himself thrice appeared to fall asleep. Whilst the General was in one of these slumbers the President of the day gave the toast 'To the Immortal Memory of the lamented Sir John Moore, to be drunk in silence.' Whether the word silence or a jog of the General's elbow roused him is not clear, but he attempted to rise from his chair and said 'Mr President, I rise' but, instead of rising, fell to his knees and then continued, 'Mr President, this is a toast I always drink kneeling.' Tremendous applause followed from everyone present sober enough to witness this drama. I left as the clock struck seven, followed separately by the Colonel and then the Captain. A few remnants of this extraordinary banquet remained until breakfast time when the General washed himself, had his boots cleaned and apparently sober appeared on parade at ten o'clock, when as a joke he put an officer under arrest for being drunk.

JOHN BUTTER

135.

An illustration of the hazards of writing military history while veterans of a campaign still live and breathe. John Keegan and others including myself had more recent experiences, when survivors of the 1944–5 northwest Europe campaign sought to defend their units against alleged slights on their honour and performance, though the duels that we faced were merely verbal.

London, October 1832

To Colonel W. Napier, C.B.

Sir,

I have just received a letter from my brother-in-law, dated Missouree [Mussourie, India], the 17th April last, in which he informs me that it was only on the day before he had read a paragraph in your third volume of the 'Peninsular War' reflecting upon his conduct at the battle of Barosa. He desires me to inform you without delay, that immediately on the termination of his military duties in India (which will take place in the latter part of the next year) he will return to this country and require from you that satisfaction which is due from one officer to another for this most unfounded calumny.

I have the honour to be, &c. &c.

[Signature deleted from Napier's published correspondence. The outraged veteran was Lt.-Gen. Sir Samuel Whittingham (1772– 1841). A brother officer contrived an uneasy reconciliation with the historian, without bloodshed.]

JOHN KEEGAN

136.

The novelist Fyodor Dostoevsky (1821–81) passed an undistinguished period in 1837–43 as a cadet at an army engineer academy.

Dostoevsky may have held most of his fellow students in contempt, but he obviously could not endure the idea of being considered by them both personally odd and socially inferior; and the struggle to maintain his social status and self-esteem is quite naively evident in his letters. He writes his father in June 1838 that all of his money had been spent, explaining that, for the May parade of the Academy before the royal family, he had bought himself a new shako. 'Absolutely all my new comrades acquired their own shakos; and my government issue might have caught the eye of the Tsar.' Since, a bit earlier, he notes proudly that 140,000 troops had participated in the spectacle, this eventuality hardly seems a likelihood.

JOSEPH FRANK

137.

On 23 December 1849, Dostovesky and eight companions were taken to Semyonov Place in St Petersburg to be shot by firing squad for alleged subversive activities.

When Dostoevsky was tied to a stake, blindfolded, waiting for execution, he heard the drums beating Retreat. Having been in the Army he
knew that he was not going to be shot. If he had heard Reveille, he
would have known that he had been shot [he was instead reprieved
and exiled].

ANTHONY POWELL

138.

*Cecil Woodham-Smith's devastating portrait of Lord Cardigan
(1797–1868), one of the most pernicious boobies in the history of the
British Army. In 1836 he assumed command of the 11th Light
Dragoons, later restyled the 11th Hussars. Four years later, while the
regiment was quartered at Canterbury, he became exasperated by the
continued presence in his mess of several socially inconsequential
officers who had served with the regiment in India.*

He was now in his glory. When he went to London it was his practice
to give a number of his smartest men a day's leave and five shillings,
and each posted himself at some point which he intended to pass.
People ran to stare as Lord Cardigan sauntered down St James's Street,
saluted at every few yards by his Hussars, brilliant as parakeets, if only
he could have got rid of the 'Indian' officers – it was incredible to him,
he used to remark to his friends, that any of them could be so thick-
skinned as to stay. No 'Indian' officer ever received an invitation to his
own house, and when cards of invitation for dinners and balls were
sent to the mess by gentlemen living in the neighbourhood, he had
made it a rule that they were not to be given to those officers whom,
he said, he had 'found sticking to the regiment in the East Indies'.

Take their drinking habits. In India it had been the custom for officers
to drink porter – it was healthier and cheaper. To this the Lieutenant-Colonel furiously objected. Porter was the drink of factory hands
and labourers, and he wished to make the 11th famous for its splendid
hospitality. He forbade bottled porter to appear on the mess table.

On May 18th, 1840, Major-General Sleigh, the Inspector-General
of Cavalry, and his staff were to dine in the mess of the 11th after an
inspection. Arrangements were made on a magnificent scale, and the

Lieutenant-Colonel gave orders that nothing but champagne was to be served. The result of the inspection was most gratifying: Lord Cardigan was highly complimented on the brilliant appearance, the magnificent mounts and the fine performance of the 11th, and as he entered the mess he was seen to be in high good humour. At dinner one of General Sleigh's aides was sitting next to a certain Captain John Reynolds, an 'Indian' officer and son of a distinguished 'Indian' officer. General Sleigh's aide asked if he might have Moselle instead of champagne, and John Reynolds gave the order to a mess waiter, who placed it on the table in its bottle.

At this moment Lord Cardigan looked down the table, and there, among the silver, the glass, the piles of hot-house fruit, he saw a black bottle – it must be porter! He was transported with rage. John Reynolds was drinking porter under his very nose, desecrating the splendour of his dinner-table. When it was explained to him that the black bottle contained Moselle, he refused to be appeased; gentlemen, he said, decanted their wine. Next day he sent a message to John Reynolds through the president of the Mess Committee, a Captain Jones, who was one of his favourites. Captain Jones found him with two other officers, one of whom did not belong to the regiment. 'The Colonel has desired me to tell you', said Captain Jones, 'that you were wrong in having a black bottle placed on the table at a great dinner like last night. The mess should be conducted like a gentleman's table and not like a pot-house.'

John Reynolds was 'utterly astonished', especially as the message was delivered before an audience, but, controlling himself, he told the other 'in a quiet manner, that he had no right to bring him an offensive message, and as a brother captain it would have been better taste if he had declined to deliver it'. Almost at once he was summoned to the orderly-room, where, before Captain Jones and the Adjutant, Lord Cardigan attacked him in furious rage. 'If you cannot behave quietly, sir, why don't you leave the regiment? That is just the way with you Indian officers; you think you know everything, but I tell you, sir, you neither know your duty nor discipline … Oh yes! I believe you do know your duty, but you have no idea whatever of discipline. I put you under arrest.'

John Reynolds remained silent. Captain Jones then offered his hand, but Reynolds refused to shake it. 'I have no quarrel with you', he said, 'and nothing has passed that makes shaking hands necessary.' Lord Cardigan burst out in a loud voice, 'You have insulted Captain Jones.' John Reynolds quietly repeated, 'I have not, my lord.' Lord Cardigan shouted, 'I say you have. You are under arrest, and I shall report the matter to the Horse Guards.' John Reynolds replied, 'I am sorry for it,' and retired. He was then placed under close arrest, but brought up from time to time, to be examined by Lord Cardigan, who railed at him, taunted him with being an 'Indian' officer, and ordered him to explain himself.

These interviews lasted as long as two hours, and John Reynolds stated, 'I never can describe the mental torture I underwent during the probing and cross-examination of my feelings, lest I should say something that might afterwards be used against me, especially as Lieutenant-Colonel the Earl of Cardigan condescended to assure me that he waived the consideration of being my commanding officer, and afterwards resumed it, so that I had great difficulty in knowing when I was addressing his Lordship as a private gentleman and when in his capacity as Lieutenant-Colonel.'

After three days he received a memorandum from Lord Hill, the Commander-in-Chief, recommending him to admit the impropriety of his conduct towards his commanding officer and to resume friendly relations with Captain Jones. He obeyed the first instruction, but refused to drink wine with Captain Jones or to shake hands with him, and remained under arrest. On June 9th Major-General Sleigh came once more to Canterbury, summoned the officers of the 11th to appear before him, and without holding an investigation, read aloud a letter from headquarters, condemning John Reynolds in the strongest possible language and approving and supporting Lord Cardigan. Reynolds's behaviour was described as 'pernicious and vindictive', and an enquiry was 'absolutely refused' on the ground that 'many things would come to light which are not for the good of the service'.

John Reynolds then asked that he might be court-martialled for the offences he was alleged to have committed, and at this General Sleigh flew into a rage. There was to be no court-martial, no enquiry, no further discussion of the affair; the Commander-in-Chief had made up his mind once and for all that the matter was to be considered as settled. And, turning angrily on John Reynolds, General Sleigh told him that he had 'forfeited the sympathy of every officer of rank in the service'. General Sleigh and Lord Cardigan then left the room together, and Captain John Reynolds resumed his regimental duties with the 11th. The following week the regiment left Canterbury for duty at Brighton Pavilion.

The Army authorities had found themselves in a dilemma; since Lord Cardigan had been reinstated [after an earlier absurdity], for better or worse he must be supported. It was too late to draw back, and the best policy seemed to be firmness: the officers of the 11th must be shown that it was useless to oppose Lord Cardigan. Unfortunately the 'Black Bottle' affair became a nine days' wonder, the phrase caught the public fancy, and 'black bottle' became a catchword. Jokes about the 11th appeared in newspapers and mock reports were circulated of 'The Battle of the Moselle, in which His Royal Highness Prince Albert's Regiment has severely suffered'. A private of the 11th was arrested for assaulting a Guardsman in the street; when reprimanded by Lieutenant-Colonel Lord Cardigan, the man stammered out, 'But, my Lord, he called me a black bottle.'

Meanwhile Captain John Reynolds's guardian pestered Lord Hill for an explanation of General Sleigh's reprimand, for production of the correspondence, and for a court-martial. When he got no satisfaction he sent an account of the affair and copies of his letters to Lord Hill to every leading newspaper in London, and in almost every instance they were printed in full. To Lord Cardigan, however, 'Black Bottle' brought unmixed satisfaction. Once more he had been supported, once more he had been proved right.

CECIL WOODHAM-SMITH

139.

Like so many famous quips of famous men, Gen. Napier's alleged message from India was apocryphal, being instead the inspiration of a woman in Britain named Catherine Winkworth, who submitted it to

Punch. *Byron Farwell (1921–99), who penned this and three other entries below, was a Chrysler executive and three-time mayor of his Virginia community, who also became an exceptionally entertaining and erudite historian of Queen Victoria's army.*

Sir Charles Napier (1782–1853) is said to have sent back a one-word announcement of his 1844 conquest of Sind: '*Peccavi.*' Of course he did not. No more than did General Pershing [arriving in France in 1917] cry 'Lafayette we are here!' But if generals have the wit to win battles there will always be wits to put words in their mouths later.

BYRON FARWELL

140.

At the Battle of Chillianwala on 13 January 1849 in the Second Sikh War, confused orders by Brigadier Pope, commanding the cavalry, caused them to turn and flee.

The flight of the cavalrymen was halted by a chaplain, the Rev. W. Whiting. He was attending the sick and wounded at a field hospital when he saw some frightened dragoons fleeing the battlefield. The chaplain stopped them, and demanded to know what had happened. 'The day is lost!' cried a dragoon. 'All our army is cut up, and the Sikhs have taken our guns and everything.'

'No, sir!' the chaplain said. 'The Almighty God would never will it that a Christian army should be cut up by a pagan host. Halt, sir! or as I am a minister of the word of God, I'll shoot you!' Whiting's knowledge of history may have been faulty, but he stopped the rout … [Gen. Hugh 1779–1869] Gough, with a soldier's view of the hierarchy of the Church of England, proposed that the chaplain be made a brevet-bishop.

BYRON FARWELL

141.

One of West Point Military Academy's more notable failures was the attempt to make a soldier of James Whistler (1834–1903), later a well-known artist but in 1853 a mere gentleman cadet.

It was not unusual at cavalry drill for Whistler, a sorry horseman, to go sliding over his horse's head. On such occasions Major Sacket ... would call: 'Mr Whistler, aren't you a little ahead of the squad?' In the first mounted drill in the riding academy in which Whistler took part, he had a hard horse. The instructor ... gave the command: 'Trot out!' At this command Whistler, who had journeyed from the withers to his croup and back again several times, tumbled in a bundle into the tanbark. He lay for a moment without movement. The dragoon soldiers, who imagined him seriously injured, ran to carry him to the hospital; but he told them to let him down. Major Porter ... called to him from his horse: 'Mr Whistler, are you hurt?' Whistler, leisurely drawing off his gauntlet and brushing the tanbark away from his hips, replied, 'No, Major! but I do not understand how any man can keep a horse for his own amusement!'

There was in the squad a horse named Quaker, which had thrown the most experienced riders in the cavalry detachment. One day this horse fell to Whistler, who coming up blinking with his myopia, said: 'Dragoon, what horse is this?' The soldier answered: 'Quaker, sir,' and Whistler replied: 'My God! He's no friend.' The most famous moment in Whistler's West Point career occurred during his chemistry examination. Asked to discuss silicon, he supposedly asserted, 'Silicon is a gas,' causing his interrogator to end the questioning by saying, 'That will do, Mr Whistler.' Years later Whistler declared, 'If silicon had been a gas, I might have become a general.'

GORDON FLEMING

142.

*John Nicholson (1821–57) was one of a handful of British officers
ruling India before the Mutiny who made himself a legend before his
early death in action.*

One night the officers of the Movable Column were waiting for their
dinner, which was overdue. A messenger was sent to the cooking tent
bringing back word that dinner would soon be ready. About half an
hour after the appointed time Nicholson stalked in, saying abruptly:
'I am sorry, gentlemen, to have kept you waiting for your dinner, but
I have been hanging your cooks.' Nicholson had heard from one of his
spies that the soup had been poisoned with aconite, so just before
dinner he sent for the soup and arrested the cooks, who denied the
accusation; but as they refused to taste it on the ground of caste, he
gave some to a monkey, which died. A few minutes later, reported an
officer who was present, 'our regimental cooks were ornamenting a
neighbouring tree'.

HESKETH PEARSON

143.

*The Battle of Balaclava on 25 October 1854 gave birth to two British
legends, of which this was the first.*

The Russians had now advanced far enough to bring Sir Colin Campbell's force within range of their guns, and they opened fire with
considerable effect. Sir Colin had drawn up his force on a hillock at
the entrance to the gorge leading to Balaclava, and he ordered his men
to lie on their faces in a line two deep on the far slope. Lying helpless
under artillery fire is notoriously a strain, and at this moment the four
squadrons came into view, bearing rapidly down from the Causeway
Heights, while behind them, just becoming visible, was the main body
of the Russian cavalry. The sight was once more too much for the
Turks; they leapt to their feet and officers and men fled for the port,
again crying 'Ship! ship! ship!' As they passed the camp of the Argylls
a soldier's wife rushed out and fell upon them, belabouring them with

a stick, kicking them, cursing them for cowards, pulling their hair and boxing their ears, and so pursued them down to the harbour. Five hundred and fifty men of the Argylls and 100 invalids were now left to stand between the Russian army and Balaclava, and Sir Colin rode down the line telling them, 'Men, remember there is no retreat from here. You must die where you stand.'

To the Russian cavalry as they came on, the hillock appeared unoccupied, when suddenly, as if out of the earth, there sprang up a line two deep of Highlanders in red coats – immortalized in British history as 'the thin red line'. Every man expected to be killed and, determined to sell his life as dearly as possible, faced the enemy with stern steadiness. The Russian cavalry checked, halted, and from the thin red line came a volley of deadly musket-fire, every bullet aimed. The Russians wavered, steadied, advanced, and a second volley was fired. Once more the Russians wavered, and there was a movement forward: the men wanted to dash out and engage the cavalry hand-to-hand, and Sir Colin Campbell was heard shouting sternly, 'Ninety-third! Ninety-third! Damn all that eagerness.' The British steadied, a third volley was fired, and the Russians wheeled and withdrew. The Highlanders burst into hurrahs.

CECIL WOODHAM-SMITH

144.

This account of the 25 October 1854 Charge of the Light Brigade by Captain Edwin Cook (1827–72), a Kent squire's son who took part with the 11th Hussars, has the merit of being much less familiar than William Howard Russell's epic of war correspondence.

Heights of Sebastopol
     December 1st (1854)
     My dear Father,
     I have abstained from making a report about this affair, except just that I had a most wonderful escape and got slightly wounded with a spent shot, as I should have unnecessarily alarmed you because it is impossible to describe it otherwise than as the most downright useless, ridiculous, except to those in it, sacrifice. *The Times* account is so good

For centuries, popular images of soldiers were dominated by sentiment and romance. (*Above*) David slaying Goliath, the most famous military anecdote in the Bible, would find a resonance with the modern Taliban, albeit not with the Pentagon. (*Below*) Classical literature's vision of war emphasized the importance of the individual hero, contrasted with a reality that big battalions were more likely to be decisive. Here, Achilles flaunts his triumph over the fallen Trojan, Hector.

This Greek hoplite rode to war. The horse served as the warrior's principal aid to mobility until the twentieth century. He was handicapped, however, by lack of stirrups, which reached Europe from the Asian steppes only in the second century BC.

(*Above*) A classic image of late fourteenth-century warfare in France, which offered little hint of its squalor and barbarism. (*Below*) Peter Paul Rubens did better with a vivid depiction of victimhood, especially that of women, in his 1639 painting *Consequences of War*.

(*Above*) Diego Velázquez portrayed the 1624 surrender of Breda by the Dutch to the Spanish as a courtly ceremony rather than an incident in one of the most destructive conflicts in European history, during the final century in which most soldiers wore armour. (*Below*) Many painters glorified heroes' battlefield deaths, as did Benjamin West with this 1770 view of the fall of a Christ-like Wolfe at the seizure of Quebec a decade earlier. Common to all such images was that they made soldiers at war appear much too clean.

Marshal Ney – 'bravest of the brave' among Napoleon's marshals – epitomized a vision of warfare as an honourable pursuit of glory, though Ney himself, least intellectual of generals, ended his career shot in a ditch near the Luxemborg Gardens in December 1815 for his betrayal of France's King Louis XVIII.

(*Clockwise*) Queen Boudicca;
William Waller; Julius Caesar;
Joan of Arc; James Graham,
1st Marquess of Montrose.

(*Clockwise*) William Cobbett;
Nadezhda Durova; Peter the
Great; James Brudenell, 7th Earl
of Cardigan; Cavalié Mercer.

(*Above*) Wellington was probably the greatest military commander in Britain's history. His survival at Waterloo represented a miracle in an age when most generals faced at least as much personal risk as their men. Hillingford's painting hints at the shrouded and confused battlefields until the passing of black powder three generations later. (*Below*) In the nineteenth century military uniforms attained a zenith of beauty as fighting men on battlefields still sought to be seen, feared, admired. The twentieth century would transform this culture, with soldiers aspiring to become invisible to their foes. This 1830 painting is from my own collection.

that I shall only say what happened to myself. Our order to charge was brought by a half-madman Capt. Nolan, the order was very difficult to understand rightly. On Lord Lucan asking what he was to charge, the only information was, there is the enemy and there is your order.

The enemy consisted of 15,000 infantry, 4,000 cavalry protected by 10 guns, to reach them we had to go down a ravine between two hills with 10 guns on each besides a host of riflemen. Down we went very steadily, the fire was terrific, it seemed impossible to escape, we were well in range of grape shot on each side besides the barkers in front. I got through safe up to the guns, cut down all that came within reach and then at the cavalry behind, but to our horror the heavy brigade had not followed in support and there was an alarm that we were cut off in rear, which was true. There was nothing left for it, but to cut our way back the same way we came, the Lancers who cut us off made a very mild resistance, they seemed to be astonished at our audacity at charging them in the wretched confusion we were in, we got through with very little loss.

Just after getting through these beggars, I thought I heard a rattle behind and, by Jove, I was only just in time, we were pursued and on looking behind a Muscovite had his sword up just in my range and in the act of cutting me down, I showed him the point of my sword instantly close to his throat, he pulled his horse almost backwards and gave me an opportunity of getting more forward. I now had nothing to fear being on a good horse except going through those infernal guns again – about a quarter of a mile from the batteries I felt a tremendous blow on the calf of the leg, and instantly my poor old horse was hit on the offside and was going to fall.

I jumped off in a second with the pistol in one hand and sword in the other, and had the satisfaction of seeing my old friends only ten yards behind. Things looked very bad indeed, but in spite of my game leg, I ran faster than ever I did before in my life and kept pace with my friends behind till their own fire got so thick that they thought fit to drop it. I was now between these two batteries and could only raise a walk. I must have been a quarter of an hour under fire on foot, and much to the astonishment of my real friends, the few who escaped, I turned up about half an hour after the rest of the stragglers. I have finished my romance, for really my escape was almost romantic, but don't on that account put it in print as some of my brave companions like.

Enough of war, in spite of living in a tent with 3 inches of mud inside and 12 out, I am not down in the mouth. My fortnight at sea put my leg to rights, and those two fools who command us Lucan and C[ardigan]. have been kind enough to place us such a distance from our supplies, the roads being very deep, that they can't feed the wretched little wreck of a brigade, the consequence is that they are dying 8 and 10 a day, now it requires no mathematician to see that this can't last, besides the horses are so weak that I don't think any one of them could trot a mile. They can't bring us into action again.

Tomorrow we move down to Balaclava to be near our supplies and if we have luck get some sort of a roof for man and beast. The Russians are worse, if possible, off than ourselves, they are beginning to starve inside and outside of Sebastopol, still they fire away as briskly as ever, we can't assault it without large reinforcements, the army outside would instantly attack us, I fancy it will not fall for some time yet, and I have also I think a chance of going home to command the depot in about two months' time. So things are looking up. The newspapers arrive very irregularly, please to speak about it and have via Marseilles put upon them. I want also the Illustrated News. Don't put any more money at the bank, but on account, and put it into the 15 per cents.

With best love,

Your affectionate son,

E. A. Cook [whose eldest son was mortally wounded commanding a Household Cavalry regiment at Ypres in October 1914]

145.

Conditions for the British Army in the Crimea were somewhat alleviated in the spring of 1855.

Food had been miraculously improved by Alexis Soyer, the famous chef of the Reform Club, who arrived in March 1855 with full authority from Lord Panmure. Soyer came out at his own expense attended

by a 'gentleman of colour' as his secretary. In manner and appearance
he was a comic opera Frenchman, but Miss [Florence] Nightingale
recognized his genius and became his friend. 'Others,' she wrote, 'have
studied cookery for the purpose of gormandizing, some for show. But
none but he for the purpose of cooking large quantities of food in the
most nutritive and economical manner for great quantities of people.'
Though the authorities received him 'very coolly', Soyer proceeded to
attack the kitchens of the Barrack Hospital which Miss Nightingale
never entered. He composed recipes for using the army rations to
make excellent soup and stews. He put an end to the frightful system
of boiling. He insisted on having permanently allocated to the kitchens soldiers who could be trained as cooks. He invented ovens to bake
bread and biscuits and a Scutari teapot which made and kept tea hot
for fifty men. As he walked the wards with his tureens of soup the men
cheered him with three times three. Finally, he gave a luncheon
attended by Lord and Lady Stratford and their suite, at which he
served delicious dishes made from army rations. In one thing Soyer
failed. Like Miss Nightingale, he strongly objected to the way the meat
was divided; since weight was the only criterion one man might get all
bone; why should not the meat be boned, and each man receive a
boneless portion, with the bones being used for broth? The answer
from Dr Cumming was that it would need a new Regulation of the
Service to bone the meat.

CECIL WOODHAM-SMITH

146.

*Count Leo Tolstoy (1828–1910) experienced the Crimea under siege
as a Russian artillery officer. The event contributed to making him a
pacifist, and to his intimate understanding of soldiers. Here, he writes
of the officers' quarters in August 1855, in his wonderful portrait,*
Sebastopol Sketches.

In the large room of the barracks there was a great number of men;
naval, artillery, and infantry officers. Some were sleeping, others were
conversing, seated on the shot-chests and gun-carriages of the cannons
of the fortifications; others still, who formed a very numerous and

noisy group behind the arch, were seated upon two felt rugs, which had been spread on the floor, and were drinking porter and playing cards.

'Ah! Kozeltzoff, Kozeltzoff! Capital! it's a good thing that he has come! He's a brave fellow! ... How's your wound?' rang out from various quarters. Here also it was evident that they loved him and were rejoiced at his coming.

After shaking hands with his friends, Kozeltzoff joined the noisy group of officers engaged in playing cards. A slender, handsome, dark-complexioned man, with a long, sharp nose and a huge moustache, which began on his cheeks, was dealing the cards with his thin, white, taper fingers, on one of which there was a heavy gold seal ring. He was dealing straight on, and carelessly, being evidently excited by something, and merely desirous of making a show of heedlessness. On his right, and beside him, lay a grey-haired major, supporting himself on his elbow, and playing for half a ruble with affected coolness, and settling up immediately. On his left squatted an officer with a red, perspiring face, who was laughing and jesting in a constrained way. When his cards won, he moved one hand about incessantly in his empty trousers pocket. He was playing high, and evidently no longer for ready money, which displeased the handsome, dark-complexioned man. A thin and pallid officer with a bald head, and a huge nose and mouth, was walking about the room, holding a large package of banknotes in his hand, staking ready money on the bank, and winning.

Kozeltzoff took a drink of vodka, and sat down by the players.

'Take a hand, Mikhaïl Semyónitch!' said the dealer to him, 'you have brought lots of money, I suppose.'

'Where should I get any money! On the contrary, I got rid of the last I had in town.'

'The idea! Someone certainly must have fleeced you in Simpferopol.'

'I really have but very little,' said Kozeltzoff, but he was evidently desirous that they should not believe him; then he unbuttoned his coat, and took the old cards in his hand.

'I don't care if I do try; there's no knowing what the Evil One will do! Queer things do come about at times. But I must have a drink, to get up my courage.'

And within a very short space of time he had drunk another glass of vodka and several of porter, and had lost his last three rubles. A hundred and fifty rubles were written down against the little, perspiring officer.

'Try to send it,' said the dealer to him, pausing a moment in laying out the cards, and glancing at him.

'Permit me to send it tomorrow,' repeated the perspiring officer, rising, and moving his hand about vigorously in his empty pocket.

'Hm!' growled the dealer, and, throwing the cards angrily to the right and left, he completed the deal. 'But this won't do,' said he. 'I'm going to stop. It won't do, Zakhár Ivánitch,' he added, 'we have been playing for ready money and not on credit.'

'What, do you doubt me? That's strange, truly!'

'From whom is one to get anything?' muttered the major, who had won about eight rubles. 'I have lost over twenty rubles, but when I have won – I get nothing.'

'How am I to pay,' said the dealer, 'when there is no money on the table?'

'I won't listen to you!' shouted the major, jumping up, 'I am playing with you, but not with him.'

All at once the perspiring officer flew into a rage.

'I tell you that I will pay tomorrow; how dare you say such impertinent things to me?'

'I shall say what I please! This is not the way to do – that's the truth!' shouted the major.

'That will do, Feódor Feodoritch!' all chimed in, holding back the major.

But let us draw a veil over this scene. Tomorrow, today, it may be, each one of these men will go cheerfully and proudly to meet his death, and he will die with firmness and composure; but the one consolation of life in these conditions, which terrify even the coldest imagination in the absence of all that is human, and the hopelessness of any escape, the one consolation is forgetfulness, the annihilation of consciousness. At the bottom of the soul of each lies that noble spark, which makes of him a hero; but this spark wearies of burning clearly – when the fateful moment comes it flashes up into a flame, and illuminates great deeds.

LEO TOLSTOY

147.

The 1857 Mutiny of the East India Company's Bengal army is today known in the subcontinent as the First War of Independence. The Indian principal of this story of its outbreak commands less respect among his modern descendants than in sentimental British annals.

In 1857, the 1st Bengal Irregular Cavalry, Skinner's Horse, were stationed at Multan under the command of Captain Neville Chamberlain. He sent for help; there was no hope of British troops but the 1st Punjab Cavalry and the 2nd Punjab Infantry were ordered to march. Until they arrived, Chamberlain must rely implicitly on his own men; he told them he trusted them and he showed it by the tasks he gave them to do. Skinner's had always recruited near Delhi; their men were counted Hindustanis by the Punjabis. They were Hindustani Muslims, Rajputs and Jats and the people known to the army as Ranghars, that is, Muslims who claim Rajput origin. At this anxious juncture, the native officer commanding the squadron of Ranghars came to Chamberlain and asked for a private interview. He was unhappy because something had given the Ranghars the impression that the captain did not trust them as completely as he trusted the other squadrons. Chamberlain heard him out, then sent his orderly to the bank, with a note requesting the banker to send him a sword he had deposited for safe custody. It was a jewelled sword, valuable to anyone, but doubly so to Chamberlain because, as everyone in the regiment knew, it had been given him by a close friend who had captured it in battle in Sind. He handed this sword to Shaidad Khan, *rissaldar* of the Ranghar squadron. 'Give me this back', he said, 'when this war is over.' The *rissaldar*'s eyes filled with tears; he knelt and touched the captain's knees. Nothing but death could sever the bond between them. Skinner's were loyal throughout the Mutiny and became the senior regiment of Indian cavalry.

PHILIP MASON

148.

*Lakshmibai, warrior ruler of Jhansi, born in 1828, is today hailed as
a martyr of the Indian independence movement. This account of her
has the merit of being based upon the testimony of a Hindu priest
present at the siege of Jhansi, rather than written by an English pen.*

In June 1857, rebels of the 12th Bengal Native Infantry seized the Star
Fort of Jhansi containing the treasure and magazine, and after persuading the British to lay down their arms on promise of quarter, massacred
around fifty European officers along with their wives and children.
Thereafter the widowed Rani of Jhansi, Lakshmibai, whose role in the
massacre was equivocal, assumed direction of the state on behalf of her
adopted son. Her forces frustrated an attempt by Mutineers to assert
the claim to the throne of a rival prince, who was captured and imprisoned. There was then an invasion of Jhansi by local allies of the East
India Company. The Rani appealed to the British for aid but the
governor-general by then believed Lakshmibai responsible for the
massacre of European officers, and ignored her request.

In August 1857, her forces nonetheless defeated the invaders. For
the ensuing six months Lakshmibai ruled unchallenged, becoming
increasingly hostile to the British. Her daily routine was most unusual
for a Hindu woman: she spent several hours practising equitation,
sword-fighting, and other martial exercises. After breakfast followed an
hour's sleep or a lengthy bath; then religious rituals, including acts of
atonement for not cutting her hair in widowhood. At noon she ate
lunch, then returned to work, and distributed gifts to the poor. She
held court at 3 p.m., sometimes dressed as a man. Her chosen personal
ornaments were few: she wore only golden bangles, a pearl necklace
and a diamond ring. In the last hours before sunset, she gave orders for
the administration of the state's departments of justice, revenue and
accounts. In the evenings, she listened to readings from religious books.

In March 1858 a British force commanded by Gen. Hugh Rose
arrived and summoned the city to surrender. The Rani responded: 'We
fight for independence. In the words of Lord Krishna, we will if we are
victorious, enjoy the fruits of victory, if defeated and killed on the field

of battle, we shall surely earn eternal glory and salvation.' Rose reported that Rani Lakshmibai is 'personable, clever and beautiful' – and 'the most dangerous of all Indian leaders'. When Rose began to bombard Jhansi, 'the Rani girded her sword and personally supervised counter-battery fire'. She rewarded her gunners handsomely and they repaired silenced guns. The British concentrated their heaviest fire on the royal palace and one shell exploded in the special apartment reserved for the Ganesh festival. This was a spacious hall, beautifully furnished with Lucknow glass, chandeliers and precious works of art. The shell shattered all the glass which gave forth a strangely sweet smell: 'That day the enemy never stopped pouring fire on us.'

Rose's army eventually stormed the city, meeting fierce resistance in every street and in every room of the palace. Fighting continued into the following day and no quarter was given, even to women and children. 'No maudlin clemency was to mark the fall of the city,' wrote a British witness without apology. Yet Vishnu Bhatt Godse wrote: 'How cruel and ruthless were these white soldiers, I thought; they were killing people for crimes they had not committed.' The Rani withdrew from the palace to the fort and eventually resigned herself to flight with her young son and a small party.

Godse, himself a fugitive, was drinking at a well when Lakshmibai and her escort pulled up: 'The Rani was dressed like a Pathan and completely covered in dust and dirt. She was very thirsty, and dismounting asked the travellers who we were.' Godse introduced them as Brahmins from Jhansi at which the Rani recognized him. He started drawing water from the well for her but she stopped him, saying that a learned Brahmin should not be drawing water for her. She drew water for herself. Then the Rani spoke despairingly to Godse, saying she had no clue what sins she committed in her previous lives to suffer so much in this one, but she recognized that there was no escape from destiny.

She reached a rebel stronghold at Kalpi very late one night. 'Next morning while they were preparing to meet the King, she realized that her periods were beginning. This depressed her much since in spite of all her efforts, the very fact of being a woman was disrupting all her plans. Besides she did not have any clothes, nor any money to buy any.

But apparently Rao Saheb got to know about the situation and sent Tatya Tope to get everything that the Rani would need.' Thereafter Lakshmibai's dwindling forces were pursued, repeatedly engaged and defeated. On 17 June 1858 in Kotah-ki-Serai near the Phool Bagh of Gwalior, a squadron of the 8th Hussars surprised the Rani as she sought yet again to flee with her remaining troops. In this engagement, one legend holds that she assumed a sowar's uniform and attacked one of the hussars; she was unhorsed and also wounded, probably by his sabre.

Shortly afterwards, as she sat bleeding by the roadside, she recognized the soldier and fired at him with a pistol, whereupon he 'dispatched the young lady with his carbine'. According to Godse's account, however, the Rani was shot almost immediately, then struck by a heavy sword blow on the head, which caused her to collapse in the saddle. Another rebel leader guided her horse away from the battlefield with its dying rider, and later presided over her cremation and burial. Hugh Rose reported that she had been buried 'with great ceremony under a tamarind tree under the Rock of Gwalior'.

VISHNU BHATT GODSE AND TRADITIONAL ACCOUNTS

149.

By 9 November 1857, Sir Colin Campbell's relief force was poised to move upon Lucknow, but the garrison possessed no means of communicating with him about the best line of advance. At last, a 36-year-old Irish civil administrator named Thomas Kavanagh, married with fourteen children, volunteered to carry a message from Lucknow's commander, Sir James Outram. Without a word to his wife, he slipped out of the Residency disguised as an Indian and made his way safely through the enemy lines. For his deed he became the first civilian to be awarded the Victoria Cross. Campbell followed [Lt. Gen. Sir James 1803–63] Outram's suggested route to the triumphant relief of Lucknow. Lacking ciphers, the general was compelled to improvise a classically Victorian method of disguising his dispatch. Kavanagh carried the following message.

Sir,

I σενδ un σκετch du γρουνδ ιντερυενινγ βετωην αλυμ βαγ et cette ποσιτιον et βεγ à συγεστ θη φολοωινγ μοδε d'οπερατιονς as that whereby vous may εφεκτ une jυνκτιον avec nous avec le λεαστ διφικυλτη. φρομ αλυμ βαγ πασινγ ρουνδ le σουθερν φασε de l'ενκλοσυρε et βετωην les υιλαγες de ὑσῆτνυγ et πορωα et προςηδινγ αλμοστ δυε εαστ pour αβουt trois μιλες pour un λευελ κουντρη de γρας λανδ et κλτιυατιον avec un σῃαλοω jῃηλ à κρος̧ς σῃορτλη après ληυινγ αλυμ βαγ (προβαβλη pas plus θαν ανκλε δηπ νοω et νο οβσταχλε aux γυνς) vous will αριωε οποσιτε à la υιλαγε de jαμαιτα sur votre λεφτ—σλαντινγ παστ wʰ pour αβουτ une μιλε à νορθ εαστ vous αριωε à la δικυσῃαρ παλασε—rien mais le παρκ ωαλ ιντερυενινγ αβουτ ειτ φητ í, wʰ est βροκεν δοων dans μανη πλασες et κουλδ βη νοκεδ δοων ανηωῃερε par a κουπλε δες πιονηρς. Le παλασε ἀωυινγ λαργε ωινδοως en ευροπεαν στιλε n'est pas λικλη être δεφενδεδ, mais ιφ σο, a φευ κανον σῃοτ wᵈ σοον εμτη ιτ et ινδηδ j'αντισιπατε λιτλε ορ νο οποσιτιον à votre οκυπατιον de δικυσῃαρ παλασε et παρκ ορ des νειγῃβουρινγ μαρτινιέρ̧ς or βιβιαπορε maisons shᵈ vous θινκ νεσεσαρη, l'ενεμή'ς τρωπς βεινγ cηφῆλη sur θις σιδε du κανal. L'υνιον jακ ὀιστεδ au τοπ de la παλασε et un ρογαλ σαλυτε φρομ υος γυνς à δραω notre ατεντιον το ιτ, shᵈ vous ἀυε ἀδ νο πρευιους φιρινγ wᵈ ινφορμ nous de votre αριυαλ et notre υνιον wʰ nous wᵈ θεν ὀιστ sur la cῃυτρ μανℨιλ παλασε (δισταντ deux à trois μιλες) will show vous que nous sommes ινφορμεδ. A la δικυσῃαρ vous avez un οπεν μαιδαν pour ενκαμπμεντ οφ νῃρλη une μιλε βετωην le παλασε & un δηπ κανal βετωην vous et la ville, les βριδγες sur wʰ sont βροκεν δοων. Par ενκαμπινγ avec votre φροντ to θη κανal avec vos γυνς en votre φροντ et φλανκς vous wᵈ κηπ δοων ανη φιρε wʰ l'ενεμη cᵈ βρινγ αγαινστ vous φρομ λε ville σιδε pour ils ont ονλη sept ορ huit γυνς en διφερεντ ποσιτιονς sur cette σιδε de la γοομτη γυαρδινγ les διφερεντ ῆγρεσσες τοωαρδς la δικυσῃαρ φρομ νοτρε ποσιτιον wʰ sont οφ διφερεντ καλιβρε et σο βαδλη φουνδ εν καριαγες ils wᵈ have σομε διφικυλτη ιν τακινγ θεμ αωαι à τυρν αγαινστ vous et ιφ ιλς διδ ρεμουε θεμ πουρ θατ πυρποσε it wᵈ φασιλιτατε notre δασῃ ουτ à μητ vous quand vous δο αδυανσε à θις σιδε du κανal where however ils ne sont pas λικελη à στανδ εξποσεδ as ils θεν wᵈ be à αταk φρομ φροντ et ρεαρ. Under κουερ de vos γυνς vous n'aurez pas de διφικυλτη εν σλοπινγ πασαγες pour votre αρτιλερη δοων votre σιδε ιντο le κανal et υπ νοτρε σιδε δυρινγ θη φιρστ νιτε ρεαδη à κρος̧ς ερλη νεξτ μορνινγ. φυρθερ δελαι, je pense, wᵈ be ιμπολιτικ, as it wᵈ give l'ενεμη τιμε à βρινγ γυνς φρομ δισταντ πλασες. λεστ μεσενγερς shᵈ μισκαρη votre σιγναλ pour ιντενδινγ à κρος̧ς le κανal ιν θη μορνινγ might be τρως γυνς φολοωεδ par trois ροκετς la nuit βεφορε, après une ρεκοναισανσε ἀδ σατισφιεδ vous de la φησιβιλτη de πρεπαρινγ les σλοπες δυρινγ λα νυιτ. Les βανκς du κανal sont φρομ vingt à vingt cinq φητ í, περαπς λες̧ς τοωαρδς votre droit avec λιτλε ορ νο eau et σουνδ βοτομ. Vous wᵈ οφ κουρσε ἀυε παρτης en οκυπατιον de la δικυσῃαρ παλασε, et après πασινγ le κανal en σομε des πρινσιπαλ βυιλδινγς κομᵛ votre λινε de κομᵛ· mais nous shᵈ μητ vous ἀλφ ωαι avec un πρετη στρονγ κολυμν d'ευροπεανς et γυνς et wᵈ θεν αρανγε τογεθερ le μοδε de μαινταινινγ la κομᵛ· βετωην votre καμπ sur le κανal et notre εντρενcῃμεντ. Vous wᵈ περαπς ἀλτ deux trois jours à αλυμ βαγ et may κοντριυε à donner nous νοτισε du jour de votre αδυανσε. Of course tous les τρωπς à αλυμ βαγ seront υνδερ vos κομανδ et un petit γυαρδ ινκλυδινγ λες κονυαλεσεντς will συφισε à μαινταιν θατ πλασε, θυς πλασινγ σομε cinq ou σιξ cents ευροπεανς à votre δισποσαλ—βεσιδες γυνς.

True Copy. J. Outram., George Couper, Seçy. Chief Comr.

150.

The Times *correspondent William Howard Russell vividly described to his editor, John Thaddeus Delane, the palace plunder that fell to Campbell's force when it finally stormed Lucknow.*

Russell had no moral objection to looting, which was universally accepted as one of the perquisites of being with the vanguard. His only regret was that he was not able to take full advantage of his own chances. In one of the palaces, the Kaiserbagh, a soldier found a boxful of diamonds, emeralds and pearls and offered them all to Russell for a hundred rupees. But he had no money on him: 'I might have made my fortune if I had had a little ready money. As it was I loaded myself with jade and got a diamond drop etc. I might have secured a small sackful. I could not believe these things were real which I saw.' Delane responded with commiserations about his correspondent's missed opportunities, 'You have done so admirably well that everyone admits that your story of Lucknow equals the very best of your Crimea achievements. Pray draw £10 on my account and carry it all in gold about you when you next accompany a storming party. It makes one blasphemous to think that you got nothing out of the Kaiserbagh for the want of a few rupees.'

ALAN HANKINSON

151.

John Keegan, writing in 1975, explored the significance of drill to the armies of the nineteenth century.

Christopher Duffy, who was lucky enough to spend some weeks teaching Yugoslav militia the elements of Napoleonic drill for a film enactment of *War and Peace*, described to me the thrill of comprehension he experienced in failing to manoeuvre his troops successfully across country 'in line' and of the corresponding ease with which he managed it 'in column', thus proving to his own satisfaction that Napoleon preferred the latter not because it more effectively harnessed the revolutionary ardour of his troops (the traditional 'glamorous' explanation), but because anything more complicated was simply impracticable. I myself recall a similar archaeological pang in catching a glimpse of a Guards sergeant marching backwards before his squad who were learning the slow-march on the Sandhurst drill square; the angle of his outstretched arms and upraised stick, his perfectly practised disregard for any obstacle in his backward path, the exhortatory rictus of his expression exactly mirrored the image, sketched from life by Rowlandson, of a Guards sergeant drilling his recruits on Horse Guards Parade a hundred and seventy years before; and through that reflection I suddenly understood the function – choreographic, ritualistic, perhaps even aesthetic, certainly much more than tactical – which drill plays in the life of long-service armies.

JOHN KEEGAN

152.

Henri Dunant (1828–1910) was a Swiss businessman who chanced to arrive at Solferino in Lombardy on the evening of 24 June 1859, after the French and Austrian armies had fought a devastating battle which left 23,000 dead and wounded men without care on the field. His experience, at the age of thirty-two, caused him first to write a memoir bearing witness to the horrors, then to become the principal inspiration for the founding of the International Committee of the Red Cross.

The sun on the twenty-fifth of June blazes down upon one of the most frightful sights imaginable. The battlefield is everywhere covered with corpses, of men and horses. They appear as if sown along the roads, in hollows, thickets and fields. Fields ready for harvesting are ruined, corn trampled, fences overturned, orchards destroyed. Here and there one finds pools of blood. The villages are deserted. They bear traces of bullets, bombs, shells and grenades. Houses whose walls have been pierced by gunfire are battered and ruined. The inhabitants, most of whom have passed almost twenty hours in their cellars without light or food, start to emerge. The stupefied expressions of these poor peasants testifies to the long terror they have endured. The ground is strewn with debris, broken weapons, articles of equipment and bloodstained clothing.

The miserable wounded gathered up during the day are pale, livid and inert. Some, the most serious cases, have a vacant look, seeming not to understand what is said to them. They turn staring eyes toward those who bring help. Others, in a dangerous state of shock, tremble convulsively. Still others, with unbandaged wounds already inflamed, seem maddened by their pain; they beg that someone may end their sufferings; with drawn faces, they writhe in the final torments of agony. Elsewhere poor fellows lie immobile, their arms and legs broken by gun wheels that have passed over them.

Shells and caseshot create unendurably painful fractures as well as terrible internal injuries. Every kind of fragment of bone, ball of earth, lead, fabric, equipment, shoe leather, aggravate wounds and increase suffering. Those who cross this vast battlefield encounter at every step inexpressible despair. Some of the units which during the action had discarded their knapsacks now find them again, but pillaged of their contents, a grave loss to poor men whose linen and uniforms are stained and torn. Not only are they deprived of apparel, but even of their pathetic savings, their tiny fortunes as well as family treasures dear to them; mementoes of mothers, sisters and sweethearts. In several places the dead are stripped naked by looters, who do not always spare the wounded.

Here old General Le Breton wanders, seeking his son-in-law, the wounded General Douay. There, Colonel de Maleville, shot at Casa Nova, expires. Here, Colonel de Genlis' dangerous wound causes a burning fever. There, Lieutenant de Selve of the artillery, only a few weeks out of Saint Cyr, has his right arm amputated. I help to care for a poor sergeant-major of the Vincennes Chasseurs, both of whose legs are pierced through with balls. I meet him again in the Brescia Hospital, but he will die crossing Mount Cenis. Lieutenant de Guiseul, who was believed dead, is found unconscious beside the standard he bore.

The courageous sub-lieutenant Fournier, gravely wounded, concludes in his twentieth year a military career commenced in his tenth by voluntary enlistment in the foreign legion. They bury Commandant de Pontgibaud, who died during the night, and young

Count de Saint Paer, who was promoted major barely a week ago. General Auger, of the artillery, is carried to the field hospital of Casa Morino with his left shoulder shattered by a six-inch shell. A fragment remained imbedded for twenty-four hours inside the muscle of his armpit. Carried to Castiglione he is attacked by gangrene, which kills him. General de Ladmirault and General Dieu, both gravely wounded, also arrived at Castiglione. The lack of water becomes ever more acute. The sun burns, ditches dry up. Soldiers have only brackish, filthy water to appease thirst. Where even the least little stream or trickling spring is found, armed guards struggle to save this water for those who need it most.

Wounded horses which lost their riders and have wandered all night, drag themselves towards other animals, from which they seem to beg help. They are put out of their agony by a bullet. One noble charger trots alone into the midst of a French company. Ornate saddle-bags show that it belongs to Prince von Isenberg, who is himself later found among the wounded. Careful nursing through a grave illness will allow him to return to Germany, where his family, in ignorance, mourned him for dead. The faces of some corpses look serene; these are men who died instantly, after being hit. But those who lingered before expiring have limbs stiffened and twisted in agony, bodies caked with filth; hands clawing at the earth; eyes open and staring; convulsive contraction exposes clenched teeth.

Three days and nights are passed in burying the dead. On so large a field, many corpses lie out of sight in ditches, thickets or irregularities of the ground, and are only belatedly discovered. They, like the dead horses, emit a fetid stench. French troops are detailed to identify and bury the dead. As far as possible men are detailed to locate comrades from their own units. They note the numbers stamped on the clothing of the dead. Then, aided in this painful duty by paid Lombardy peasants, they lay the corpses in a common grave. In their inevitable haste, and because of the carelessness of the paid hands, more than one living man is, alas, interred with the dead.

153.

Sir Harry Smith's Arab charger Aliwal, having carried him through
all the battles of the Sutlej campaign, followed him home to England
in 1847 and thence to South Africa and back again in 1852. In
1859 the old general concluded that the old warhorse's day was done.
His aide-de-camp's daughter describes the parting.

My sister and I have a vivid recollection of the lovely horse, and how,
when we used to meet Sir Harry when we were out walking and he was
riding, he would call out, 'Stand still, children,' and then come galloping up at full speed, and Aliwal would stop at our very feet; and my
mother used to tell us that on the anniversary of the battle of Aliwal
(28 January 1846), when there was always a full-dress dinner at the
General's house, someone would propose Aliwal's health, and Sir
Harry would order him to be sent for. The groom would lead the
beautiful creature all round the dinner-table, glittering with plate,
lights, uniforms, and brilliant dresses, and he would be quite quiet,
only giving a snort now and then, though, when his health had been
drunk and the groom had led him out, you could hear him on the
gravel outside, prancing and capering. The horse was now old, and Sir
Harry, in his new house in London, would not be able to keep him;
and though Sir Robert Gerard kindly offered him a home, Sir Harry
feared that his old age would perhaps be an unhappy one, and he
resolved to shoot him. My father and the faithful groom were with Sir
Harry when he did so, and I believe they all shed tears.

MISS PAYNE

154.

Lord Roberts (1832–1914) recalls a frustrating experience in 1860,
when he was an ambitious young quartermaster-general serving in
India.

For some time I had been indulging a hope that I might be sent to
China with my old general, Hope Grant, who had been nominated to
command the expedition which, in co-operation with the French, was

being prepared to wipe out the disgrace of the repulse experienced early in the year, by the combined French and English naval squadrons in their attack on the Taku forts. My hope, however, was doomed to disappointment. A day or two afterwards we dined with the Cannings, and Lord Clyde took my wife in to dinner. His first remark to her was: 'I think I have earned your gratitude, if I have not managed to satisfy everyone by these China appointments.' On my wife asking for what she was expected to be grateful, he said: 'Why, for not sending your husband with the expedition, of course. I suppose you would rather not be left in a foreign country alone a few months after your marriage? If Roberts had not been a newly-married man, I would have sent him.' This was too much for my wife, who sympathized greatly with my disappointment, and she could not help retorting: 'I am afraid I cannot be very grateful to you for making my husband feel I am ruining his career by standing in the way of his being sent on service. You have done your best to make him regret his marriage.' The poor old Chief was greatly astonished, and burst out in his not too refined way: 'Well, I'll be hanged if I can understand you women! I have done the very thing I thought you would like, and have only succeeded in making you angry. I will never try to help a woman again.'

LORD ROBERTS

155.

The Third China War of 1860, which Roberts was so miserable to miss, was one of a long series of nineteenth-century clashes in which the European powers – in this case, Britain and France – sought to assert the will of the 'Western barbarians' against the Chinese on the most spurious pretexts. Having stormed the Taku forts at the mouth of the Pai Ho river, the army proceeded to Peking where the French looted the Summer Palace, a Chinese Versailles. The British, not to be outdone, then burned down what was left, to 'teach the Chinese a lesson'.

As regards the burning of the palace, we have never at any time hesitated to assume the responsibility of the deed, but not from the motives attributed by General de Montauban, who now justifies an act in which at the time he refused to participate. The general states: 'We hoped, therefore, that this vigorous demonstration would have a good result. Unfortunately, our hopes were not realized soon enough to save some unfortunate European envoys from frightful tortures; still more, the correspondent of the "*Times*" was put to death; and it was when the bloody fragments of his body were found that the English resolved to burn the palace in revenge for the murder of their countryman. I perfectly remember that I then made some observations to Lord Elgin [the British Envoy], who replied in a form of discreet confidence, "What would the '*Times*' say of me if I did not avenge its correspondent?"'

AN UNNAMED BRITISH MEDICAL OFFICER

156.

From the seventeenth until the mid-nineteenth century, the West African kingdom of Dahomey, in what is now Benin, employed women soldiers who fought alongside their male counterparts. The first Europeans who encountered them dubbed the Dahomean soldiers 'black Amazons'. The Dahomeans called them abosi *– 'the king's wives' – or* minos *– 'our mothers'.*

The female regiments lived in the palace, which was off limits to all men except the king. Unlike other palace women, such as concubines, they were required to remain celibate. In compensation, they enjoyed more autonomy than most women in Dahomey. In their time off they swaggered like any other band of elite soldiers with a three-day pass: drinking, dancing and singing rowdy songs, many of them to the effect that the men could stay behind and plant crops, a job Dahomeans considered women's work, while the *minos* headed out to eviscerate their enemies. They not only disdained 'women's work' but claimed that by proving themselves equal to, or better than, their male counterparts on the battlefield, they had become men.

Most early references to Dahomey's women warriors describe them carrying muskets and comment on their prowess with the weapon. British traveller John Duncan was impressed: 'I was certainly surprised to see the certainty of their deadly aim ...' Dahomean soldiers also carried machetes, a weapon that French trader Edmond Chardoin reported they 'wield with much skill and with which they lop off a limb or a head with a single blow as if it were an ordinary cane of bamboo'. The *minos* fought not in units alongside men but in separate units commanded by women. The units were not easily distinguished from each other in the field because their uniforms were similar: sleeveless, kilt-length tunics and shorts. Oral history traces the origin of the *minos* to a group of women who hunted elephants for King Wegabja (ca. 1645–1680). Known as the *gheto*, they were presumably responsible for supplying him with ivory and meat. While the jump from women hunting elephants to women fighting the enemy seems logical enough, there is no direct link to prove it is true. The closest tie we have is a probably apocryphal story that when King Gezo (ruled 1818–1858) praised his female elephant-hunters for their courage, they answered 'a nice manhunt would suit them even better.'

Veterans of Dahomey's all-female regiments survived into the twentieth century. Several travellers reported encounters with elderly women who claimed to be former *minos*. The most touching of these comes from a man who told the story of an ancient woman he and his friends used to see on the street of Cotonou when they were boys, around 1930. On one occasion, they startled her with a sharp noise. As they watched, she straightened to attention, fired an imaginary rifle, pounced on an imaginary enemy, and stabbed him with an imaginary dagger. She performed a victory dance, holding up an imaginary trophy of severed genitals, then shrank, confused, into old age again. An adult explained to the shaken boys: 'She is a former warrior … In the time of our former kings, there were women soldiers. Their battles ended long ago, but she continues the war in her head.' As what soldier does not?

PAMELA TOLER

157.

In the summer of 1861 Ulysses S. Grant (1822–85) was appointed Colonel of the 21st Illinois Volunteers, and soon afterwards ordered to march against a Confederate force at the little town of Florida, commanded by Col. Thomas Harris.

In the twenty-five miles we had to march we did not see a person, old or young, male or female, except two horsemen who were on a road that crossed ours. As soon as they saw us they decamped as fast as their horses could carry them. I kept my men in the ranks and forbade their entering any of the deserted houses. We halted at night on the road and proceeded the next morning at an early hour. Harris had been encamped in a creek bottom for the sake of being near water. As we approached the brow of the hill from which it was expected we could see Harris' camp, and possibly find his men ready formed to meet us, my heart kept getting higher and higher until it felt to me it was in my throat. I would have given anything then to have been back in Illinois, but I had not the moral courage to halt and consider what to do; I kept right on. When we reached a point from which the valley below was in full view I halted. The place where Harris had been encamped a few days before was still there, but the troops were gone. My heart resumed its place. It occurred to me at once that Harris had been as much afraid of me as I had been of him. This was a view of the question I had never taken before, but it was one I never forgot afterwards. From that event to the close of the war, I never experienced trepidation upon confronting an enemy, though I always felt more or less anxiety. I never forgot that he had as much reason to fear my forces as I had his.

ULYSSES S. GRANT

158.

Through the ages more than a few civilians have rejected Samuel Johnson's view that 'every man thinks meanly of himself for not having been a soldier', and have disdained warriors. One such was the novelist Anthony Trollope (1815–82), who during a tour of North America in 1861 visited the Army of the Potomac, and did not like what he saw.

To me the soldiers seemed to be innumerable, hanging like locusts over the whole country – a swarm desolating everything around them. Those pomps and circumstances are not glorious in my eyes. They affect me with a melancholy which I cannot avoid. Soldiers gathered together in a camp are uncouth and ugly when they are idle; and when they are at work their work is worse than idleness. When I have seen a thousand men together, moving their feet hither at one sound and thither at another, throwing their muskets about awkwardly, prodding at the air with their bayonets, trotting twenty paces here and backing ten paces there, wheeling round in uneven lines, and looking, as they did so, miserably conscious of the absurdity of their own performances, I have always been inclined to think how little the world can have advanced in civilization, while grown-up men are still forced to spend their days in such grotesque performances. Those to whom the 'pomps and circumstances' are dear – nay, those by whom they are considered simply necessary – will be able to confute me by a thousand arguments. I readily own myself confuted. There must be soldiers, and soldiers must be taught. But not the less pitiful is it to see men of thirty undergoing the goose-step, and tortured by orders as to the proper mode of handling a long instrument which is half gun and half spear. In the days of Hector and Ajax, the thing was done in a more picturesque manner; and the songs of battle should, I think, be confined to those ages.

ANTHONY TROLLOPE

159.

Lew Wallace (1827–1905), who later became a Union general and the author of Ben Hur, *as a boy was disappointed in a yearning to attend West Point, and also failed in an early quest to be a novelist. Instead he became a lawyer and local politician.*

When the Civil War began he was made Colonel of the 11th Indiana. He was sent to Paducah soon after the place was occupied by Union troops, and – his political connections being first-rate – it was not long before he learned that he was being made a brigadier-general. This unsettled him a bit, and he went to General [Charles F.] Smith to ask advice. Smith had taken over a big residence for headquarters, and Wallace found him sitting by the fire after dinner, taking his ease, his long legs stretched out, a decanter on the table. Smith was, said Wallace, 'by all odds the handsomest, stateliest, most commanding figure I had ever seen'. Somewhat hesitantly Wallace showed him his notice of promotion and asked if he should accept.

Smith had worked thirty-five years to get his own commission as a brigadier, and the idea that any officer might hesitate to accept such a thing stumped him. Why on earth, he asked, should Wallace not take it? 'Because,' confessed Wallace, 'I don't know anything about the duties of a brigadier.' Smith blinked at him. 'This,' he said at last, 'is extraordinary. Here I have been spending a long life to get an appointment like this one about which you are hesitating. And yet – that isn't it. That you should confess your ignorance – good God!' Then Smith reached for the decanter, poured Wallace a drink, and told him to accept the promotion and stop worrying. He dug into a table drawer, got out a copy of the United States Army Regulations, and declared that a general should know these rules 'as the preacher knows his Bible'.

Then he went on to sum up his own soldierly philosophy in words which Wallace remembered: 'Battle is the ultimate to which the whole life's labour of an officer should be directed. He may live to the age of retirement without seeing a battle; still, he must always be getting ready for it as if he knew the hour of the day it is to break upon him. And then, whether it come late or early, he must be willing to fight – he must fight.'

BRUCE CATTON

160.

General Thomas J. Jackson (1824–63) at Bull Run, 21 July 1861, where he earned the sobriquet 'Stonewall', forever afterwards attached to his name.

Jackson had been ordered to the Stone Bridge. Hearing the heavy fire to his left increasing in intensity, he had turned the head of his column towards the most pressing danger, and had sent a messenger to Bee to announce his coming. As he pushed rapidly forward, part of the troops he intended to support swept by in disorder to the rear. Imboden's battery came dashing back, and this officer, meeting Jackson, expressed with a profanity which was evidently displeasing to the general his disgust at being left without support. 'I'll support your battery,' was the brief reply; 'unlimber right here.' At this moment appeared General

Bee, approaching at full gallop, and he and Jackson met face to face.
The latter was cool and composed; Bee covered with dust and sweat,
his sword in his hand, and his horse foaming. 'General,' he said, 'they
are beating us back!' 'Then, sir, we will give them the bayonet'; the thin
lips closed like a vice, and the First Brigade, pressing up the slope,
formed into line.

Jackson's determined bearing inspired Bee with renewed confidence.
He turned bridle and galloped back to the ravine where his officers
were attempting to reform their broken companies. Riding into the
midst of the throng, he pointed with his sword to the Virginia regiments, deployed in well-ordered array on the height above. 'Look!' he
shouted, 'there is Jackson standing like a stone wall! Rally behind the
Virginians!' The men took up the cry; and the happy augury of the
expression, applied at a time when defeat seemed imminent and hearts
were failing, was remembered when the danger had passed away.

G. F. R. HENDERSON

161.

*Jackson's subordinates were frequently exasperated by his passion for
secrecy – security, as we should call it – concerning his intentions.*

When Jackson was informed of the irritation of his generals he merely
smiled, and said, 'If I can deceive my own friends I can make certain
of deceiving the enemy.' Nothing shook his faith in Frederick the
Great's maxim, which he was fond of quoting: 'If I thought my coat
knew my plans, I would take it off and burn it.' An anecdote told by
one of his brigadiers illustrates his reluctance to say more than necessary. Previous to the march to Richmond this officer met Jackson.
'Colonel,' said the general, 'have you received the order?' 'No, sir.'
'Want you to march.' 'When, sir?' 'Now.' 'Which way?' 'Get in the
cars. Go with Lawton.' 'How must I send my train and battery?' 'By
the road.' 'Well, general, I hate to ask questions, but it is impossible to
send my wagons off without knowing which road to send them.' 'Oh!'
– laughing – 'send them by the road the others go.'

G. F. R. HENDERSON

162.

Gettysburg, 1 July 1863, from the Confederate ranks.

Late in the afternoon of this first day's battle, when the firing had greatly decreased along most of the lines, General Richard S. Ewell and I were riding through the streets of Gettysburg. In a previous battle he had lost one of his legs, but prided himself on the efficiency of the wooden one which he used in its place. As we rode together, a body of Union soldiers, posted behind some buildings and fences on the outskirts of the town, suddenly opened a brisk fire. A number of Confederates were killed or wounded, and I heard the ominous thud of a Minié ball as it struck General Ewell at my side. I quickly asked: 'Are you hurt, sir?' 'No, no,' he replied; 'I'm not hurt. But suppose that ball had struck you, we would have had the trouble of carrying you off the field, sir. You see how much better fixed for a fight I am than you are. It don't hurt a bit to be shot in a wooden leg.'

JOHN B. GORDON

163.

The death of Gen. John Sedgwick (1813–64), commanding general of the Union's VI Corps, Spotsylvania Court House, Virginia, 9 May 1864.

After a conference with Grant, Sedgwick rode forward to an elevation near the centre of his position, found that his men were a little nervous because of the fire of Confederate sharpshooters, assured them that there was nothing to worry about because 'they couldn't hit an elephant at this distance …', and then himself fell dead with a sharpshooter's bullet in his brain.

BRUCE CATTON

164.

After the Spotsylvania battles of 10–12 May 1864. The author of this tale had run away from home at Buffalo, New York, two months earlier at the age of fourteen to enlist as an artilleryman with the Union army.

I breakfasted about 3 p.m., and then, feeling frisky, volunteered to go to a spring a quarter of a mile to the rear, the first portion of the path to which was commanded by Confederate rifles. The crew of the gun I belonged to loaded me down with their empty canteens, and I ran, to avoid the sharpshooters' fire, to the protection of the forest behind us. There I saw many soldiers. Hollow-eyed, tired-looking men they were too, they had sought the comparative safety of the forest to sleep. Near the spring, which rose in a dense thicket, the shade was thick and the forest gloomy. The water in the spring had been roiled, so I searched for another higher up the run.

While searching for it I saw a colonel of infantry put on his war paint. It was a howling farce in one act – one brief act of not more than twenty seconds' duration, but the fun of the world was crowded into it. This blonde, bewhiskered brave sat safely behind a large oak tree. He looked around quickly; his face hardened with resolution. He took a cartridge out of his vest pocket, tore the paper with his strong white teeth, spilled the powder into his right palm, spat on it, and then, first casting a quick glance around to see if he was observed, he rubbed the moistened powder on his face and hands, and then dust-coated the war paint.

Instantly he was transformed from a trembling coward who lurked behind a tree into an exhausted brave taking a little well-earned repose. I laughed silently at the spectacle, and filled my canteens, then rejoined my comrades, and together we laughed at and then drank to the health of the blonde warrior. That night I slept and dreamed of comic plays and extravagant burlesques; but in the wildest of dream vagaries there was no picture that at all compared with the actual one I had seen in the forest. That colonel is yet alive.

165.

Treatment of a journalist in the Union lines that many another general on many another battlefield would have relished an opportunity to match. Frank Wilkeson's own father was war correspondent for the New York Times.

On one of these six Cold Harbour days, when my battery was in action, I saw a party of horsemen riding towards us from the left. I smiled as the absurdity of men riding along a battle-line for pleasure filled my sense of the ridiculous; but as I looked I saw that the party consisted of a civilian under escort. The party passed close behind our guns, and in passing the civilian exposed a large placard, which was fastened to his back, and which bore the words, 'Libeller of the Press'. We all agreed that he had been guilty of some dreadful deed, and were pleased to see him ride the battle-line. He was howled at, and the wish to tear him limb from limb and strew him over the ground was fiercely expressed. This man escaped death from the shot and shells and bullets that filled the air. I afterwards met him [Edward Crapsey of the *Philadelphia Inquirer*] in Washington, and he told me that he was a newspaper war correspondent, and that his offence was in writing, as he thought, truthfully, to his journal, that General Meade advised General Grant to retreat to the north of the Rapidan after the battle of the Wilderness.

FRANK WILKESON

166.

Leonidas Polk (1806–64), who was killed at Pine Mountain in June 1864 during Sherman's advance on Atlanta, was distinguished by being both a Confederate general and a bishop.

Like his opponent, General Sherman, Johnston had been aware of the unwieldy length of his position. He could shorten and strengthen it by withdrawing from Pine Mountain, and anchoring his line on Kennesaw. On the morning of June 14 he rode with Hardee and Polk to the crest of Pine Mountain to look the situation over. Three quarters of a mile below, a Union battery, looking like toy cannon tended by toy soldiers, lobbed two shells in their direction. Johnston and Hardee took cover, as did other officers who had joined the group from curiosity. Polk remained unruffled, viewing through his glass the scene below. A third shell whistled from the Union guns. It struck the bishop-general squarely in the chest, tearing his lungs out. For a second or two he remained erect – even in death he would not be hurried – then slipped to the ground without a sound. Ignoring the Federal cannon, Johnston ran to his side, raising the head of the fallen general on his arm. Weeping unrestrainedly he whispered, 'I would rather anything than this.' An ambulance was summoned by the signal station. While waiting, his fellow officers discovered in Polk's bloodied tunic a copy of Dr Quintard's poems, 'Balm for the Weary and Wounded', with the corner of a page turned down to mark the stanza: 'There is an unseen battlefield/In every human breast/Where two opposing forces meet/ And where they seldom rest.

Later a sad and silent cavalcade wound down the mountain, with Johnston riding bareheaded beside the body of his general. Polk's remains were carried to the Marietta depot to be taken to Atlanta, and Johnston returned to the front to break the news to his assembled army. 'In this most distinguished leader', he told the troops, 'we have lost the most courteous of gentlemen, the most gallant of soldiers. The Christian patriot has neither lived nor died in vain. His example is before you; his mantle rests with you.' Polk's loss was not a major military tragedy to the defending army. His fellow officers agreed that he was 'more theoretical than practical', and even the troops considered him a little ineffectual. Yet they loved him withal, as even hardened troops can love a man for his humanity rather than his skill in combat. Perhaps the most eloquent tribute was found by Federal troops who occupied Pine Mountain two days later. Greeting them was a crudely lettered sign: YOU YANKEE SONS OF BITCHES HAVE KILLED OUR OLD GEN. POLK.

167.

After being expelled from Atlanta by Sherman in September 1864,
Hood's Confederate army swung in a wide arc towards Tennessee
along the route by which Sherman had advanced, hoping to induce
him to follow them. Along their line of march, Union positions that
they encountered were summoned to surrender, provoking such
exchanges of civilities as those that follow.

Around Allatoona, October 5, 1864
TO THE Commanding Officer, United States Forces, Allatoona.

I have placed the forces under my command in such positions that
you are surrounded, and to avoid a needless effusion of blood I call
on you to surrender your forces at once, and unconditionally.

Five minutes will be allowed you to decide. Should you accede to
this, you will be treated in the most honourable manner as prisoners
of war,

I have the honour to be, very respectfully yours,

S. G. FRENCH

Major-General commanding forces Confederate States.

General Corse replied immediately:

Allatoona, Georgia, 8.30 a.m. October 5 1864
Major-General S. G. French, Confederate States etc.

Your communication demanding surrender of my command I
acknowledge receipt of, and respectfully reply that we are prepared
for the 'needless effusion of blood' whenever it is agreeable to you.

I am, very respectfully, your obedient servant,

JOHN M. CORSE

Brigadier-General commanding forces United States.

In the Field, October 12, 1864
To the Officer commanding the United States forces at Resaca,
Georgia.

Sir, I demand the immediate and unconditional surrender of the post
and garrison under your command, and, should this be acceded to,
all white officers and soldiers will be paroled in a few days. If the
place is carried by assault, no prisoners will be taken. Most
respectfully, your obedient servant,

    J. B. HOOD, General

Resaca, Georgia, October 12, 1864
To General J. B. Hood

Your communication of this date just received. In reply, I have to
state that I am somewhat surprised at the concluding paragraph to
the effect that, if the place is carried by assault, no prisoners will be
taken. In my opinion I can hold this post. If you want it, come and
take it.

    I am, general, very respectfully, your most obedient servant,

    CLARK R. WEAVER, Commanding Officer.

W. T. SHERMAN

168.

*Gen. William Tecumseh Sherman (1820–91) tells a sanctimonious
tale of the manner in which the memory of his friendship saved a
Confederate matron from being despoiled by his men.*

Toward evening of February 17th [1865], the mayor, Dr Goodwin,
came to my quarters at Duncan's house, and remarked that there was
a lady in Columbia who professed to be a special friend of mine. On
his giving her name, I could not recall it, but inquired as to her
maiden or family name. He answered Poyas. It so happened that,
when I was a lieutenant at Fort Moultrie, in 1842–'46, I used very
often to visit a family of that name on the east branch of Cooper
River, about forty miles from Fort Moultrie, and to hunt with the

son, Mr James Poyas, an elegant young fellow and a fine sportsman. His father, mother, and several sisters, composed the family, and were extremely hospitable.

One of the ladies was very fond of painting in watercolors, which was one of my weaknesses, and on one occasion I had presented her with a volume treating of watercolors. Of course, I was glad to renew the acquaintance, and proposed to Dr Goodwin that we should walk to her house and visit this lady, which we did. The house stood beyond the Charlotte depot, in a large lot, was of frame, with a high porch, which was reached by a set of steps outside. Entering this yard, I noticed ducks and chickens, and a general air of peace and comfort that was really pleasant to behold at that time of universal desolation; the lady in question met us at the head of the steps and invited us into a parlor which was perfectly neat and well furnished.

After inquiring about her father, mother, sisters, and especially her brother James, my special friend, I could not help saying that I was pleased to notice that our men had not handled her house and premises as roughly as was their wont. 'I owe it to you, general,' she answered. 'Not at all. I did not know you were here till a few minutes ago.' She reiterated that she was indebted to me for the perfect safety of her house and property, and added, 'You remember, when you were at our house on Cooper River in 1845, you gave me a book'; and she handed me the book in question, on the flyleaf of which was written: 'To Miss – Poyas, with the compliments of W. T. Sherman, First-lieutenant Third Artillery.' She then explained that, as our army approached Columbia, there was a doubt in her mind whether the terrible Sherman who was devastating the land were W. T. Sherman or T. W. Sherman, both known to be generals in the Northern army; but, on the supposition that he was her old acquaintance, when Wade Hampton's cavalry drew out of the city, calling out that the Yankees were coming, she armed herself with this book, and awaited the crisis.

Soon the shouts about the market-house announced that the Yankees had come; very soon men were seen running up and down the streets; a parcel of them poured over the fence, began to chase the chickens and ducks, and to enter her house. She observed one large man, with full beard, who exercised some authority, and to him she appealed in the name of 'his general'. 'What do you know of Uncle Billy?' 'Why,' she said, 'when he was a young man he used to be our friend in Charleston, and here is a book he gave me.' The officer or soldier took the book, looked at the inscription, and, turning to his fellows, said: 'Boys, that's so; that's Uncle Billy's writing, for I have seen it often before.' He at once commanded the party to stop pillaging, and left a man in charge of the house, to protect her until the regular provost-guard should be established. I then asked her if the regular guard or sentinel had been as good to her. She assured me that he was a very nice young man; that he had been telling her all about his family in Iowa; and that at that very instant of time he was in another room minding her baby. Now, this lady had good sense and tact, and had thus turned aside a party who, in five minutes more, would have rifled her premises of all that was good to eat or wear.

W. T. SHERMAN

169.

The French Foreign Legion's annual parade of the wooden hand of Captain Jean Danjou (1828–63) is one of history's odder military traditions. This was the only relic of the officer to be recovered after the Mexican action in which he died, and it remains today the Legion's proudest treasure.

Camerone is the most evocative name in Legion history, even though it will not be found on any but the largest-scale maps. Yet it is important to know what happened on 30th April 1863 at the farmhouse whose ruins inspired this inscription on a marble plaque placed in Les Invalides in Paris:

QUOS HIC NON PLUS LX
   ADVERSI TOTIUS AGMINIS
   MOLES CONSTRAVIT
   VITA PRIUS QUAM VIRTUS
   MILITES DESERVIT GALLICOS
   DIE XXX MENSI APR. ANNI MDCCCLXIII
   (Those who lie here, though less than sixty in number, fought an
entire army before being overwhelmed by sheer weight. Life

abandoned these French soldiers before honour did on the 30 of
April, 1863.)

These few words epitomize the story of Captain Jean Danjou, thirty-
five years old, veteran of the Crimea, Italy, and North Africa, and his
company of sixty-four men – men with names like Bartolotto, Katau,
Wenzel, Kunassek, Gorski. Captain Danjou's orders were to keep open
the highway connecting Vera Cruz and Mexico City so that the French
could send through a convoy of 60 carts and 150 mules carrying arms,
ammunition, and three million francs in gold to headquarters in the
capital.

For this purpose he volunteered to lead a depleted company of sixty-
two men together with two officers from another unit on a
reconnaissance mission along the Vera Cruz–Puebla road. Danjou left
camp at one o'clock in the morning and, marching all night, reached
high ground at seven in the morning when it was decided to brew up
coffee. The Legion troop had, in the meantime, been observed by the
Mexicans whose leader, Colonel Milan, now decided to wipe them out
before attacking the convoy coming from Vera Cruz. As soon as he was
harassed by the Mexican cavalry, Captain Danjou withdrew his men to
a farmhouse in the village of Camerone, no doubt hoping to be able to
hold off sporadic attacks by small groups of horsemen until the main
body of the Legion came to his relief. But he was mistaken; within a
matter of hours, his company was besieged by a small army of at least
2,000 men – 300 regular cavalry, 350 guerillas, and three battalions of
infantry. Of his own company of sixty-four, sixteen were already dead,
wounded, or missing, leaving only three officers and forty-six legionnaires to defend the farmhouse. They could not hope to survive against
such odds, which may explain why Captain Danjou, before he died,
demanded that each of them take an oath to fight to the end.

The end came in the evening after a day of non-stop fighting during
which the defenders had had nothing to eat and, worse, nothing to
drink. By six o'clock the original company was reduced to one officer,
second-lieutenant Maudet, and eleven legionnaires. The others were
either dead or badly wounded, among them Captain Danjou, shot in
the head, and his second-in-command, Lieutenant Vilain, mortally

wounded. Soon after six the Mexicans decided on an all-out attack, urged on by Colonel Milan who realized that to lose this battle would be a lasting disgrace to the Mexican army and the cause of liberation. This time the assault was overwhelming. But as the Mexicans swarmed in through the now undefended windows and doors of the farmhouse, Lieutenant Maudet ordered a bayonet charge. At the head of his four remaining legionnaires, he rushed out into the courtyard to meet the cross-fire of the besiegers and fell, hit in the face and body. One of his men was shot dead in this last charge; the other three were taken prisoner.

The battle in the farmhouse at Camerone had lasted nine hours, and when it was over, two officers and twenty legionnaires were dead; one officer and twenty-two legionnaires were wounded; and twenty legionnaires had been taken prisoner. The Mexican casualties were around three hundred. The Third Company of the Second Battalion of the Legion had been wiped out, but the convoy of arms and money reached Mexico City safely. Camerone was symbolic – 'a glorious defeat' on the one hand, a 'glorious victory' on the other.

JAMES WELLARD

170.

Cathy Williams (c. 1844–92) was the first African-American woman known to have served in the United States Army – a two-year stint in which she passed as a man.

Born a slave near Independence, Missouri, she was a 'house girl' on the Johnson plantation in Cole County, near the Missouri capital of Jefferson City, when the Civil War began. After General Nathaniel Lyons's troops captured Jefferson City, which had become a rebel stronghold, the Eighth Indiana Volunteer Infantry claimed Williams and other escaped or displaced slaves as 'contrabands.' She travelled with the regiment for the rest of the war, working as a laundress. When the war was over, she was free for the first time, but without family, home, or job. We can only speculate as to why she chose to enlist. It is probable that, like many women who walked a similar path before her, her motivation was as basic as economic security. She could earn more as a soldier than as a laundress, or even as a cook, which was the highest paid, most prestigious job available to black women in the United States.

In November 1866, she enlisted for a three-year term of service as 'William Cathay'. After what must have been a cursory medical examination, she was assigned to the newly formed Thirty-Eighth United States Infantry Regiment – one of six all-black regiments of 'Buffalo Soldiers' created by Congress in August 1866 with a view towards filling the need for soldiers created by westward expansion. She spent most of her military career on sick call: she was hospitalized five times in four different hospitals over the two years that she served. Apparently no one discovered she was a woman during any of these hospital visits – which raises questions about the quality of the medical care black soldiers received at the time. Or, perhaps, doctors repeatedly discovered the truth about her gender and didn't bother to report it.

On October 14, 1868, Private William Cathay was discharged from the army for medical reasons. In June 1891, she filed an application for an invalid pension based on her military service. In February 1892, the Pension Bureau rejected her claim on the grounds that no disability existed, not on the grounds that she was a woman and therefore her enlistment in the army was illegal. Cathy Williams was not a military hero. She probably never faced an enemy in the field. But she earned a place in history.

PAMELA TOLER

171.

An episode during white America's notorious wars against its native inhabitants.

Late in December 1867 the survivors of Black Kettle's band began arriving at Fort Cobb, Oklahoma. They had to come on foot, because [Col. George Armstrong] Custer had killed all of their ponies. Little Robe was now the nominal leader of the tribe, and when he was taken to see General Philip Sheridan (1831–88) he told the bearlike soldier chief that his people were starving. Custer had burned their winter meat supply; they could find no buffalo along the Washita; they had eaten all their dogs. Sheridan replied that the Cheyennes would be fed if they all came into Fort Cobb and surrendered unconditionally. 'You cannot make peace now and commence killing whites again in the spring,' Sheridan added. 'If you are not willing to make a complete peace, you can go back and we will fight this thing out.' Little Robe knew there was but one answer he could give. 'It is for you to say what we have to do,' he said. Yellow Bear of the Arapahos also agreed to bring his people to Fort Cobb. A few days later, Tosawi brought in the first band of Comanches to surrender. When he was presented to Sheridan, Tosawi's eyes brightened. He spoke his own name and added two words of broken English. 'Tosawi, good Indian,' he said. It was then that General Sheridan uttered the immortal words: 'The only good Indians I ever saw were dead.' Lieutenant Charles Nordstrom, who was present, remembered the words and passed them on, until in time they were honed into an American aphorism: The only good Indian is a dead Indian.

DEE BROWN

172.

Not all military music is heroic.

Queen Victoria, listening to an army band at Windsor, was captivated by a certain tune and sent a messenger to ascertain the title of it. He returned in some embarrassment and said it was called 'Come Where the Booze is Cheaper'.

CHRISTOPHER PULLING

173.

Career French officer Léonce Patry (1841–1917), Paris-born son of a language teacher, carried in his wallet throughout the 1870 Franco-Prussian war this letter from his adored father, expressing sentiments shared by millions of parents throughout the history of conflict.

Grenoble, 20 July 1870

My dear Boy,

Here you are, exposed to fire for the first time. It is your duty. I have nothing else to say. But you must understand how much we suffer, your mother and me. So be careful. Remember that between the timidity which fears everything and the rashness which fears nothing, there is a nobility which confronts necessary dangers and a caution which knows how to protect itself from unnecessary risk. So think of your mother, your sister and me. We all say 'March Forward like the brave man you are'. But we add 'Don't be rash. Don't put our lives at hazard by needlessly risking your own'. Put yourself in the hands of God who demands from us and you this supreme sacrifice for the defence of the country. Go, my fine son, face the fire covered with God's protection, our kisses and our blessing.

Your father who dotes on you, E. PATRY

LÉONCE PATRY

174.

Sedan, on 1 September 1870, proved the decisive clash of the Franco-Prussian War. The 130,000-strong army of Emperor Napoleon III (1808–73) was trapped amid overwhelming shell and small-arms fire from a much larger host commanded by the greatest Prussian field general of the nineteenth century, Helmuth von Moltke the elder (1800–91).

It was now a superb day, and Moltke's staff had found for the King [of Prussia] a vantage-point from which a view of the battle could be obtained such as no commander of an army in Western Europe was ever to see again. In a clearing on the wooded hills above Frénois, south of the Meuse, there gathered a glittering concourse of uniformed notabilities more suitable to an opera-house or a racecourse than to a climactic battle which was to decide the destinies of Europe and perhaps of the world. There was the King himself; there was Moltke, Roon and their staff officers watching the crown to their labours, while Bismarck, Hatzfeldt and the Foreign Office officials watched the beginnings of theirs. There was Colonel Walker from the British army and General Kutusow from the Russian; there was General Sheridan from the United States, Mr W. H. Russell of *The Times*, and a whole crowd of German princelings: Leopold of Bavaria and William of Wurttemberg, Duke Frederick of Schleswig-Holstein and the Duke of

Saxe-Coburg, the Grand Duke of Saxe-Weimar and the Grand Duke of Mecklenburg-Strelitz and half a dozen others, watching the remains of their independence dwindling hour by hour as the Prussian, Saxon, and Bavarian guns decimated the French army round Sedan.

As at Morsbach, as at Vionville, it was shown that when faced with resolute men armed with breech-loading rifles all the anachronistic splendour and courage of French chivalry was impotent. The German skirmishing-lines were overrun, but the supporting formations stood immovable and poured their volleys into the advancing mass. At no point was the German line broken. The cavalry torrent divided and swept by it to either side, northward towards Illy to return to their own ranks, southward to crash into the quarries of Gaulier or to be rounded up in the valley towards Glaire, leaving the carcasses of horses and the bodies of their riders lying thick in front of the German lines.

As the survivors of the charge rallied, Ducrot sought out their commander, General de Gallifet, and asked him whether they could try again. 'As often as you like, *mon général*,' replied Gallifet cheerfully, 'so long as there's one of us left.' So the scattered squadrons were rallied and once more the watchers above Frénois saw them plunging down the hill to certain destruction. King William was stirred to exclaim at their courage in words still carved on their memorial above Floing: '*Ah! Les braves gens!*' but it was not for him to lament that it was courage tragically wasted. Even now the cavalry were not exhausted. At 3 p.m. Ducrot, his front everywhere crumbling, threw them in yet again, while he and his staff rode along the ranks of the infantry trying in vain to rouse them to advance in the wake of the horse. This last attack, its cohesion gone, was repulsed as decisively as the rest, and with the greatest bloodshed of all. A pleasing legend speaks of Gallifet and his last followers passing exhausted within a few feet of the German infantry regiment. The Germans ceased fire; their officers saluted; and the Frenchmen were allowed to ride slowly away, honoured and unharmed.

With the business of the day safely over, Bismarck considered it safe to send for [the captive French emperor to meet] the King of Prussia. The interview between the sovereigns was brief and embarrassed. There was little for Napoleon to say, except to compliment William on his army – above all on his artillery – and lament the inadequacy of his own. He asked only one favour – that he might go into captivity, not by the same road as his army, but through Belgium, which would avoid an embarrassing passage through the French countryside. Bismarck approved. Napoleon might still be useful. Peace would eventually have to be made. So on 3rd September Napoleon with his suite, his powdered postilions, and the train of waggons which had so encumbered the movements of his army, drove into captivity, bound for the palace of Wilhelmshöhe above Cassel. [His wife Eugenie demanded in anger and bewilderment: 'Why didn't he kill himself?'] His troops, marching through pouring rain to the makeshift internment camp which the Germans had improvised for them in the loop of the Meuse round Iges – *le camp de la misère* as they called it after a week of starvation under pelting rain – watched his departure with indifference punctuated by abuse. Moltke wondered, a little tortuously, whether Napoleon might not have devised the whole operation to secure his untroubled retreat from his responsibilities. Bismarck merely remarked reflectively, 'There is a dynasty on its way out.' Then both returned to the gigantic problems which their victory had set them to solve.

MICHAEL HOWARD

175.

Captain Henry 'Charlie' Harford (1850–1937), an ardent collector of beetles, butterflies and moths, was also one of the very few men taking part in Chelmsford's invasion of Zululand who spoke its language. Lord Chelmsford was commander-in-chief of the British Army in Zululand. At a parley in one of the chiefs' kraals, a warrior who had participated in the Zulu triumph over the 24th Foot at Isandlwana on 22 January 1879 was delighted to discover that Harford had been nearby.

He caught hold of both my hands and shook them firmly in a great state of delight, saying it was a splendid fight. 'You fought well, and we fought well,' he exclaimed, and then showed me eleven wounds that he had received, bounding off in the greatest ecstasy to show how it all happened. Rushing up towards me, he jumped, fell on his stomach, got up again, rolled over and over, crawled flat, bounded on again and so forth, until he came right up to me, his movements being applauded by the warriors squatting in the centre of the kraal with a loud '*Gee!*' I now had a look at his wounds. One bullet had gone through his hand, three had gone through his shoulder and smashed his shoulder-blade, two had cut the skin and slightly into the flesh right down the chest and stomach, and one had gone clean through the fleshy part of the thigh. The others were mere scratches in comparison, but there he was, after about eight months, as well as ever and ready for another set-to. Could anything more clearly show the splendid spirit in which the Zulus fought us? No animosity, no revengeful feeling, but sheer love of a good fight in which the courage of both sides could be tested.

HENRY HARFORD

176.

Rudyard Kipling (1865–1936), the condescending yet superlatively intimate bard of 'Tommy Atkins', lived close to the great military cantonment at Mian Mir while working in 1883–7 as a writer for the Civil & Military Gazette in Lahore. He came to know prototypes of his 'Private Simmons', hanged for killing a bullying fellow ranker, and the habitual drunk in 'Cells', who lamented 'My wife she cries at the barrack gate, my kid in the barrack yard'. Here, such realities are catalogued by a half-admirer of Kipling's 'bouncing, vulgar vitality', who also knew the sub-continent well:

The sweltering barracks in Gibraltar or Lucknow, the redcoats, the pipeclayed belts and the pillbox hats, the beer, the fights, the floggings, hangings and crucifixions, the bugle-calls, the smell of the oats and horse-piss, the bellowing sergeants with foot-long moustaches, the bloody skirmishes, invariably mishandled, the crowded troop-ships, the cholera-stricken camps, the 'native' concubines, the ultimate death in the workhouse. It is a crude, vulgar picture … but from it future generations will be able to gather some idea of what a long-term, volunteer army was like.

GEORGE ORWELL

177.

A pen portrait by Lt. Winston Churchill of John Palmer Brabazon
(1843–1922), who served 1891–6 as commanding officer of his own
regiment, the 4th Hussars. The eccentricities described here were later
enhanced, when during the Boer War Brabazon urged that cavalry
should be equipped with tomahawks for mounted action.

Colonel Brabazon was an impoverished Irish landlord [who] personi-
fied the heroes of Ouida. From his entry into the Grenadier Guards in
the early 60s he had been in the van of fashion. He was one of the
brightest military stars in London society. A close lifelong friendship
had subsisted between him and the Prince of Wales. At Court, in the
Clubs, on the racecourse, in the hunting field, he was accepted as a
most distinguished figure. Though he had always remained a bachelor,
he was by no means a misogynist. As a young man he must have been
exceptionally good-looking. He was exactly the right height for a man
to be. He was not actually six feet, but he looked it. Now, in his prime,
his appearance was magnificent. His clean-cut symmetrical features,
his bright grey eyes and strong jaw, were shown to the best advantage
by a moustache which the Kaiser might well have taken as his unat-
tainable ideal. To all this he added the airs and manners of the dandies
of the generation before his own, and an inability real or affected to
pronounce the letter 'R'. Apt and experienced in conversation, his
remarkable personality was never at a loss in any company, polite or
otherwise.

His military career had been long and varied. He had had to leave
the Grenadier Guards after six years through straitened finances, and
passed through a period of serious difficulty. He served as a gentleman
volunteer – a great privilege – in the Ashanti Campaign of 1874. Here
he so distinguished himself that there was a strong movement in high
circles to restore to him his commission. The Prince of Wales was most
anxious that he should be appointed to his own regiment – the 10th
Hussars – in those days probably the most exclusive regiment in the
Army. However, as no vacancy was immediately available he was in the
interval posted to an infantry regiment of the Line. To the question,

'What do you belong to now, Brab?' he replied, 'I never can wemember, but they have gween facings and you get at 'em from Waterloo.' Of the stationmaster at Aldershot he enquired on one occasion in later years: 'Where is the London twain?' 'It has gone, Colonel.' 'Gone! Bwing another.'

Translated at length into the 10th Hussars he served with increasing reputation through the Afghan War in 1878 and 1879 and through the fierce fighting round Suakim in 1884. As he had gained two successive brevets upon active service he was in army rank actually senior to the Colonel of his own regiment.

This produced at least one embarrassing situation conceivable only in the British Army of those days. The Colonel of the 10th had occasion to find fault with Brabazon's squadron and went so far in his displeasure as to order it home to barracks. Brabazon was deeply mortified. However, a few weeks later the 10th Hussars were brigaded for some manoeuvres with another cavalry regiment. Regimental seniority no longer ruled, and Brabazon's army rank gave him automatically the command of the brigade. Face to face with his own commanding officer, now for the moment his subordinate, Brabazon had repeated the same remarks and cutting sentences so recently addressed to him, and finished by the harsh order, 'Take your wegiment home, Sir!'

The fashionable part of the army had been agog with this episode. That Brabazon had the law on his side could not be gainsaid. In those days men were accustomed to assert their rights in a rigid manner which would now be thought unsuitable. There were, however, two opinions upon the matter. As it was clear that his regimental seniority would never enable him to command the 10th, the War Office had offered him in 1893 the command of the 4th Hussars. This was in itself an inevitable reflection upon the senior officers of that regiment. No regiment relishes the arrival of a stranger with the idea of 'smartening them up'; and there must have been a great deal of tension when this terrific Colonel, blazing with medals and clasps, and clad in all his social and military prestige, first assumed command of a regiment which had even longer traditions than the 10th Hussars. Brabazon made little attempt to conciliate. On the contrary he displayed a masterful confidence which won not only unquestioning obedience from all, but intense admiration, at any rate from the captains and subalterns. Some of the seniors, however, were made to feel their position. 'And what chemist do you get this champagne fwom?' he enquired one evening of an irascible Mess president.

To me, apart from service matters in which he was a strict disciplinarian, he was always charming. But I soon discovered that behind all his talk of war and sport, which together with questions of religion or irreligion and one or two other topics formed the staple of Mess conversation, there lay in the Colonel's mind a very wide reading. When, for instance, on one occasion I quoted, 'God tempers the wind to the shorn lamb', and Brabazon asked 'Where do you get that fwom?' I had replied with some complacency that, though it was attributed often to the Bible, it really occurred in Sterne's *Sentimental Journey*. 'Have you ever wead it?' he asked, in the most innocent manner. Luckily I was not only naturally truthful, but also on my guard. I admitted that I had not. It was, it seemed, one of the Colonel's special favourites.

The Colonel, however, had his own rebuffs. Shortly before I joined the regiment he came into sharp collision with no less a personage than Sir Evelyn Wood who then commanded at Aldershot. Brabazon had not only introduced a number of minor irregularities, mostly extremely sensible, into the working uniform of the regiment – as for instance chrome yellow stripes for drill instead of gold lace – but he had worn for more than thirty years a small 'imperial' beard under his lower lip. This was of course contrary to the Queen's Regulations, Section VII: 'The chin and underlip are to be shaved (except by pioneers, who will wear beards).' But in thirty years of war and peace no superior authority had ever challenged Brabazon's imperial. He had established it as a recognized privilege and institution of which no doubt he was enormously proud.

No sooner had he brought his regiment into the Aldershot command than Sir Evelyn Wood was eager to show himself no respecter of persons. Away went the chrome yellow stripes on the pantaloons, away went the comfortable serge jumpers in which the regiment was accustomed to drill; back came the gold lace stripes and the tight-fitting cloth stable-jackets of the old regime. Forced to obey, the Colonel carried his complaints unofficially to the War Office. There was no doubt he had reason on his side. In fact within a year these sensible and economical innovations were imposed compulsorily upon the whole army. But no one at the War Office or in London dared override Sir Evelyn Wood, armed as he was with the text of the Queen's Regulations.

As soon as Sir Evelyn Wood learned that Brabazon had criticized his decisions, he resolved upon a bold stroke. He sent the Colonel a written order to appear upon his next parade 'shaved in accordance with the regulations'. This was of course a mortal insult. Brabazon had no choice but to obey. That very night he made the sacrifice, and the next morning appeared disfigured before his men, who were aghast at the spectacle, and shocked at the tale they heard. The Colonel felt this situation so deeply that he never referred to it on any occasion. Except when obliged by military duty, he never spoke to Sir Evelyn Wood again.

WINSTON S. CHURCHILL

178.

*Marcel Proust (1871–1922) was asked in a Confession Album:
'What event in military history do you admire most?' He replied
unhesitatingly, 'My own enlistment.' In 1889 Proust, a lifelong
valetudinarian, offered himself as a 'volunteer', enjoying the privileges
of an officer cadet though messing with rankers, and entered service
with the 76th Infantry at Orléans, an experience described by his best
biographer, aided by the novelist himself.*

The following year, of the discipline and love of comrades which to certain neurotics are so welcome, was among the happiest of his life. 'It's curious', he wrote to a friend fifteen years later, 'that you should have regarded the army as a prison, I as a paradise.' He swam, rode, fenced and marched, rejoiced to be called '*mon vieux*' by the common soldiers his companions: he experienced, for one whole year, the delightful illusion of being normal and accepted. There was a new poetry in the grey autumnal landscape, in the daily scenes of life in the

barrack-room, which he likened to the *genre* paintings of the Dutch
School. 'The rural character of the places,' he wrote in *Les Plaisirs et Les
Jours*, 'the simplicity of some of my peasant-comrades, whose bodies
were more beautiful and agile, their minds more original, their character more natural than those of the young men I had known before
or knew later, the calm of a life in which occupations are more regu-
lated and the imagination less trammelled than in any other, in which
pleasure is constantly with us because we have no time to run about
looking for it and so miss it altogether, all these things concur to make
this period of my life a series of little pictures full of happy reality and
a charm on which time has since shed its delicious sadness and poetry.'

GEORGE PAINTER

179.

*Anthony Powell (1905–2000), a fellow-novelist who modelled
himself upon Proust, traces the links between the latter's great work* In
Search of Lost Time *and his military experience as a volunteer.*

Captain de Borodino, Saint-Loup's squadron commander, is one of
the characters drawn from life. His prototype was Captain Walewski,
a company commander in the 76th, grandson of Bonaparte by a Polish
lady, an affair well-known to history. As it happened the Captain's
mother, in addition to his grandmother's imperial connections, had
been mistress to Napoleon III. That such a figure, with origins, appearance and behaviour all crying out for chronicling, should turn up in
Proust's regiment illustrates one of those peculiar pieces of literary luck
which sometimes attend novelists.

General de Froberville's prototype in real life was General de
Gallifet, a well-known personality in the world with which Proust
deals. Gallifet, who had led the cavalry charge at Sedan, suppressed the
Commune with an almost Communist savagery, and (though not
Dreyfusard) insisted on a revisionist approach to the Dreyfus case, was
also a wit and a womanizer. Mr Painter mentions several stories about
him: the silver plate covering the wound in his abdomen (received in
the Mexican campaign) alleged to lend physical subtlety to his many
love affairs; the distinguished lady archaeologist, rather masculine in

dress, who insisted on joining the men after dinner, at which the General took her by the arm with the words 'Come along, my dear fellow, let's go and have a pee.'

In *Jean Santeuil* [Proust devotes] a good deal of space to Colonel Georges Picquart (1854–1914), another good instance of Proust's approach to army matters, and also his technique of absorbing 'real people' into his writing. Picquart's story should be briefly recalled. An Alsatian, sixteen years old when Alsace was annexed by Germany, he was regarded as an ambitious and very promising officer, he had served on Gallifet's staff, been present at Dreyfus's court-martial, and, in due course, put in charge of the Secret Service Section – an outstandingly ramshackle one – at the French War Office.

On taking over, Picquart re-examined the Dreyfus file held by his Section, coming to the conclusion that something had gone badly wrong in the Court's examination of evidence. He drew this fact to the attention of his superiors, with the consequence that he was himself posted to North Africa (stationed where there was a good chance of death in action) then, when he persisted in making further representations about Dreyfus, put under arrest, imprisoned, and placed in the running for condemnation to five years in a fortress.

All this is striking enough; but when it is added that Picquart, if not a rabid anti-semite, was decidedly unfriendly towards Jews, he will at once be seen to be building up the sort of character upon which a writer likes to get to work. When Dreyfus was cleared, Picquart refused to meet him; and when, in due course, Picquart rose in rank and was in a position to be of some assistance in Dreyfus's professional rehabil-itation in the army, he would take no step to make things easier. Clemenceau, in a slapstick mood, appointed him his Minister of War, a post Picquart filled without great distinction, behaving rather badly to officers who had merely been carrying out orders issued by former anti-Dreyfusard superiors. Picquart remained unmarried all his life; dying, in consequence of being thrown from his horse, when in command of an Army Corps, about six months before the outbreak of war in 1914.

Proust's own health [at the outbreak of war] was naturally far too precarious for there to be any question of serving again in the army.

That did not prevent the routine requirements of medical boards, which he accepted – one recalls the great to-do D. H. Lawrence made in similar circumstances – as inevitable consequences of a world war. All the same, there was one aspect of them that was exceedingly troublesome to Proust – the time the boards took place. He dreaded these orders to present himself, merely because they threatened the hour or two's sleep he could achieve only during daytime. By one of those clerical errors endemic to military administration, certainly a classical one, he was ordered on one occasion to report to the Invalides for medical examination at 3.30 a.m., instead of the same hour in the afternoon. To many people such an instruction would have been disturbing. Proust was charmed. This nocturnal summons seemed just another example of how accommodating the military authorities could sometimes show themselves.

ANTHONY POWELL

180.

Major-General Sir Charles Callwell (1859–1928), as he eventually became, was an Ulsterman, a lifelong bachelor, who proved himself one of the British Army's most brilliant and original thinkers. His 1896 book Small Wars: Their Principles and Practice *became a classic – Douglas Porch characterizes its author as 'the Clausewitz of colonial wars'. Britain and other Western nations might have engaged in fewer doomed twentieth- and twenty-first-century overseas interventions had modern commanders heeded Callwell's precepts, but his ruthlessness must shock our own, professedly more humane generation. Here he discusses the importance, but chronic difficulty, of identifying an objective when fighting tribesmen – non-state enemies – in faraway places.*

When there is no king to conquer, no capital to seize, no organized army to overthrow, when there are no celebrated strongholds to capture, and no great centres of population to occupy, the objective is not so easy to select. 'In planning a war against an uncivilized nation who has, perhaps, no capital,' says Lord Wolseley, 'your first object should be the capture of whatever they prize most, and the destruction or deprivation of which will probably bring the war most rapidly to a conclusion.' This goes to the root of the whole matter. If the enemy cannot be touched in his patriotism or his honour, he can be touched through his pocket. Fighting the Kirghiz and other nomads of the steppes the Russians have always trusted largely to carrying off the camels and flocks of the enemy. In Algeria the French, adopting the methods of Abd-el-Kader and his followers, made sudden raids or 'razzias', carrying off the livestock and property of their wandering opponents. In the [South African] wars, especially in 1852, this mode of procedure has been very common. The United States troops used to retaliate upon the Red Indians in similar fashion.

The destruction of the crops and stores of grain of the enemy is another way of carrying on hostilities. This method of warfare is more exasperating to the adversary than carrying off livestock; for while they appreciate the principle that the victor is entitled to the spoils, wanton damage tends to embitter their feeling of enmity. The same applies to the destruction of villages so often resorted to in punitive expeditions, but hardly to the same extent, since the dwellings of these races can be reconstructed easily while their food supplies, if destroyed, cannot be replaced.

It is so often the case that the power which undertakes a small war desires to acquire the friendship of the people which its armies are chastising, that the system of what is called 'military execution' is ill-adapted to the end in view. However much it may shock humanitarian susceptibilities it is useless to conceal the fact that the most satisfactory way of bringing such foes to reason is by the rifle and the sword, for they understand this mode of warfare and respect it. Sometimes the circumstances do not admit of it, and then their villages must be demolished and their crops and granaries destroyed; but it is unfortunate when it is so.

When, however, the campaign takes the form of quelling an insurrection, the object is not only to prove unmistakably to the opposing force which is the stronger, but also to inflict punishment on those who have taken up arms. In this case it is often necessary to injure property. 'A war,' wrote Sir G. Cathcart from [South Africa] in 1852, 'may be terminated by the surrender or capitulation of the hostile sovereign or chief, who answers for his people; but in the suppression of a rebellion the refractory subjects of the ruling power must all be chastised and subdued.'

Still, there is a limit to the licence in destruction which is expedient. [Gen. Lazare] Hoche, whose conduct of the [1793] campaign against the Chouans and insurgents from La Vendée will ever remain a model of operations of this kind, achieved success as much by his happy combination of clemency with firmness as by his masterly dispositions in the theatre of war. Expeditions to put down revolt are not put in motion merely to bring about a temporary cessation of hostility, their purpose is to ensure a lasting peace. Therefore, in choosing the objective the overawing and not the exasperation of the enemy is the aim to keep in view.

The main points of difference between small wars and regular campaigns are that in the former the beating of the hostile armies is not necessarily the main object even if such exists, that normal effect is often far more important than material success, and that the operations are sometimes limited to committing havoc which the laws of regular warfare do not sanction. The crushing of an insurrectionary movement or the settlement of a conquered country are undertakings so distinct from enterprises entered upon to overawe a semi-civilized state, that what may present itself as the obvious objective under the former set of circumstances may be non-existent in the latter.

CHARLES CALLWELL

181.

Few military reputations have been won as briskly as that of future US President Theodore Roosevelt (1858–1919). In April 1898 when the US declared war on Spain, he was granted, at the age of thirty-nine, command of a regiment of volunteer 'Rough Riders' which landed in Cuba on 15 June. Within two months he was back in New York, a national hero, having led and survived on 1 July a reckless charge up San Juan hill under heavy enemy fire, as he later described in his autobiography.

I had not enjoyed the Guasimas fight at all, because I had been so uncertain as to what I ought to do. But the San Juan fight was entirely different. The Spaniards had a hard position [for us] to attack, it is true, but we could see them, and I knew exactly how to proceed. I kept on horseback, merely because I found it difficult to convey orders along the line, as the men were lying down; and it is always hard to get men to start when they cannot see whether their comrades are also going. So I rode up and down the lines, keeping them straightened out, and gradually worked through line after line until I found myself at the head of the regiment. By the time I had reached the lines of the regulars of the first brigade I had come to the conclusion that it was silly to stay in the valley firing at the hills, because that was really where we were most exposed, and that the thing to do was to try to rush the intrenchments. Where I struck the regulars there was no one of superior rank to mine, and after asking why they did not charge, and being answered that they had no orders, I said I would give the order. There was naturally a little reluctance shown by the elderly officer in command to accept my order, so I said, 'Then let my men through, sir,' and I marched through, followed by my grinning men. The younger officers and the enlisted men of the regulars jumped up and joined us. I waved my hat, and we went up the hill with a rush. Having taken it, we looked across at the Spaniards in the trenches under the San Juan blockhouse to our left, which Hawkins's brigade was assaulting. I ordered our men to open fire on the Spaniards in the trenches.

Memory plays funny tricks in such a fight, where things happen quickly, and all kinds of mental images succeed one another in a detached kind of way, while the work goes on. As I gave the order in question there slipped through my mind Mahan's account of Nelson's orders that each ship as it sailed forward, if it saw another ship engaged with an enemy's ship, should rake the latter as it passed. When Hawkins's soldiers captured the blockhouse, I, very much elated, ordered a charge on my own hook to a line of hills still farther on. Hardly anybody heard this order, however; only four men started with me, three of whom were shot. I gave one of them, who was only wounded, my canteen of water, and ran back, much irritated that I had not been followed – which was quite unjustifiable, because I found that nobody had heard my orders.

General Sumner had come up by this time, and I asked his permission to lead the charge. He ordered me to do so, and this time away we went, and stormed the Spanish intrenchments. In the final fighting at San Juan, when we captured one of the trenches, Jack Greenway had seized a Spaniard, and shortly afterwards I found Jack leading his captive round with a string. I told him to turn him over to a man who had two or three other captives, so that they should all be taken to the rear. It was the only time I ever saw Jack look aggrieved. 'Why, Colonel, can't I keep him for myself?' he asked, plaintively. I think he had an idea that as a trophy of his bow and spear the Spaniard would make a fine body servant.

THEODORE ROOSEVELT

182.

By a curious symmetry, less than three months later and seven thousand miles to the east, another future national leader, Winston Churchill (1874–1965), participated in another equally reckless action. His description of the 2 September 1898 charge of the 21st Lancers at Omdurman – Kerreri, as the battle is known to the Sudanese – is a classic of narrative, though the cavalry action was the epitome of futility. This account appeared in Churchill's 1930 memoir My Early Life, *his literary masterpiece.*

Everyone expected that we were going to make a charge. That was the one idea that had been in all minds since we had started from Cairo. Of course there would be a charge. In those days, British cavalry had been taught little else. Here was clearly the occasion for a charge. But against what body of enemy, over what ground, in which direction or with what purpose, were matters hidden from the rank and file. We continued to pace forward over the hard sand, peering into the mirage-twisted plain in a high state of suppressed excitement.

Presently I noticed, 300 yards away on our flank and parallel to the line on which we were advancing, a long row of blue-black objects, two or three yards apart. I thought there were about a hundred and fifty. Then I became sure that these were men – enemy men – squatting on the ground. Almost at the same moment the trumpet sounded 'Trot', and the whole long column of cavalry began to jingle and clatter across the front of these crouching figures. We were in the lull of the battle and there was perfect silence. Forthwith from every blue-black blob came a white puff of smoke, and a loud volley of musketry broke the odd stillness. Such a target at such a distance could scarcely be missed, and all along the column here and there horses bounded and a few men fell.

The intentions of our Colonel had no doubt been to move round the flank of the body of Dervishes he had now located, and who, concealed in a fold of the ground behind their riflemen, were invisible to us, and then to attack them from a more advantageous quarter; but once the fire was opened and losses began to grow, he must have judged it inexpedient to prolong his procession across the open plain. The trumpet sounded 'Right wheel into line', and all the sixteen troops swung round towards the blue-black riflemen. Almost immediately the regiment broke into a gallop, and the 21st Lancers were committed to their first charge in war!

I propose to describe exactly what happened to me: what I saw and what I felt. The troop I commanded was, when we wheeled into line, the second from the right of the regiment. I was riding a handy, sure-footed, grey Arab polo pony. Before we wheeled and began to gallop, the officers had been marching with drawn swords. On account of my [polo-injured] shoulder I had always decided that if I were involved in hand-to-hand fighting, I must use a pistol and not a sword. I had purchased in London a Mauser automatic, then the newest and the latest design. I had practised carefully with this during our march and journey up the river.

This then was the weapon with which I determined to fight. I had first of all to return my sword into its scabbard, which is not the easiest thing to do at a gallop. I had then to draw my pistol from its wooden holster and bring it to full cock. This dual operation took an appreciable time, and until it was finished, apart from a few glances to my left to see what effect the fire was producing, I did not look up at the general scene.

Then I saw immediately before me, and now only half the length of a polo ground away, the row of crouching blue figures firing frantically, wreathed in white smoke. On my right and left my neighbouring troop leaders made a good line. Immediately behind was a long dancing row of lances couched for the charge. We were going at a fast but steady gallop. There was too much trampling and rifle fire to hear any bullets. I looked again towards the enemy. The scene appeared to be suddenly transformed. The blue-black men were still firing, but behind them there now came into view a depression like a shallow sunken road. This was crowded and crammed with men rising up from the ground where they had hidden. Bright flags appeared as if by magic, and I saw arriving from nowhere Emirs on horseback among and around the mass of the enemy.

The Dervishes appeared to be ten or twelve deep at the thickest, a great grey mass gleaming with steel, filling the dry watercourse. In the same twinkling of an eye I saw also that our right overlapped their left, that my troop would just strike the edge of their array, and that the troop on my right would charge into air. My subaltern comrade on the right, Wormald of the 7th Hussars, could see the situation too; and we both increased our speed to the very fastest gallop and curved inwards like the horns of the moon. One really had not time to be frightened or to think of anything else but these particular necessary actions which I have described. They completely occupied mind and senses.

The collision was now very near. I saw immediately before me, not ten yards away, the two blue men who lay in my path. They were perhaps a couple of yards apart. I rode at the interval between them. They both fired. I passed through the smoke conscious that I was unhurt. The trooper immediately behind me was killed at this place and at this moment, whether by these shots or not I do not know. I checked my pony as the ground began to fall away beneath his feet. The clever animal dropped like a cat four or five feet down on to the sandy bed of the watercourse, and I found myself surrounded by what seemed to be dozens of men. They were not thickly-packed enough at this point for me to experience any actual collision with them. Whereas Grenfell's troop next but one on my left was brought to a complete standstill and suffered very heavy losses, we seemed to push our way through as one has sometimes seen mounted policemen break up a crowd. In less time than it takes to relate, my pony had scrambled up the other side of the ditch. I looked round.

Once again I was on the hard, crisp desert, my horse at a trot. I had the impression of scattered Dervishes running to and fro in all directions. Straight before me a man threw himself on the ground. The reader must remember that I had been trained to believe that if ever cavalry broke into a mass of infantry, the latter would be at their mercy. My first idea therefore was that the man was terrified. But simultaneously I saw the gleam of his curved sword as he drew it back for a ham-stringing cut. I had room and time enough to turn my pony out of his reach, and leaning over on the off side I fired two shots into him at about three yards. As I straightened myself in the saddle, I saw before me another figure with uplifted sword. I raised my pistol and fired. So close were we that the pistol itself actually struck him. Man and sword disappeared below and behind me. On my left, ten yards away, was an Arab horseman in a bright-coloured tunic and steel helmet, with chain-mail hangings. I fired at him. He turned aside. I pulled my horse into a walk and looked around again.

In one respect a cavalry charge is very like ordinary life. So long as you are all right, firmly in your saddle, your horse in hand, and well-armed, lots of enemies will give you a wide berth. But as soon as you have lost a stirrup, have a rein cut, have dropped your weapon, are wounded, or your horse is wounded, then is the moment when from all quarters enemies rush upon you. Such was the fate of not a few of my comrades in the troops immediately on my left. Brought to an actual standstill in the enemy's mass, clutched at from every side, stabbed at and hacked at by spear and sword, they were dragged from their horses and cut to pieces by the infuriated foe. But this I did not at the time see or understand.

My impressions continued to be sanguine. I thought we were masters of the situation, riding the enemy down, scattering them and killing them. I pulled my horse up and looked about me. There was a mass of Dervishes about forty or fifty yards away on my left. They were huddling and clumping themselves together, rallying for mutual protection. They seemed wild with excitement, dancing about on their feet, shaking their spears up and down. The whole scene seemed to flicker. I have an impression, but it is too fleeting to define, of brown-clad Lancers mixed up here and there with this surging mob. The scattered individuals in my immediate neighbourhood made no attempt to molest me. Where was my troop? Where were the other troops of the squadron? Within a hundred yards of me I could not see a single officer or man. I looked back at the Dervish mass.

I saw two or three riflemen crouching and aiming their rifles at me from the fringe of it. Then for the first time that morning I experienced a sudden sensation of fear. I felt myself absolutely alone. I thought these riflemen would hit me and the rest devour me like wolves. What a fool I was to loiter like this in the midst of the enemy! I crouched over the saddle, spurred my horse into a gallop and drew clear of the melee. Two or three hundred yards away I found my troop already faced about and partly formed up.

The other three troops of the squadron were re-forming close by. Suddenly in the midst of the troop up sprung a Dervish. How he got there I do not know. He must have leaped out of some scrub or hole. All the troopers turned upon him thrusting with their lances: but he darted to and fro causing for the moment a frantic commotion. Wounded several times, he staggered towards me raising his spear. I shot him at less than a yard. He fell on the sand, and lay there dead. How easy to kill a man! But I did not worry about it. I found I had fired the whole magazine of my Mauser, so I put in a new clip of ten cartridges before thinking of anything else.

I was still prepossessed with the idea that we had inflicted great slaughter on the enemy and had scarcely suffered at all ourselves. Three or four men were missing from my troop. Six men and nine or ten horses were bleeding from spear thrusts or sword cuts. We all expected to be ordered immediately to charge back again. The men were ready, though they all looked serious. Several asked to be allowed to throw away their lances and draw their swords. I asked my second sergeant if he had enjoyed himself. His answer was 'Well, I don't exactly say I enjoyed it, Sir; but I think I'll get more used to it next time.' [Of 400 men who charged, 70 were killed or wounded. Omdurman became the regiment's sole, highly equivocal, battle honour.]

183.

No episode in the life of Field-Marshal Earl Kitchener (1850–1916)
took him longer to live down than his treatment of the remains of
Dervish leader the Mahdi, Muhammad Ahmad bin Abd Allah
(1844–85).

On 6 September 1898, four days after Omdurman, he issued orders that the Mahdi's tomb should be razed to the ground, and that the bones of [Gen. Charles] Gordon's great enemy [killed by the Dervishes at their January 1885 capture of Khartoum] should be cast into the Nile. Gordon's nephew, Major S. W. Gordon, R.E., was entrusted with the execution of that order; and the Madhi's skull, which was unusually large and shapely, was saved from destruction and presented to Kitchener as a trophy. Some members of the 'band of boys' with whom it amused him occasionally to relax, suggested that he should cause the skull to be mounted in silver or gold, and that he should use it as an inkstand or as a drinking-cup.

Kitchener played with that idea and with the skull for a short time; and he acquired somehow the idea that Napoleon's intestines had found their way from St Helena to the museum of the Royal College of Surgeons in London. Accordingly, he announced incautiously to some of his staff that he proposed to send the Mahdi's skull to the College of Surgeons with a request that it should be placed on exhibition alongside the guts of Napoleon. That story of the Mahdi's skull obtained a wide currency; and it caused, in February 1899, a great howl of rage against Kitchener, which was compounded, in approximately equal parts, of frothy but sincere sentiment and of jealousy. Radical and intellectual circles hated Kitchener at that time; the Army was intensely jealous of him; and he had gone out of his way to insult the Press. For a few weeks, therefore, while unfriendly questions were being asked in Parliament and elsewhere, Kitchener felt extremely uncomfortable.

On 27 February 1899 Salisbury [the prime minister] telegraphed to Cromer [consul-general and effective viceroy of Egypt]: 'The Queen is shocked by the treatment the Mahdi's body has received, and thinks the head ought to be buried. Putting it in a museum, she thinks, will do great harm.' On 2 March Cromer replied: 'The dead set against Kitchener was sure to come, sooner or later. Apart from the natural reaction, he has not the faculty of making friends. The soldiers are furiously jealous of him, and many of the newspaper correspondents, whom he took no pains to conciliate, have long been waiting for an opportunity to attack him. He has his faults. No one is more aware of them than myself. But for all that, he is the most able of the English soldiers I have come across in my time. He was quite right in destroying the Mahdi's tomb, but the details of the destruction were obviously open to objection. Kitchener is himself responsible for the rather unwise course of sending the skull to the College of Surgeons.'

Kitchener's relations with the Press had been bad from the start of the campaign. He made two exceptions among the newspaper correspondents in favour of *The Times* and the *Daily Mail* (Hubert Howard and G. W. Steevens); but that favouritism caused trouble; and he seldom let slip an opportunity of demonstrating the contempt in which he held the profession as a whole. Only a day or two before Omdurman he was informed that a group of correspondents had been waiting outside his tent for some time in the belief that he had a statement to make. He let them wait, until he was ready to emerge, and then, as he strode angrily through their midst he made a statement, which consisted only of the words: 'Get out of my way, you drunken swabs!'

In those circumstances the commotion in the British and American press about Kitchener and the Mahdi's skull was prolonged maliciously for several weeks. It was combined with charges [entirely accurate, and later supported by Lt. Churchill] that Kitchener had left all the Dervish wounded to die without succour on the battlefield of Omdurman, and that he had personally ordered a massacre of civilians in Omdurman after the battle [not true]. Those attacks worried Kitchener, who wrote (7 March 1899) to the Queen: 'Lord Kitchener is much distressed that Your Majesty should think that the destruction of the Mahdi's tomb, and the disposal of his bones was improperly carried out. He is very sorry that anything he has done should have caused Your Majesty a moment's uneasiness.

'A few days after the battle, I consulted with some native officers of the Sudanese troops, and spoke on the matter with some influential natives here; and they told me that, although no educated person believed in the Mahdi being anything but an impostor who had attempted to change the Mohammedan religion, ... some of the soldiers in our ranks still believed in the Mahdi; and they recommended the destruction of the tomb, and that the bones should be thrown into the Nile, which would entirely dissipate any such belief. Nothing in the matter was done in a hurry, but four days after the battle I gave the order for the destruction, thinking it was the safest and wisest course; and this was carried out in my absence. There was no coffin, and when the bones were found the soldiers seemed all astonished, and exclaimed – "By God! This was not the Mahdi after all he told us!" They had previously believed that the Mahdi had been translated bodily to heaven. When I returned from Fashoda, the Mahdi's skull, in a box, was brought to me, and I did not know what to do with it. I had thought of sending it to the College of Surgeons where, I believe, such things are kept. It has now been buried in a Moslem cemetery.'

PHILIP MAGNUS

184.

The elephantine sense of fun displayed by Colonel Sir Robert Baden-Powell (1857–1941) as garrison commander of besieged Mafeking (October 1899–May 1900) in South Africa enchanted his Victorian contemporaries, but suggests to a modern reader that enemy action was the least of the miseries his comrades endured. He acted in character when he later became founder of the Boy Scout movement.

The garrison, in the face of increasing losses and decreasing food, lost none of the high spirits which it reflected from its commander. The programme of a single day of jubilee – Heaven only knows what they had to hold jubilee over – shows a cricket match in the morning, sports in the afternoon, a concert in the evening, and a dance, given by the bachelor officers, to wind up. Baden-Powell himself seems to have descended from the eyrie from which, like a captain on the bridge, he rang bells and telephoned orders, to bring the house down with a comic song and a humorous recitation. The ball went admirably, save that there was an interval to repel an attack which disarranged the programme. Sports were zealously cultivated, and the grimy inhabitants of casemates and trenches were pitted against each other at cricket or football. Sunday cricket so shocked Snyman [the Boer commander] that he threatened to fire upon it if it were continued.

ARTHUR CONAN DOYLE

185.

General Sir Redvers Buller's (1839–1908) sybaritic habits were the source of caustic humour among contemporaries during his campaign for the relief of Ladysmith, November 1899–February 1900.

Finding his supply of champagne was getting very low, he telegraphed home to his wine-merchants to send out fifty cases of the usual brand, with strict injunctions that the cases were to be marked 'Castor Oil'. About the time the wine was due, the general wrote to the base and informed the officer in charge that he expected fifty cases of castor oil, which he wished despatched to his headquarters without delay. The reply from the base came in a few days, and was as follows: 'Regret exceedingly no cases as described have yet reached us, but this day we have procured all the castor oil possible (twenty cases), and have despatched it without delay, as you desired. We trust this unavoidable delay has caused no serious inconvenience.'

MACCARTHY O'MOORE

186.

Few modern writers about soldiers have entertained so large a readership and stirred such controversy as psychologist Norman Dixon (1922–2013), with his 1976 book On the Psychology of Military Incompetence, *identifying flaws in such commanders as Douglas Haig that allegedly derived from anal obsession and authoritarianism. Dixon found matching characteristics in Lord Raglan of the Crimean era and Sir Redvers Buller in the Boer War, who remained inexplicably popular with their men despite displaying absolute unfitness for command.*

Both men were genial, courteous and kind. Both were inexperienced, irresolute and lacking moral courage. Both were rich and well-connected, but both were only too ready to divest themselves of all responsibility for the errors which they had made. Certain characteristics of the incompetence just described include:

1. An underestimation of the enemy
2. An equating of war with sport
3. An inability to profit from past experience
4. A resistance to adopting and exploiting available technology and novel tactics
5. An aversion to reconnaissance, coupled with a dislike of intelligence (in both senses of the word)
6. Great physical bravery but little moral courage
7. An apparent imperviousness to loss of life and human suffering
8. A love of the frontal assault
9. A love of 'bull', smartness, precision
10. A high regard for tradition and other aspects of conservatism
11. A tendency to eschew moderate risks in favour of tasks so difficult the failure might seem excusable
12. Procrastination

NORMAN DIXON

187.

Adrian Carton de Wiart, VC (1880–1963), became a legend for bombastic courage of a kind associated with Evelyn Waugh's fictional Ben Ritchie-Hook. Here he makes an early appearance as a young ADC to the commander-in-chief in South Africa, Sir Henry Hildyard, soon after the end of the Boer War.

Lady Hildyard was a most charming hostess but an inveterate gambler, and South Africa with its fortunes won and lost overnight was a dangerous centre for the unstable. One day she came to me in great distress. She had gambled and lost an enormous sum, practically all Sir Henry's capital, and what should she do? I advised her to confess at once. All Sir Henry said was: 'Never mind, my dear, I might have done much worse myself.' I was always a reluctant card player, but bridge was considered as an essential part of an A.D.C.'s equipment. One night Lady Hildyard, who was my partner, had committed what I considered to be several enormities and as she got up to leave the room at the end of our game I shook my fist after her retreating back. Sir Henry entered the room at that unfortunate moment, and I thought I was for home. Instead, he turned to Major Winwood the military secretary and said: 'De Wiart's a very patient man, isn't he?'

ADRIAN CARTON DE WIART

188.

The Royal Military College at Sandhurst was the scene of several mutinies. Here, a participant recalled in old age that of 1902, which followed disciplinary action against the entire cadet body for a series of mysterious fires.

The cadets had a burning sense of grievance when all leave was stopped on account of something they knew nothing about; so, about half-past nine, after mess, they collected on the Main Entrance steps of the College and cheered Kruger, Smuts, De Wet and every Boer general they could think of, and there were hoots about 'Bobs' (Field-Marshal Earl Roberts) and the War Office. While this was going on some suggested that we should go to the Fête, which was on in Camberley. Some started to roll the [parade-ground trophy] guns down into the lake. As soon as the Fete was suggested, the cadets moved en masse down the drive. Every lamp post was bent double; everything that was breakable was broken; and everything movable was thrown into the lake.

The gate-keeper wisely went into his house, and the cadets trooped into Camberley singing the well-known songs of the Sandhurst of that day. One small body went off to the Governor's House, where he was having a dinner party, and serenaded him. When we got to the Fête ground, the gate-keeper showed great pleasure and thought we were going to pay to go in. Instead, with one wild rush down went the money-collectors, and in rushed the cadets. I was knocked down and got through on my hands and knees – we were all in red mess kit.

Some of us got on the roundabouts, some on the swings, and the Fair authorities thought it was a bit risky to stop the roundabouts in case we broke them up. Then the rumour came that they had sent officers and senior corporals to round us up; so we asked them to stop the merry-go-round and all got off. We found the gates closed, so wandered round until we found a barbed-wire fence and crawled over it into the Staff College grounds, from where we got into the RMC. The orderly on duty took our names.

The next morning there were headlines in the Press of mutiny at Sandhurst. However everything quietened down and leave was re-opened. Two days after, however, the fifth fire occurred; and that put the lid on it. A telegram came from the War Office, to say that all the cadets of 'C' Company were to be rusticated if they could not prove an alibi, and twenty-nine, of whom I was one, were for it. We thought this very unfair but all went off and packed and were driven, with our luggage, to Camberley station in four-wheeler cabs amidst the cheers of the College. Every window had a cadet cheering.

At my home in Eastbourne was a friend called Cavendish, and his father knew Winston Churchill; so he suggested that his boy and I should go up to London and see Churchill, a backbench MP. In the meanwhile a number of letters had appeared in *The Times* and questions had been asked in the House.

I am not sure whether Churchill ever did ask his question, but a few days later we got a communication from the War Office to say that the Commander-in-Chief wanted to interview all twenty-nine of us. He shook hands with me, sat me down in an armchair and checked my name. He asked if I was any relation of Colonel Hadow's of the 15th Sikhs. I said, 'Yes, I'm his son.'

'Oh indeed,' he said, 'I remember him in Kabul and Kandahar in '79. He was wounded wasn't he, in the Tirah Show? How is he? Remember me to him.'

We chatted for a bit and then at the end of about five minutes he said, 'Oh, by the way, do you know anything about these fires ... who did it ... or who was likely to have done it? And if you know anything perhaps you would tell me in confidence.' But I told him that I had no more idea than he had who had started them. With that he thanked

me and I got up and shook hands and went out. Shortly after I got a
letter from the War Office saying that as the Commander-in-Chief
had interviewed me personally and satisfied himself that I knew nothing about the fires and couldn't help in any way, that I was completely
exonerated; I could return to Sandhurst at the beginning of the next
term and count my exams of the term before as passed.

H. R. HADOW

189.

*Private Frank Richards – real name Francis Woodruff DCM, MM
(1883–1961) – was among the best-ever ranker memoirists, perhaps
because he received literary assistance with his two books, published in
1933 and 1936, from the poet Robert Graves, a wartime officer of his
regiment, the Royal Welch Fusiliers. Born in Monmouthshire,
Richards was orphaned at the age of nine and worked as a miner
before enlisting in 1901. Amazingly, he served unwounded
throughout World War I on the Western Front, refusing promotion.
Here, he recalls a pre-war comrade driven to desperate measures to
secure a discharge.*

He had joined the Army in a fit of despair over the young lady with
whom he had been walking out, who had chucked him. It was Archie's
idea to be sent to the South African War and win a posthumous V.C.,
so that she would be sorry for the manner in which she had jilted
him. But after he had been in the Army for a time he forgot his
broken heart, and the South African War ended, and he wanted to
return to civil life. It was while we were in Jersey that Archie began to
'work his ticket', as it was called. An Adjutant's Parade was the first
occasion. His company had already fallen in, the roll had been called
and the orderly-sergeant had reported Archie absent, when he came
strolling out of his room, trailing his rifle behind him, with a far-away
look in his eyes.

He fell in on the left of his company, just as the company officer
began to inspect it. The Adjutant, who spotted him, rubbed his eyes
in amazement and certainly the way Archie was dressed would have
made a cat laugh. On his red jacket he had stitched a dozen lids of

Day and Martin's Soldier's Friend, together with metal-polish tins, all
of them highly polished. Tied to the back of his braces and hanging
over his backside was a frying-pan. The Adjutant was too astonished
to say a word until Archie was about to be marched to the Guard-
room under escort. Then he roared: 'Bring that damned lunatic in
front of me.'

When questioned as to why he had appeared on parade improperly
dressed, Archie assured the Adjutant that he was properly dressed. He
said that he was entitled to the decorations and medals, having won
them during the years he had served with the Emperor of Abyssinia's
army. He said that the large decoration he wore on his backside was
the most coveted honour; when the Emperor decorated him with it he
had also promoted him to full general. He said that the generals and
the princes became jealous, and if he hadn't left the country he would
have been dead meat.

The Adjutant ordered the escort to take him to hospital, but after
he had been there a week the medical officer came to the conclusion
that he was perfectly sane. For making a laughing-stock of the King's
uniform and pretending he was balmy he was lucky enough to get the
light sentence of fourteen days' cells. He still acted strangely after he
came out. He was determined to leave the Army and, like other men
who tried to work their tickets, did not have the necessary twenty-one
pounds to buy himself out. He would have deserted, but he knew that
if he went back to his relatives he would soon be arrested, which would
mean six months' imprisonment. He decided to stick to his original
plan. In India his manner became stranger than ever. He used to have
long interesting conversations with himself, mostly about love or
Abyssinia, and was twice sent to hospital for observation. He was not
punished any more, because the doctors were undecided as to whether
he had lost his mental balance. A lot of us believed that he was really
up the loop from having played at it so long.

On our march back to Meerut we stayed one day at a place where
there was a magnificent temple on the bank of a large, deep lake. That
afternoon quite a number of us were enjoying a swim and Archie, who
could not swim a stroke, sat watching. Some time later, I heard a man
exclaim. I looked around and was surprised to see Archie stripped and

standing on the top of a high pillar of stone on the edge of the lake. I shouted to him not to be a fool, but at that moment he made a wonderful dive, going in so straight and making so little of a splash that the men who did not know him uttered a cry of amazement and said: 'That fellow Archie must have been a professional high-diver!' But the rest of us, who knew him better, dived in and fished him out half-drowned.

Archie's final stunt was a masterpiece. The Divisional Sports were being held in a few weeks' time and he entered for every running event from the hundred yards to the mile. He refused to be assisted in his training, which he did, so he said, about half an hour before twilight every evening. Late on Sunday evening, on the day before the Sports, Archie left the tent, saying that he was going out for a final spin on his secret training-ground. We had not followed him before, but that evening we thought we would. He stuck to the main road after leaving the Camp until he was about twenty yards from the entrance to the Protestant Church. There we were surprised to see him cut across country and disappear.

The shadows of twilight were falling at the entrance to the church, where all the best society of Meerut attended evensong. Just as we opened the gate the congregation began to file out. Suddenly a man with a pair of running pumps on his feet but otherwise as naked as the day he was born jumped out from behind the plants and began running round and round the church with the speed of a hare. It was Archie. Some of the ladies screamed, others did their best to close their eyes. I expect that the full-blooded ones who had old and decrepit husbands closed only one eye and gazed with the other in rapturous admiration. Archie was physically handsome in feature and limb and old Mother Nature had been kind to him in many ways. For a few moments the ladies' esquires were too astonished to do anything, and it was the same with us. He had completed two laps and was halfway on a third before we burst into the grounds, shouting that the man was a lunatic.

We caught him and rushed him behind the plants. The three of us now thought that he was really up the loop. He did not seem to realize that he had done anything out of the ordinary and said: 'Well, boys, do you think any man has a ghost of a chance against me tomorrow?

You'll see, I'll cake-walk every event.' An officer who had been in church, ordered us to conduct him to hospital and in less than a fortnight he was on his way to be interned in an asylum. His last words to the escort were: 'Well, so long, boys. I'll be thinking of you when I'm back in Blighty. I am supposed to be balmy, and so I was to join the Army. But, one thing, I'm not half so balmy, and never have been, as those balmy bastards who still have to do six or seven years in this God-damned country.' Within twelve months we had news that he had been discharged from the Army, that he had an excellent job in his home town and was happily married to a young lady who, he said, was worth a hundred of the one for whose sake he had behaved in such a rash manner.

FRANK RICHARDS

190.

Before World War I, 4th Gurkha Rifles occupied a remote station on India's North-West Frontier, where one of its officers took pride in his success as a 'ladies' man'.

He never failed to attend the regimental ordeal known as the Tuesday Bunfight. The ladies of the station, perhaps five or six in number, were invited to the mess for tea, cakes, and tennis. Every available officer had to be present to entertain them – all except one, a man who could hardly be induced to speak at all to any woman, and never spoke to one politely. This man had succeeded, through his known misogyny and addiction to work, in getting permanent permission to absent himself from the Tuesday Bunfights. What few people ever knew – and none at the time – was that as soon as the ladykiller left for the Bunfight the misogynist pedalled furiously down the steep road to the former's Indian mistress, and returned late at night with ardour quelled but misogyny unabated.

JOHN MASTERS

191.

*Having failed the Sandhurst entrance exam, to his profound dismay
the aesthete Osbert Sitwell (1892–1969) found that his family had
secured him a Yeomanry commission, attached to a Hussar regiment
stationed at Aldershot. There, in 1912, he spent a miserable spring
and summer.*

From time to time I still tried to reach London. Though only thirty-six
miles away, the capital seemed infinitely distant. There were many
things that I wanted to see; the second Post-Impressionist Exhibition
and a small show of drawings and paintings by Augustus John; there
were operas to hear, and concerts. I wanted, also, to keep in touch with
the few friends I possessed, and from whom my incarceration at Aldershot cut me off no less effectually than banishment to Siberia. In June,
therefore, I asked for the two or three days to which I had become
entitled. But when the Commanding Officer enquired where I wished
to spend it, and received the reply 'London', I could see the look of
genuine consternation that passed over his face. 'London!' he plainly
said to himself. 'Imagine wishing to leave Aldershot, earthly paradise
that it is, for so mean a city!

'But what can you do there; what can you want to do? There's nothing to do,' he reiterated in a tortured voice, and with a soldier's simple
vocabulary. When he had recovered sufficiently from the shock, he
refused permission. But I think his story of 'the Young Officer who
wanted to go to London!' went the rounds: for Generals, when they
visited us, surveyed me carefully, as if I were a dangerous wild beast,
and the senior regimental officers seemed to regard me with increased
distaste. 'What can be the state of mind', their eyes clearly goggled the
message, 'of a young man who wishes to leave Aldershot to spend a few
days in London!'

London! Why, you could not even kill anything there!

(It was tantalizing, too, to see all those living creatures behind their
bars, walking, pacing, climbing, swinging about in the Zoological
Gardens, and not be able to get at them, not be able to fire a single
shot!) No huntin': no shootin': no polo, even. Of course, there was

always Tattersall's, that they admitted, but it need not occupy more than a single afternoon. You could be back in the dear old Mess in time for dinner.

[By the summer of 1913, Sitwell had relinquished his Yeomanry commission and joined the Grenadier Guards to appease his father's wishes, without any access of military enthusiasm.] One day while I was Ensign on King's Guard at St James's Palace, the Captain of it – an awe-inspiring individual, with a heavy, but regular profile, and moustachios left over from the drawing-rooms of George du Maurier – enquired, after an immense effort that resembled the wheezing of an old clock about to strike, but was none the less born of a kindly intention to try to lessen the tedium of long hours spent in the red-papered guard-room, 'Do you like horses?', and I replied, 'No, but I like giraffes – they have such a beautiful line,' he took the answer unexpectedly well, and even attempted to smile. It had been my turn to take the Early Parade at 6 a.m. After breakfast, the Adjutant sent for me to the Orderly Room. I obeyed the intimidating summons. I entered, and saluted the great man. He said, looking up, 'Mr Sitwell, it is reported to me that you were late this morning for Early Parade.' I expressed dissent. On this, he enquired, 'Were the men on parade when you arrived?' I replied, 'I didn't take any notice, sir. I did not look.' I can see now that my answer, which was quite genuine and unaffected, must have been disconcerting to a mind of such excellent military punctuality and precision.

OSBERT SITWELL

192.

On the morning of 1 August 1914, the German ambassador in London, Prince Lichnowsky, telegraphed Kaiser Wilhelm II (1859–1941) to declare his belief that if Germany did not attack France, Britain would agree to remain neutral, and to keep France neutral, in a Russo-German war. Barbara Tuchman's account is disputed by some modern scholars, but makes gripping reading.

The Kaiser clutched at Lichnowsky's passport to a one-front war. Minutes counted. Already mobilization was rolling inexorably toward the French frontier. The first hostile act, seizure of a railway junction in Luxembourg, whose neutrality the five Great Powers, including Germany, had guaranteed, was scheduled within an hour. It must be stopped, stopped at once. But how? Where was Moltke [Helmuth von Moltke the Younger (1848–1916), German chief of staff]? Moltke had left the palace. An aide was sent off, with siren screaming, to intercept him. He was brought back. The Kaiser was himself again, the All-Highest, the War Lord, blazing with a new idea, planning, proposing, disposing. He read Moltke the telegram and said in triumph: 'Now we can go to war against Russia only. We simply march the whole of our Army to the East!'

Aghast at the thought of his marvellous machinery of mobilization wrenched into reverse, Moltke refused point-blank. For the past ten years, Moltke's job had been planning for this day, The Day, *Der Tag*, for which all Germany's energies were gathered, on which the march to final mastery of Europe would begin. It weighed upon him with an oppressive, almost unbearable responsibility.

Tall, heavy, bald, and sixty-six years old, Moltke habitually wore an expression of profound distress which led the Kaiser to call him *der traurige Julius* (or what might be rendered 'Gloomy Gus'). Poor health, for which he took an annual cure at Carlsbad, and the shadow of a great uncle were perhaps cause for gloom. From his window in the red brick General Staff building on the Königplatz where he lived as well as worked, he looked out every day on the equestrian statue of his namesake, the hero of 1870 and, together with Bismarck, the architect of the German Empire. The nephew was a poor horseman with a habit of falling off on staff rides and, worse, a follower of Christian Science with a side interest in anthroposophism and other cults.

For this unbecoming weakness in a Prussian officer he was considered 'soft'; what is more, he painted, played the cello, carried Goethe's *Faust* in his pocket, and had begun a translation of Maeterlinck's *Pelléas et Mélisande*. Introspective and a doubter by nature, he had said to the Kaiser upon his appointment in 1906: 'I do not know how I shall get on in the event of a campaign. I am very critical of myself.' Yet he was neither personally nor politically timid. In 1911, disgusted by Germany's retreat in the Agadir crisis, he wrote to [Austria's chief of staff] Conrad von Hotzendorff that if things got worse he would resign, propose to disband the army and 'place ourselves under the protection of Japan; then we can make money undisturbed and turn into imbeciles'.

He did not hesitate to talk back to the Kaiser, told him 'quite brutally' in 1900 that his Peking expedition was a 'crazy adventure', and when offered the appointment as Chief of Staff, asked the Kaiser if he expected 'to win the big prize twice in the same lottery' – a thought that had certainly influenced William's choice [of the great Moltke's nephew]. He refused to take the post unless the Kaiser stopped his habit of winning all the war games which was making nonsense of manoeuvres. Surprisingly, the Kaiser meekly obeyed.

Now, on the climactic night of August 1, Moltke was in no mood for any more of the Kaiser's meddling with serious military matters, or with meddling of any kind with the fixed arrangements. To turn around the deployment of a million men from west to east at the very moment of departure would have taken a more iron nerve than Moltke disposed of. He saw a vision of the deployment crumbling apart in confusion, supplies here, soldiers there, ammunition lost in the middle, companies without officers, divisions without staffs, and those 11,000 trains, each exquisitely scheduled to click over specified tracks at specified intervals of ten minutes, tangled in a grotesque ruin of the most perfectly planned military movement in history.

'Your Majesty,' Moltke said to him now, 'it cannot be done. The deployment of millions cannot be improvised. If Your Majesty insists on leading the whole army to the East it will not be an army ready for battle but a disorganized mob of armed men with no arrangements for supply. Those arrangements took a whole year of intricate labour to complete' – and Moltke closed upon that rigid phrase, the basis for every major German mistake, the phrase that launched the invasion of Belgium and the submarine war against the United States, the inevitable phrase when military plans dictate policy – 'and once settled, it cannot be altered'.

In fact it could have been altered. The German General Staff, though committed since 1905 to a plan of attack upon France first, had in their files, revised each year until 1913, an alternative plan against Russia with all the trains running eastward. On the night of August 1, Moltke, clinging to the fixed plan, lacked the necessary nerve. 'Your uncle would have given me a different answer,' the Kaiser said to him bitterly.

BARBARA TUCHMAN

193.

One of the first cavalry encounters of the war took place when two Russian hussar patrols encountered a German picket in East Prussia.

Suddenly there was a clatter of hooves, and seven horsemen trotted up. Lieutenant Stepanov had started at the same time as Genishta's patrol. Genishta rode up to Stepanov. 'Stepa, let's have a go at them!' he said. Now they could clearly see the advancing Germans. Sixteen to eighteen riders were approaching the village in a deployed formation, shooting their carbines as they rode. The distance was about 600 yards; the Russian troopers tensed as they watched their quarry draw near. Because of the buildings it was impossible to align the men; they clustered in small groups behind houses and sheds. Lieutenant Genishta turned to give a command, but Stepanov anticipated him. Spurring his horse and yelling at the top of his voice, he dashed forward. Pell-mell through the narrow alleys the Horse Grenadiers rushed forward shouting 'Hurrah!'

Out in the open a semblance of a line was formed, some of the men galloping with lances couched, others with raised swords. Genishta and Stepanov, both good horsemen and well mounted, had drawn ahead, Stepanov leading. The German patrol stopped, their leader gave a command, and wheeling about they galloped off in good order. The German lieutenant, however, restrained his horse and stood for a moment or two facing the oncoming Russians. Then he too wheeled and followed his men. Now the three Russian officers were galloping abreast – Genishta on the right, Stepanov in the middle and Egerstrom on the left. Several lengths behind came their men. Thus they galloped for about a mile. The Germans came to a wide ditch with rather boggy sides. Everyone made it over except the lieutenant. His horse refused suddenly, sinking hock-deep in mud.

Lieutenant von Lütken was catapulted over its head and he landed on the other side of the ditch. He jumped quickly to his feet and looked back. His men were galloping away. No one turned back to help him. Undeterred, he drew his revolver and stood alone facing the enemy.

The first to reach the ditch was Lieutenant Stepanov who headed directly at the German standing on the other side. As the horse rose for the jump, von Lütken fired. The bullet hit Stepanov's horse in the head, killing it outright. It crashed into the ditch, sending its rider flying. Stepanov fell at von Lütken's feet, and lost his sword. The German fired and missed. At this moment Lieutenant Genishta jumped the ditch somewhat to the right. Von Lütken wheeled left and fired, but Genishta galloped right by, too far to reach him with his sword. Stepanov had jumped up and seized the German; they fell struggling to the ground. Von Lütken tried to use his revolver, but Stepanov pinned his arm to the ground. In desperation von Lütken bit Stepanov's finger to the bone.

Several Horse Grenadiers had, in the meanwhile, jumped the ditch. One of the troopers, Semikopenko, hurled himself down from his horse and while still in the air, slashed at the German with his sword. The blow was accurate and deadly: it severed the wrist holding the revolver. Another trooper thrust his lance into the brave von Lütken, killing him. Knowing the German squadrons to be close at hand, Genishta stopped the pursuit. Scarlet faces dripping with sweat, grey coats torn and spattered with blood, blood dripping too from sword blades and lance heads, the Horse Grenadiers rode in.

Someone brought back the dead German lieutenant's dispatch case. Genishta opened it. Inside were two papers. One was the report to his regiment. The other was a letter to his parents, written, addressed, but never sent. That evening Lieutenant Genishta sat down and wrote a letter to the family of the late Lieutenant von Lütken. He told them about their son's brave death and said how sorry he was. He sent the letter c/o the Red Cross.

Several years passed; the war ended – for the Germans in defeat, for the Russians in revolution. The year 1924 found the ex-cavalry officer Genishta an exile in Paris. Living was hard, jobs were scarce and it was difficult to start a new life. The former Guards officer was eking out an existence driving a taxi. One evening, returning to his dimly-lit garret, he found a letter postmarked Germany. It read:

'Dear Lieutenant Genishta, Please forgive this tardy (10 years) answer, but only now have we learned quite by accident of your whereabouts. My parents, now deceased, and I were deeply touched by your letter describing my brother's death; we always wanted to thank you for your kind words and consideration. Your letter hangs framed below the portrait of my late brother.'

Enclosed in the letter was a photograph of a large castle somewhere in Germany, the residence of the von Lütken family. A long correspondence followed between Genishta and von Lütken's sister. Knowing how difficult life had become for the former cavalry officer, the German lady invited him to come and stay with her family. Poor but proud, the Russian officer thanked her and refused.

ALEXIS WRANGEL

194.

The British Expeditionary Force boards transports for France in August 1914, described by a staff officer, J. F. C. 'Boney' Fuller (1878–1966). Fuller afterwards became a general, distinguished war historian and exponent of armoured warfare, though his later life was badly tarnished by avowed fascist sympathies.

When the Oxfordshire Hussars embarked, they brought with them a vast quantity of kit: tin uniform boxes, suitcases and cabin trunks. Someone questioned the loading of this baggage, whereupon a red-faced Major burst into my office in a towering rage: 'This is simply damnable!' he shouted. 'Winston [Churchill, a former officer of the regiment] said we could take 'em, and now one of your prize B.F.s says we can't ...' 'All right! All right!' I cut him short. 'What is the trouble about?' And having ascertained what the First Lord of the Admiralty had sanctioned, I telephoned down [orders] to load the officers' trousseaux – a word which did not seem to please my furious friend. All were loaded, and, I believe, a week later were unpacked by German hands.

No sooner had he left the room than in burst a Hussar captain. He also was boiling over with anger. He stuttered and had a high-pitched voice: 'Do you expect that I am going to get on that old barge?' (the *Archimedes*, a cattle-ship, later torpedoed). 'Why,' he continued, 'there is no notepaper on board.' 'Yes,' I replied, 'that is so; for, since the outbreak of war, Argentine bullocks have been considerate enough to do without it. May I, however, give you a tip – well, they have dispensed with toilet paper also.'

There was a strange incident in which Major Maclean played the part of fairy godmother. As was often the case, a howling crowd of friends and relatives collected outside the dock gates. An elderly woman was demanding to be let in, and by her side was standing a girl literally dripping tears. Maclean, always good-natured, allowed the two women to enter, and then discovered that the girl was expecting a baby, and that the culprit, a sergeant, was somewhere in the crowded docks. Pacifying her, Maclean said: 'You leave it to me: it will be all right,' and off he went. Three-quarters of an hour later he came back with the sergeant: how he found him remains a mystery, for there must have been some twelve to fifteen thousand troops embarking. Then he took the sergeant and the girl into the door-keeper's hutch, which stands just inside the gates, and making them sign some document which he hastily concocted, he married them in Scottish fashion. When later on someone chipped him and said: 'Well, that's not a legal marriage,' he replied: 'No – but now it is a very good case for breach of promise.'

J. F. C. FULLER

195.

*The 17 August meeting between Field-Marshal Sir John French
(1852–1925), commander-in-chief of the British Expeditionary
Force, and General Charles Lanrezac (1852–1925), commanding the
French Fifth Army on his flank, was described by Lt. Edward Louis
Spears (1886–1974), British liaison officer. Although then only
twenty-eight years old, and in some eyes an intolerably cocky
mountebank, Spears became one of the most influential figures in
Anglo-French relations during two world wars.*

Sir John stepped out of his car looking very spick and span. He was a good deal shorter than General Lanrezac, who came out to greet him. The two walked into Lanrezac's sanctum together, the one short, brisk, taking long strides out of proportion to his size, the other big, bulky, heavy, moving with short steps as if his body were too heavy for his legs. We knew that Lanrezac spoke no English, and Sir John, though he understood a little French, at that time could hardly speak it at all. Sir John, stepping up to a map in the *3ieme Bureau*, took out his glasses, located a place with his finger, and said to Lanrezac: '*Mon Général, est-ce-que –*' His French then gave out, and turning to one of his staff, he asked: 'How do you say "to cross the river" in French?' He was told, and proceeded: '*Est-ce-que les Allemands vont traverser la Meuse à – à –.*' Then he fumbled over the pronunciation of the name. 'Huy' was the place, unfortunately one of the most difficult words imaginable to pronounce, the 'u' having practically to be whistled. It was quite beyond Sir John. 'Hoy,' he said at last, triumphantly. 'What does he say? What does he say?' exclaimed Lanrezac. Somebody explained that the Marshal wanted to know whether in his opinion the Germans were going to cross the river at Huy? Lanrezac shrugged his shoulders impatiently. 'Tell the Marshal', he said curtly, 'that in my opinion the Germans have merely gone to the Meuse to fish.'

EDWARD SPEARS

196.

An episode on 27 August during the retreat from Mons, recalled by a British medical officer.

As we turned into the Grande Place at St Quentin on that late August afternoon, the whole square was thronged with British infantrymen. Scores had gone to sleep on the pavement, their backs against the fronts of the shops. Some few, obviously intoxicated, wandered about firing in the air at real or imaginary German aeroplanes. The great majority were not only without their arms but had apparently either lost or thrown away their belts, water-bottles and other equipment. Apparently they were without officers – anyway, no officers were to be seen. On the road down to the station we found Major Tom Bridges with part of his squadron and a few Lancers, horse-gunners and other stragglers who had attached themselves to his command.

We followed him down to the station. Apparently some hours before our arrival the last train that was to leave St Quentin Pariswards for several years, had steamed out, carrying with it most of the British General Staff. A mob of disorganized soldiery had collected, and I was told some had booed and cheered ironically as the Staff train steamed out. Certainly many appeared to be in a queer, rather truculent, mood. Bridges, who had sized up the situation, harangued this disorganized mob that only a few hours before had represented at least two famous regiments. Dismounted and standing far back in the crowd I could not hear what he said, but his words of encouragement and exhortation were received with sullen disapproval and murmurs by the bulk of those around him. One man shouted out: 'Our oldman [Colonel] has surrendered, and we'll stick to him. We don't want any bloody cavalry interfering!' and he pointed his rifle at Bridges. I failed at first to understand how all these soldiers could have surrendered to the Germans whom we had left several miles outside the city. But I was tired and hungry and I didn't much care what happened.

I rode back up to the Grande Place, hoping I should find some food and a sofa on which I could lie down. Many of the men in the street stared at me disdainfully, their arms folded; scarcely one saluted – I was for them only 'one of the bloody interfering cavalry officers'. When I awoke it was dusk. Bridges was having an interview with some official – I believe, the Mayor of St Quentin – urging him to provide horses and carts to take those of our men who were too sore-footed to be able to march out of town. I walked over to listen. The official – Mayor or whoever he was – was very indignant; he kept saying: 'You understand, m'sieu le Majeur, it is now too late. These men have surrendered.'

'How? The Germans are not here!'

'Their colonel and officers have signed a paper giving me the numbers of the men of each regiment and the names of the officers who are prepared to surrender, and I have sent a copy of this out under a white flag to the Commander of the approaching German Army!'

'But you have no business, m'sieu, as a loyal Frenchman, to assist allied troops to surrender!'

'What else?' urged the Mayor. 'Consider, m'sieu le Majeur, the alternatives. The German Army is at Gricourt? Very well, I, representing the inhabitants of St Quentin, who do not want our beautiful town unnecessarily destroyed by shell-fire because it happens to be full of English troops, have said to your colonels, and your men: "Will you please go out and fight the German Army outside St Quentin", but your men they say: "No! We cannot fight! We have lost nearly all our officers, our Staff have gone away by train, we do not know where to. Also, we have no artillery, most of us have neither rifles nor ammunition, and we are all so very tired!"

'Then, m'sieu le Majeur, I say to them: "please if you will not fight will you go right away, and presently the Germans will enter St Quentin peacefully; so the inhabitants will be glad to be tranquil, and not killed, and all our good shops not burned." But they reply to me: "No, we cannot go away! We are terribly, terribly tired. We have had no proper food nor rest for many days, and yesterday we fought a great battle. We have not got any maps, and we do not even know where to go to. So we will stay in St Quentin and have a little rest!" Then I say to them: "Since you will neither fight nor go away, then please you must surrender." And now all is properly arranged!'

Arranged! Yet the logic of this argument was irresistible – but for one point, which Bridges had quickly seized upon. The men could be got away if every horse and cart in St Quentin was collected for those men too tired to march; his cavalrymen would escort them out of the town. So the shops and streets would be cleared of tired and drunken men, and there would be no more firing off of rifles; but there was to be no more of this wine, only tea or coffee and bread. So eventually it was arranged; Bridges had saved the situation.

Disorganized stragglers had arrived by the hundred. They had tramped beneath the blazing August sun with empty stomachs, dispirited and utterly weary; many had received quantities of wine from kindly French peasants to revive them in those dusty lanes. Literally, in many cases, their bellies were full of wine and their boots half full of blood; that I saw myself. The English soldier's feet like his head, but unlike his heart, are not his strong point. Bridges asked me to count the men and get them into fours. I counted one hundred and ten fours. A few had whistles and Jew's harps, perhaps they had them in their haversacks as soldiers often do, and they formed a sort of band. We persuaded one of the colonels to march in front. He looked very pale, entirely dazed, had no Sam Browne belt, and leant heavily on his stick, apparently so exhausted that he could scarcely have known what he was doing. Some of his men called to him encouraging words, affectionate and familiar, but not meant insolently – such as: 'Buck up, sir! Cheer up, Daddy! Now we shan't be long! We are all going back to "Hang-le-Tear"!'

I saw him saluting one of his own corporals who did not even look surprised. What with fatigue, heat, drink and the demoralization of defeat, many hardly knew what they were doing. I was so tired myself that I went to sleep on my horse almost immediately after I remounted, and nearly fell off, much to the amusement of the infantry who supposed I was as drunk as some of their comrades. It was nearly half-past twelve before we left St Quentin. The sultry August day had passed to leave a thick summer mist. Every kind of vehicle had been filled with men with blistered feet. In front of them, on foot, were several hundred infantry and behind, to form the rearguard to this extraordinary cavalcade, Tom Bridges' mounted column – the gallant little band of 4th Dragoon Guards with driblets of Lancers, Hussars, Irish Horse, signallers and the rest of the stragglers.

In front of all rode a liaison officer and a guide sent by the Mayor, and I think Tom Bridges. By his side, armed with a walking stick, was one of the two colonels – a thick-set man – who had surrendered. (The other had disappeared.) And immediately behind them the miscellaneous 'band' of Jew's harps and penny whistles. So through the darkness and the thick shrouding fog of that summer night we marched out, literally feeling our way through the countryside, so thick was the mist. I woke up to the fact that my precious map-case was missing, and I had to return to look for it in the now deserted Grande Place. As for a moment I sat on my horse alone there, taking a last look round, I heard an ominous sound – the metallic rattle on the cobbles, of cavalry in formation entering the town through one of the darkened side-streets. The Germans must have entered St Quentin but a few minutes after the tail of our queer little column disappeared westward through the fog.

[Both defaulting colonels, John Elkington of the Warwicks and Arthur Mainwaring of the Dublin Fusiliers, were cashiered for their attempted surrender. On 14 September 1914, Army Orders recorded their conviction for 'behaving in a scandalous manner unbecoming to the character of an officer and a gentleman'. Elkington, though forty-nine years old, responded in a manner worthy of fiction by joining the French Foreign Legion, with which he lost a leg and was awarded the Legion of Honour. King George V later reinstated him in the British Army and gave him a DSO.]

ARTHUR OSBURN

197.

The difficulties of the Germans invading Belgium were increased by the Dowager Duchess of Sutherland (1867–1965), who had arrived in Namur with a private volunteer medical team, which her family had established in London some weeks earlier with the intention of supporting Ulster's cause in an Irish civil war. When Namur was occupied, the duchess took it for granted that her social grandeur transcended mere issues of war and peace. She harried German commanders mercilessly, as she described in a contemporary memoir.

The doctor and I thought we had better visit the Commander, General Otto von Below. The Germans were perfectly civil to us. Some of them said that they had already invested Brussels. They seemed so absolutely sure of themselves that they still treated the English with politeness and were for the moment only terrorizing and bullying the Belgians. Herr General von Below and his smartly-uniformed officers received my card with great courtesy, and I began to see that it would be necessary to keep up this courtesy by a fixed determination on my part to get all I wanted. The Headquarters Staff was established at the Hôtel de Hollande. The Germans were being importuned by residents asking various favours and questions. One Belgian lady asked if she might follow her husband, who was a prisoner, to Germany. 'You may follow him if you like, *madame*,' was the reply, 'but you cannot accompany him.' The lady looked very sorrowful.

General von Below apologized for receiving me in his bedroom, so terribly overflowing were all the other rooms. Feld-Marschall von der Goltz, who arrived en route to take up his duties in Brussels, was kept waiting while the General spoke to me. He was buttoned up to his nose in an overcoat. Above the collar gleamed a pair of enormous glasses. He was covered with orders. He shook me by the hand, and went out. I did not discuss the situation with General von Below. I took him for granted. He said he was sure he had met me at Homburg.

'Accept my admiration for your work, Duchess,' he said. He spoke perfect English. General von Below 'did me the honour' to call the next morning at our Ambulance. He was accompanied by Baron Kessler, his aide-de-camp, who composed the scenario of *La Légende de Josephe*. He had been much connected with Russian opera in London during the past season. It was exceedingly odd to meet him under such circumstances, after having so often discussed 'art' with him in London. A message came through to me that about 20 English prisoners had passed through Namur station going to Germany. They were closely guarded and were not even given water to drink nor food, because the Germans said, the English were using 'dum-dum' bullets. I laid a complaint about this to the new German Commander of Namur. He assured me there must be some mistake, and he gave me permission to go to the station and look after any other wounded and take my nurses.

When I was at the 'Kommandantur', there appeared to be some depression among the Germans. The head doctor of the garrison, Dr Schilling, who had hitherto been most civil to me, seemed agitated. He looked at my passport and his hands trembled as he held it. He said, 'How wicked of you English and your "Mr" Grey to fight against the Germans and leave us to those devilish Russians.' I used to go every day and visit the 'Kommandantur' and quote the Convention of Geneva and do all I could to lighten the lot of our wounded. I disliked very much trying to get favours out of the Germans, but it had to be done.

On September 4 I called all my wounded, Belgian and French, together and told them they would have to go as prisoners of war to

Germany; but I put them on their word of honour that none of them would try to escape from our hospital. I pointed out the probability that if one escaped, the rest would be shot. One man told me that some of them had considered the idea of climbing over the Convent wall and making a dash for liberty, but they owed so much to my sisters' gentle nursing and to me that they gave their word of honour they would not betray us. They kept their word, too.

September 9. After the [church] service I went down to see the Commander. He and the German officers were at lunch at the Hotel St Aubain. They talked in loud voices, and seemed very hungry. I sent a message in to say that I wished to speak to the Commander, and he came out from the dining-room with his aide-de-camp. 'I must apologize, Duchess,' said the aide-de-camp in English, 'for the growth of my beard.' I felt very indifferent about his beard, but I asked him why he did not get shaved. 'Shaved by a Belgian,' he exclaimed; 'why he would cut my throat!'

The Commander was delighted, he said, to help us in any way to leave Namur. I think he was getting sick of me. Then Dr Schilling followed him, waving his serviette and smoking a large black cigar and asked if I had evacuated our Ambulance. When I told him the Commander had said we might go to Maubeuge he called out at first, '*Nein, nein,*' and then ran back into the dining-room. This was rather discouraging. He soon returned and said we could go when our Ambulance was evacuated. 'But,' I replied, 'the doctor at the college declares that he does not want our wounded until tomorrow, Monday.'

'Nonsense,' replied Schilling, 'he has nothing to do with it. They must go today. Even if it is Sunday I will send carts for them. When you have done what I order you, you can go to Maubeuge tomorrow.'

'How are we to go to Maubeuge?' I persisted. 'I have no motor.'

He reflected for a moment. 'I will get you and your nurses a carriage on a German military train. Here is an order. You had better visit the station and see to it.' I determined, as Maubeuge was invested by the Germans, to report myself to the Commander, Major Abercron, at Headquarters. He asked me to bring in all the nurses and the doctor, and informed me that the French Red Cross were giving him a lot of trouble, that he had shot a doctor and even a nurse for being spies, and as he did not like to see a 'high-born lady in an invidious position' he thought that we had better leave Maubeuge as soon as possible.

He asked, 'What, under the circumstances, do you wish to do?' It was the first critical moment of difficulty that I had had so far with the German authorities. I had to call all my Scottish mother-wit to my aid. 'Our only object, *mein Herr*, is to nurse the wounded. Perhaps you will allow us to proceed into France, where we might find our own troops.' This amazing request caused him to reflect. He was now joined by Baron W—, whose name I will not give as he had evidently known me before the war. 'You cannot go into France,' said Abercron. 'Have you any money on you?'

'I have sufficient,' I replied. 'Then I will give you a pass,' he said, 'to enable you to go to England. If it is necessary to pay for services required, you will do so.'

DUCHESS OF SUTHERLAND

198.

An unexpected hazard of the French retirement westwards.

3rd Sept 1914. The prospect of the retreat being continued was depressing to all of us, and especially exasperating to Commandant Lamotte, the efficient little officer in charge of the distribution of maps. With the Army moving at such a rate, to supply all units with the vast quantity required was no light task. Lamotte kept dashing off to Paris to exhort the map-printing department, already working day and night, to even greater efforts. Maps of France, always more maps of France, were called for, whilst vast quantities of maps of Germany, carefully prepared for a successful offensive, filled the vaults, never to be disturbed. The special grievance of our little cartographer was that people would insist on fighting battles at the junction of two maps, thereby thoughtlessly and wastefully using two sheets where one should have sufficed.

EDWARD SPEARS

199.

During the Allied retreat drastic measures were implemented, to punish and deter desertion. Spears suggests unconvincingly that a condemned man suddenly saw the point of his own fate.

General Louis de Maud'huy (1857–1921) had just been roused from sleep on 12 September on the straw of a shed and was standing in the street, when a little group of unmistakable purport came round the corner. Twelve French soldiers and an N.C.O., a firing party, a couple of gendarmes, and between them an unarmed soldier. My heart sank and a feeling of horror overcame me. An execution was about to take place. General de Maud'huy gave a look, then held up his hand so that the party halted, and with his characteristic quick step went up to the doomed man. He asked what he had been condemned for. It was for abandoning his post.

The General then began to talk to the man. Quite simply he explained discipline. Abandoning your post was letting down your pals, more, it was letting down your country that looked to you to defend her. He spoke of the necessity of example, how some could do their duty without prompting but others, less strong, had to know and understand the supreme cost of failure. He told the condemned man that his crime was not venial, not low, and that he must die as an example, so that others should not fail.

Surprisingly the wretch agreed, nodded his head. The burden of infamy was lifted from his shoulders. He saw a glimmer of something, redemption in his own eyes, a real hope, though he knew he was to die. Maud'huy went on, carrying the man with him to comprehension that any sacrifice was worthwhile if it helped France ever so little. What did anything matter if he knew this? Finally de Maud'huy held out his hand: 'Yours also is a way of dying for France,' he said. The procession started again, but now the victim was a willing one. The sound of a volley in the distance announced that all was over. General de Maud'huy wiped the beads of perspiration from his brow, and for the first time perhaps his hand trembled as he lit his pipe.

EDWARD SPEARS

200.

*The flower painter Madame Lemaire (1845–1928) had been fin de
siècle hostess of the most famous literary and artistic salon in Paris. In
1914 she and her daughter Suzette retreated to their beloved chateau
at Reveillon, thirty miles east of the capital, which found itself in the
path of the Kaiser's armies. Marcel Proust's biographer describes their
experience.*

For a single day on 5 September, eve of the battle of the Marne, the
chateau was within the fringe of the German advance. As they walked
in the garden a German officer on horseback jumped the hedge,
clapped his monocle to his eye, cried: 'I wanted to see Madeleine
Lemaire, and now I have!' and galloped away. That night a German
detachment was billeted in the chateau.

GEORGE PAINTER

201.

*Although Lt. James Marshall-Cornwall (1887–1985) was a gunner,
in the confusion of the September advance he found himself
commanding cavalrymen of the 15th Hussars.*

As the country seemed deserted, I was riding some hundreds of yards
ahead of my squadron, accompanied only by my trumpeter. On turning the corner of a village street, I ran into a patrol of four Uhlans, not
30 yards away. Drawing my sword, I shouted 'Troop, charge!' and then
turned about quickly and fled at a gallop. The enemy patrol also
wheeled about and galloped in the opposite direction. It was the only
time that I ever drew my sword in the presence of the enemy.

JAMES MARSHALL-CORNWALL

202.

*Lt. Bernard Montgomery (1887–1976) of the Warwickshire
Regiment also drew his sword, during an attack on 13 October 1914,
but felt obliged to resort to cruder methods.*

When zero hour arrived I drew my recently sharpened sword and shouted to my platoon to follow me. We charged forward towards the village; there was considerable fire and some of my men became casualties, but we continued on our way. As we neared the objective I suddenly saw in front a trench full of Germans, one of whom was aiming his rifle at me. In my training as a young officer I had received much instruction in how to kill my enemy with a bayonet fixed to a rifle. I knew all about the various movements – right parry, left parry, forward lunge. I had been taught how to put the left foot on the corpse and extract the bayonet, giving at the same time a loud grunt. Indeed, I had been considered good on the bayonet-fighting course against sacks filled with straw, and had won prizes in man-to-man contests in the gymnasium. But now I had no rifle and bayonet; I had only a sharp sword, and I was confronted by a large German who was about to shoot me. In all my short career in the Army no one had taught me how to kill a German with a sword. The only sword exercise I knew was saluting drill, learnt under the sergeant-major on the barrack square. An immediate decision was vital. I hurled myself through the air at the German and kicked him as hard as I could in the lower part of the stomach; the blow was well-aimed at a tender spot. I had read much about the value of surprise. There is no doubt that the German was surprised and it must have seemed to him a new form of war; he fell to the ground in great pain and I took my first prisoner!

FIELD-MARSHAL VISCOUNT MONTGOMERY

203.

Indian soldiers serving alongside the British Army in France suffered terribly, having been drafted into an uncomprehended struggle, in an alien country and culture and a bitter climate. Here a British nurse notes the sorrows of wounded Indians aboard a hospital train in November 1914.

They are such pathetic babes, just as inarticulate to us and crying as if it was a crèche. I've done a great trade in Hindustani, picked up at a desperate pace from a Hindu officer today! If you write it down you can soon learn it, and I've got all the necessary medical jargon now;

you read it off, and then spout it without looking at your notebook. The awkward part is when they answer something you haven't got! They are nearly all 47th Sikhs, perfect lambs; they hold up their wounded hands and arms like babies for you to see. They behave like gentlemen, and salaam after you've dressed them. They have masses of long, fine, dark hair under their turbans done up with yellow combs, glorious teeth and melting dark eyes. One died.

Those of a different caste had to sleep on the floor of the corridors, as the others wouldn't have them in. One compartment of four lying-down ones got restless with the pain of their arms, and I found them all sitting up rocking their arms and wailing, '*Aie, Aie, Aie.*' Poor pets. They all had morphia, and subsided. This long journey from Belgium down to Havre has been a strange mixture. Glorious country, towns and valleys. Glorious British Army lying broken in the train – sleep (or the chance of it) three hours one night and four the next, with all the hours between hard working putting the British Army together again.

SISTER KATHERINE LUARD

204.

The Christmas Day truce of 1914, described in a letter home by Captain Sir Edward Hulse (1889–1915), a Guards officer who had less than three months to live.

At 8.30 a.m. I was looking out, and saw four Germans leave their trenches and come towards us; I told two of my men to go and meet them, unarmed (as the Germans were unarmed), and to see that they did not pass the half-way line. We were 350–400 yards apart at this point. My fellows were not very keen, not knowing what was up, so I went out alone, and met Barry, one of our ensigns, also coming out from another part of the line. By the time we got to them, they were three quarters of the way over, and much too near our barbed wire, so I moved them back. They were three private soldiers and a stretcher-bearer, and their spokesman started off by saying that he thought it only right to come over and wish us a happy Christmas, and trusted us implicitly to keep the truce. He came from Suffolk, where he had left his best girl and a 3½ h.p. motor-bike! He told me that he could not get a letter to the girl, and wanted to send one through me. I made him write out a post card in front of me, in English, and I sent it off that night. I told him that she probably would not be a bit keen to see him again. We then entered on a long discussion on every sort of thing. I was dressed in an old stocking-cap and a man's overcoat, and they took me for a corporal, a thing which I did not discourage, as I had an eye to going as near their lines as possible.

I asked them what orders they had from their officers as to coming over to us, and they said none; they had just come out of goodwill. I kept it up for half an hour, and then escorted them back as far as their barbed wire, having a jolly good look round all the time, and picking up various little bits of information which I had not had an opportunity of doing under fire! I left instructions with them that if any came out later they must not come over the half-way line, and appointed a ditch as the meeting-place. We parted after an exchange of Albany cigarettes and German cigars, and I went straight to H.-qrs. to report.

On my return at 10 a.m. I was surprised to hear a hell of a din going on, and not a single man left in my trenches; they were completely denuded (against my orders), and nothing lived! I heard strains of 'Tipperary' floating down the breeze, swiftly followed by a tremendous burst of '*Deutschland über Alles*', and as I got to my own Coy. H.-qrs. dug-out, I saw, to my amazement, not only a crowd of about 150 British and Germans at the half-way house, but six or seven such crowds, all the way down our lines, extending towards the 8th Division on our right. I bustled out and asked if there were any German officers in my crowd, and the noise died down (as this time I was myself in my own cap and badges of rank).

I found two, but had to talk through an interpreter, as they could neither talk English nor French. I explained that strict orders must be maintained as to meeting half-way, and everyone unarmed; and we both agreed not to fire until the other did, thereby creating a complete deadlock and armistice (if strictly observed). Meanwhile Scots and Huns were fraternizing in the most genuine possible manner. Every sort of souvenir was exchanged, addresses given and received, photos of families shown, etc. One of our fellows offered a German a cigarette; the German said, 'Virginian?' Our fellow said, 'Aye, straight-cut':

the German said, 'No thanks, I only smoke Turkish!' (Sort of 10/- a 100 me!) It gave us all a good laugh.

A German N.C.O. with the Iron Cross – gained, he told me, for conspicuous skill in sniping – started his fellows off on some marching tune. When they had done I set the note for 'The Boys of Bonnie Scotland, where the heather and the bluebells grow', and so we went on, singing everything from 'Good King Wenceslaus' down to the ordinary Tommies' songs, and ended up with 'Auld Lang Syne', which we all, English, Scots, Irish, Prussians, Wurtembergers, etc., joined in. It was absolutely astounding, and if I had seen it on a cinematograph film I should have sworn that it was faked!

Just after we had finished 'Auld Lang Syne' an old hare started up, and seeing so many of us about in an unwonted spot – did not know which way to go. I gave one loud 'View Holloa', and one and all, British and Germans, rushed about giving chase, slipping up on the frozen plough, falling about, and after a hot two minutes we killed in the open, a German and one of our fellows falling together heavily upon the completely baffled hare. Shortly afterwards we saw four more hares, and killed one again; both were good heavy weight and had evidently been out between the two rows of trenches for the last two months, well-fed on the cabbage patches, etc., many of which are untouched in the 'no-man's land'. The enemy kept one and we kept the other.

It was now 11.30 a.m. and at this moment George Paynter arrived on the scene with a hearty 'Well, my lads, a Merry Christmas to you! This is d—d comic, isn't it?' George told them that he thought it was only right that we should show that we could desist from hostilities on a day which was so important in both countries; and he then said, 'Well, my boys, I've brought you over something to celebrate this funny show with,' and he produced from his pocket a large bottle of rum (not ration rum, but the proper stuff). One large shout went up, and the nasty little spokesman uncorked it, and in a heavy ceremonious manner, drank our healths, in the name of his '*kameraden*'; the bottle was then passed on and polished off before you could say knife.

During the afternoon the same extraordinary scene was enacted, and one of the enemy told me that he was longing to get back to London: I assured him that 'So was I'. He said that he was sick of the war, and I told him that when the truce was ended, any of his friends would be welcome in our trenches, and would be well-received, fed, and given a free passage to the Isle of Man! Another coursing meeting took place, with no result, and at 4.30 p.m. we agreed to keep in our respective trenches, and told them that the truce was ended.

Giles Loder, our Adjutant, went down to see if he could come to an agreement about our dead, who were still lying out between the trenches. He found an extremely pleasant and superior stamp of German officer, who arranged to bring all our dead to the half-way line. We took them over there, and buried 29 exactly half-way between the two lines. Giles collected all personal effects, pay-books and identity discs, but was stopped by the Germans when he told some men to bring in the rifles; all rifles lying on their side of the half-way line they kept carefully! They apparently treated our prisoners well, and did all they could for our wounded. This officer kept on pointing to our dead and saying, 'Les braves, c'est bien dommage.' When George heard of it he went down to that section and talked to the nice officer and gave him a scarf. That same evening a German orderly came to the half-way line, and brought a pair of warm, woolly gloves as a present in return for George.

SIR EDWARD HULSE

205.

A staff officer's frustration, serving with Home Command in the winter of 1914.

One day the General said to me: 'The War Office is very nervous about an invasion, there are five million (or whatever the number was) sheep in Sussex, Kent and Surrey. When the enemy land, the sheep will at once be moved by route march to Salisbury Plain.' I knew that this was an impossible task, and that Sir John Moore had proclaimed it as such in 1805. But there was no arguing over it, so I spent days and days working out march-tables for sheep. One day I said to him: 'Do you realize, sir, that should all these sheep be set in movement, every road will be blocked?' 'Of course,' he answered; 'at once arrange to have a number of signposts ready and marked, "Sheep are not to use this road."'

'But,' I replied, 'what if the less well-educated sheep are unable to read them?' This brought our conversation to an end, but, unfortunately, not my tribulations; for as none of us was even an amateur farmer, no one of us had thought of the lambing season, and when it came along all our time and space factors had to be readjusted. If ever there was a wicked waste of time, it was this.

Another hare-brained idea was to destroy all intoxicants in the public-houses the moment the enemy landed. Our general summoned a meeting of the Local Emergency Committee, upon which sat a variety of ancient celebrities, among whom was a General Heath, aged about seventy-five and an ardent teetotaller. To this assembly of notables the General pointed out that, as the enemy were likely to land in Thanet, it would be better to double the liquor than remove what was there, because the drunker they got, the more time we should have in which to collect our scattered forces. Whereupon Heath, white and trembling, rose to exclaim: 'But what of my wife, General, what of my poor wife?' Considering that this good lady must have been over seventy, he quite rightly replied that she ought to be safe.

J. F. C. FULLER

206.

Captain Sir Basil Liddell Hart (1895–1970) had only a brief active service career before being wounded and posted home for staff duties. He nonetheless became an influential war historian and advocate of armoured warfare. His thinking exercised less influence upon Hitler's panzer commanders than their survivors obliged him after 1945 by pretending, but 'BLH' deserves to be remembered as a brilliant controversialist. He should be forgiven the following youthful credo, written from Cambridge Officers' Training Corps in January 1915, and mostly repudiated when he grew up.

Before the war, I, Basil Hart, was a Socialist, a Pacifist, an anti-conscriptionist anti-disciplinist ... held thinkers in greater admiration than warriors. Now having studied the principles of warfare and undergone military training and seen the effects of it on my companions the following are my opinions:

1. *I believe* (i) in the supremacy of the aristocracy of race (and birth) (ii) in the supremacy of the individual.
2. In compulsory military service because it is the only possible life for a *man* and brings out all the finest qualities of manhood.
3. I have acquired rather a contempt for mere thinkers and men of books who have not come to a full realisation of what true manhood means …
4. I exalt the great general into the highest position in the roll of great men and consider it requires higher mental qualities than any other line of life.
5. I consider the Slavs, by which I mean a greater Russia, will rule both Europe and Asia and will have world dominion, being the finest and most virile civilization …
6. Socialism and its forms are an impossibility unless human nature radically alters …
7. Many of the German militarist ideas are very sound, but I oppose the Germans because I do not consider that the German type of mind is the one to carry out their ideas. I prefer brilliance to mechanical and methodical mediocrity …

My belief in the necessary inferiority of women is more pronounced than ever.

If the war ends by Easter [1915] it will be a great thing for the virility and manhood of Europe. If it continues until Xmas, it will be a disaster.

B. H. LIDDELL HART

207.

The Duchess of Westminster (1875–1970), a former Olympic sailing competitor, established her own private hospital in a French casino, in emulation of the Duchess of Sutherland's pioneer attempt, and perhaps also in compensation for her own arid marriage to the 2nd Duke, 'Bendor', soon to be dissolved. A British nurse who joined the hospital in February 1915 describes its chatelaine.

The Duchess was a splendid lady, quite young and really beautiful. She had the most lovely uniform and always wore a very dainty cap, and wherever she went her great wolfhound followed her – into the wards, everywhere. She had a villa at Le Touquet, and very soon after war broke out she went there with some other ladies who were all friends of hers to start a hospital. But, of course, they couldn't manage on their own. Having seen the wounded coming from Mons I always thought it was rather splendid of them not to give up, and as they realized that a hospital was badly needed, they sent to the Red Cross asking for trained nurses. They stayed on themselves, of course, and were very useful because we didn't have time to do things for the men like write letters, and read to them, and do other things to keep them happy and entertained. So the ladies did that, and they also did all the administration and clerical work and saw to the medical supplies and food. The Duchess thought that one of the best things they could do was to raise the morale of the soldiers. When the convoys came in we all rushed to attend to them, and they had to have their particulars taken down. The Duchess and the other ladies used to do that and they always dressed themselves up! 'It's the least we can do to cheer up the men,' the Duchess used to say. Whenever we got word that a convoy was coming in, even if it was nine o'clock in the morning, they went upstairs and changed into full evening dress, with diamond tiaras and everything. Then they would stand at the entrance to take the names. They used to set the gramophone going too, so that they would have a welcome. They meant very well, but it did look funny, these ladies all dressed up and the men, all muddy on the stretchers, looking at them as if they couldn't believe their eyes. I remember one saying, 'We thought we were going to Hell and now it seems we are in Heaven!'

LYNETTE POWELL

208.

It may be hard to believe that anyone enjoyed World War I, but Carton de Wiart did so. After some early 1914 adventures in East Africa, he was invalided home, having lost an eye.

On my appearing before the Medical Board they seemed rather shocked at my desire to go to France. We argued, and they produced the astonishing solution that if I could wear a satisfactory glass eye they would consider me. I imagine they did not wish the Germans to think that we were reduced to sending out one-eyed officers. At my next board I appeared with a startling, excessively uncomfortable glass eye. I was passed fit for general service. On emerging I called a taxi, threw my glass eye out of the window, put on my black patch, and have never worn a glass eye since.

[Having secured a posting to France early in 1915, he was soon wounded again.] My hand was a ghastly sight; two of the fingers were hanging by a bit of skin, all the palm was shot away and most of the wrist. I asked the doctor to take my fingers off; he refused, so I pulled them off myself and felt absolutely no pain. [Once again he qualified for sick leave.] When I had nearly recovered I went along to White's. A member, known to me by sight, came up and asked if I would do him a favour. I answered cautiously that I would if it was not a financial transaction, as no good seems to come out of borrowing or lending money. He then told me there was a man paying undue attention to a lady he knew and he wanted to fight him and asked me to second him in a duel! I agreed at once, as I think duelling a most excellent solution in matters of the heart, and saw that my man was a tremendous fire-eater with only one object in view, to kill his opponent.

I went off to see his opponent, whom I knew; true to form, he found the whole idea ridiculous. I assured him that my friend was adamant and determined to fight with any suggested weapon, preferably pistols at the range of a few feet. It took some time to penetrate into the gentleman's mind that this was serious, and with a great deal of reluctance he appointed seconds. As a last resort our opponent produced what he considered a telling argument, which was that if this episode was found out we should all get into serious trouble.

My reply was that the war was on, everyone too busy to be interested, and that it would be simple to go to some secluded spot like Ashdown Forest with a can of petrol and cremate the remains of whichever was killed. This suggestion finished him off; the mere thought of his ashes scattered to the four winds, unhonoured and

unsung, was too much. He promptly sat down and wrote an affidavit
not to see the lady again. It was a tame end: it seemed to me that as he
did not like the lady enough to fight for her, he needed a thrashing.

ADRIAN CARTON DE WIART

209.

An early German gas attack on the Western Front in April 1915.

About sunset on a calm evening, slowly the brown line of trenches and
earth began to change to a dull luminous green. Looking intently I saw
great clouds of greenish-yellow vapour creeping across from the
German lines. We had heard talk of gas, and once or twice detected
strange chemical odours, but here was a gas-attack, a mile away, which
I could see with my own eyes. The men had already shown signs of
nervousness of gas, based on the wild stories that runners had brought.
The other signs of battle had filled me with a curious elation. The gas,
with its green paralysis, changed my mood. I was angry rather than
frightened, angry as the dog that snaps at the unaccustomed. Our
seniors were alarmed, for they saw that at any moment we might be
called upon to deal with a situation that neither they nor we had been
trained to meet.

Unexpectedly help came. A parcel was delivered for each company
labelled, 'Gas Masks, Type 1'. Unpacked, the parcel revealed bundles
of small squares of blue flannel, just large enough to cover the mouth,
with a tape on each side to tie round behind the head. Whatever
benign personage contrived these amiable death-traps I do not know.
But anything more futile could never have been devised. We preferred
to resort to the face-towels dipped in our own urine, which an earlier
order had suggested would be a palliative. Nor was our confidence
restored a day later by the arrival of 'Gas Masks, Type II', which
consisted of large pieces of hairy Harris tweed about three feet long
and one in width, again with tapes nattily fixed to the sides. With
much laughter the men tried to don their new masks. At the bottom
of the parcel I found a small printed label entitled 'BODY BELTS'. I
ordered my men to put them to whatever use seemed best. To a man,
they placed them round their long-suffering stomachs. I have often

wondered what inspired genius was at work back in England to give us these gifts. I have been told since that Gas Mask Type I was invented by the fertile brain of a Cabinet Minister. I feel tempted to attribute Type II to the Archbishop of Canterbury.

STANLEY CASSON

210.

Lady Dorothie Feilding (1889–1935), a daughter of the Earl of Denbigh, served for most of the war as an ambulance driver on the Western Front, chiefly in the French and Belgian sectors. She became the first woman to be awarded the Military Medal for bravery in the field, together with the Croix de Guerre. Her letters home, such as this one, are minor classics of their time and place.

1 May 1915, Tuesday
I am lying on my underneath (damn uncomfortable too) in the sand by the sea, snatching a little peace. I am fed up with the war and very weary, we really have been very rushed lately and so many late nights make one rather tired, more so mentally than physically and I find myself getting very peevish and stuffy. It must be 3 weeks now since we got to bed before 2 or 3 in the morning and many nights later. Yesterday we were at Steenstrade till 5am as there were a lot of wounded, French infantry mostly, and you can only get them in at night, otherwise things have been quieter down there lately DG [*Danke Gott*]. One of the most lovely sights I have seen 2 mornings ago was dawn at Reninghe over the inundations, and to see the sun streaming up and being caught by the reflections in the waters and the ruined churches and villages sticking up out of it all.

Last night we got to bed at 12 till 3.30am and then had to climb out because there was a call for a car and then turned in from 6 to 10am again. I got so bored dressing and undressing I shall economise labour tonight and go to sleep with my clothes and false teeth all ready on for an emergency exit. It's so nice in the sun here and the sea and so peaceful except for some Godforsaken soldiers that will shoot at a floating target. I never saw such putrid shots. They most of them seem to be going in the direction of Dover. They are cavalry and I think had better go back to their muttons and leave others to try and shoot a bit straighter.

Hély d'Oissel sent 2,000 of his Zouaves down to reinforce the asphyxiated French at Lizerne and it was they that did that very fine attack and took the village but they lost 800 men out of 2,000, a huge total, and all except a tiny handful of officers. They have a wonderful fine 'moral[e]' those Zouaves though of course it depends a lot on individual regiments. Many of the last turned-out regiments are more hotel proprietors from Dinard than Algerians. One Zouave was given the VC and made a corporal lately for doing fine things and he told his colonel he was afraid he wouldn't be much use as he could neither read or write. His Col told him not to fret as it was fighting was wanted. Hély d'Oissel saw this same Tommy in hospital yesterday, he had been shot through both eyes 'stone blind' in the attack at Lizerne and in awful pain but he never complained but said 'My General – you should tell the Colonel that he was right and that it doesn't matter now that I know not how to read nor write' [translated from the French] I think it's so sad, perhaps all these little things don't interest you, but it's little trifles of pluck and all the millions of individual efforts that count in this war.

The things that do one so much good too and make up for any nights up is the extraordinary gratitude of these men one helps. There are so many of them that think of such pathetic ways to thank you that it gives you a lump in your throat and makes one see red and want to put all these beastly Germans in a pit and chop them up with spades like the boys used to do with jellyfish at Colwyn Bay.

[Lady Dorothie eventually went home, to marry and become mother to five children.]

DOROTHIE FEILDING

211.

A truce at Anzac Beach, Gallipoli, on 24 May 1915 described by Hon. Aubrey Herbert (1880–1923), an MP, Somerset squire, multiple linguist, Middle East traveller and wartime officer who was twice offered the throne of Albania, and whose daughter Laura later

married the novelist Evelyn Waugh. Herbert was also a close friend of
T. E. Lawrence and a model for John Buchan's Sandy Arbuthnot.

We walked from the sea and passed immediately up the hill, through a field of tall corn filled with poppies, then another cornfield; then the fearful smell of death began as we came upon scattered bodies. We mounted over a plateau and down through gullies filled with thyme, where there lay about 4,000 Turkish dead. It was indescribable. One was grateful for the rain and the grey sky. There were two wounded crying in that multitude of silence. The Turkish captain with me said: 'At this spectacle even the most gentle must feel savage, and the most savage must weep.' The dead fill acres of ground, mostly killed in the one big attack. One saw the result of machine-gun fire very clearly; entire companies annihilated – not wounded, but killed, their heads doubled under them with the impetus of their rush and both hands clasping their bayonets. It was as if God had breathed in their faces.

There was a good deal of friction at first. The trenches were 10 to 15 yards apart. Each side was on the *qui vive* for treachery. In one gully the dead had got to be left. It was impossible to bury them without one side seeing the position of the other. In the Turkish parapet there were many bodies. Fahreddin told Skeen he wanted to bury them, 'but,' he said, 'we cannot take them out without putting something in their place.' I talked to the Turks, one of whom pointed to the graves. 'That's politics,' he said. Then he pointed to the bodies and said: 'That's diplomacy. God pity all of us poor soldiers.' Then Skeen came. He told me to get back as quickly as possible to Quinn's Post, as I said I was nervous at being away, and to retire the troops at 4 and the white-flag men at 4.15.

At 4 o'clock the Turks came to me for orders. I do not believe this could have happened anywhere else. I retired their troops and ours, walking along the line. At 4.7 I retired the white-flag men, making them shake hands with our men. Then I came to the upper end. About a dozen Turks came out. I chaffed them, and said that they would shoot me next day. They said, in a horrified chorus: 'God forbid!' The Albanians laughed and cheered, and said: 'We will never shoot you.' Then the Australians began coming up, and said: 'Goodbye, old chap;

good luck!' And the Turks said: '*Oghur Ola gule gule gedejek – seniz, gule gule gelejekseniz*' (Smiling may you go and smiling come again). Then I told them all to get into their trenches, and unthinkingly went up to the Turkish trench and got a deep salaam.

AUBREY HERBERT

212.

A skirmish at Gallipoli with an enemy common to all wars, narrated by a Royal Marine. Private Bruckshaw (1891–1917) was born in Hampshire, and worked as a junior manager in a Cheshire mill before enlisting in 1914. He kept a diary until shortly before his death in action at Arras, bequeathing to his widow Phoebe the tiny fortune of £242 18s 10d.

The following is a strictly true account of about the most horrible and deadly encounter I have so far had the misfortune to be in. The enemy in enormously strong numbers occupied some strong positions and simply could not be turned out. They were behind every small ridge and filled up every crevice of which there was an abundance. We had continually had encounters with them in a casual sort of way but now it had been determined to shift them at all costs. The enemy had their bite first however. One night just after we had piped down and settled ourselves for the night routine of the trenches they made an attack. We repulsed them with very considerable loss and our monkey was well up now and as soon as daylight appeared we made a determined counter-attack.

It was no end of a stiff do. Blood, blood everywhere. Even splashes of it on my face and as for my hands, well, I looked an object indeed. You know too it was Summer time and by now the sun had risen so that streams of perspiration rolled from us. We took no prisoners and the slaughter was really terrible. Even as we killed them off, fresh ones appeared to take our special attention, and so the slaughter continued. We had at last cleared the place except for sundry stragglers who would no doubt be seen off later. We had killed scores, yes hundreds of the loathed enemy. On our shirts. The next time I enlist and go on Active Service (and I hope others will take my advice) I shall always carry a plentiful supply of Keatings powder for it kills bugs, fleas, moths, beetles, etc. and ensures a peaceful night.

PRIVATE HORACE BRUCKSHAW

213.

The Anzacs – Australian and New Zealand Army Corps – won undying fame at Gallipoli, which represented a terrible coming-of-age for their nations. There the world learned the meaning of 'mateship', reflected in the suicidal Australian assault at Lone Pine in August 1915. War correspondent Charles Bean described how one man shouldered his way into the tensed line of attackers manning the forward trench, minutes before H-Hour, to share almost certain death.

'Jim here?' he asked. A voice on the firestep answered: 'Right, Bill, here.' 'Do you chaps mind shiftin' up a piece,' said the first voice. 'Him and me are mates, an' we're going over together.'

CHARLES BEAN

214.

The Anzacs won the battle, at terrible cost in casualties.

Seven Australians won the Victoria Cross, two of them posthumously. All the medals were won during the three days after the opening charge. All were about the bomb-throwing and subterranean anarchy of the second part of the battle. All were about furious little scraps around trench corners and over piled-up sandbags. The background of the seven VC winners tells us much about Australian society in 1915. Lance-Corporal Leonard Keysor was a 29-year-old Jew who had emigrated to Sydney in 1914. He returned to London after the war, lived quietly and said the war was the only adventure he ever had. Captain Alfred Shout was born in Wellington, New Zealand, and served with the New Zealand contingent in the Boer War, where he was twice wounded. He emigrated to Sydney in 1907. Three others – Lieutenant William Symons, Captain Frederick Tubb and Corporal William Dunstan, all born in country Victoria – were of Cornish stock. Before the war, Private John Hamilton worked as a butcher's boy, Dunstan as a messenger boy in a draper's shop, Symons as a clerk in a grocer's shop, Shout as a carpenter, Tubb as a farmer. Tubb and Corporal Alexander Burton, both from the Euroa area of Victoria, had joined up together.

Burton was killed by a bomb. Tubb was with him and later said: 'Just before he died he looked up at me, smiled quietly, and was then killed. His was a fine death, and I almost wish I had died, too.' Tubb was killed in Belgium in 1917. Dunstan later became general manager of the *Herald & Weekly Times* newspaper group when it was run by Keith Murdoch. He never talked about that day at Lone Pine.

LES CARLYON

215.

Siegfried Sassoon (1886–1967), an infantry officer who was to become one of the most famous chroniclers of those times, through his poetry and his Memoirs of a Foxhunting Man *and* Memoirs of An Infantry Officer, *daydreamed about his comrades at a tactical school in France, most of whom were doomed to wounds or death.*

The star turn in the schoolroom was a massive sandy-haired Highland Major whose subject was 'The Spirit of the Bayonet' – he was afterwards awarded the DSO for lecturing. He took as his text a few leading points from the *Manual of Bayonet Training*. He spoke with homicidal eloquence, keeping the game alive with genial and well-judged jokes. He had a sergeant to assist him. The sergeant, a tall sinewy machine, had been trained to such a pitch of frightfulness that at a moment's warning he could divest himself of all semblance of humanity. With rifle and bayonet he illustrated the Major's ferocious aphorisms, including facial expression. When told to 'put on the killing face', he did so, combining it with an ultra-vindictive attitude.

'To instil fear into the opponent' was one of the Major's main maxims. Man, it seemed, had been created to jab the life out of Germans. To hear the Major talk, one might have thought that he did it himself every day before breakfast. His final words were: 'Remember that every Boche you fellows kill is a point scored to our side; every

Boche you kill brings victory one minute nearer and shortens the war by one minute. Kill them! Kill them! There's only one good Boche, and that's a dead one!'

Afterwards I went up the hill to my favourite sanctuary, a wood of hazels and beeches. The evening air smelt of wet mould and wet leaves; the trees were misty-green; the church bell was tolling in the town, and smoke rose from the roofs. Peace was there in the twilight of that prophetic foreign spring. But the lecturer's voice still battered on my brain. 'The bullet and the bayonet are brother and sister.' 'If you don't kill him, he'll kill you.' 'Stick him between the eyes, in the throat, in the chest.' 'Don't waste good steel. Six inches are enough. What's the use of a foot of steel sticking out at the back of a man's neck? Three inches will do for him; when he coughs, go and look for another.'

SIEGFRIED SASSOON

216.

A Lewis gunner of 6th Royal Scots Fusiliers describes the December 1915 arrival of an unexpected commanding officer.

Charlie Broon, Company runner, came into our billet and announced: 'We're gettin' a new C.O. – some fella ca'd Churchill.' I asked him if it was the ex-First Lord of the Admiralty, the ex-Home Secretary, M.P. for my home town of Dundee. 'Aw Ah ken is that he's ca'd Churchill.' At the battalion parade on the morning following Churchill's arrival we soon sensed that a new force had come among us. In his initial address he said: 'You men have had a hard time. Now you're going to have it easy for some time, I hope.' On the previous evening, to the assembled officers he spoke these now-historic words: 'Gentlemen, we are going to make war – on the lice!'

Cleaning, scrubbing and de-lousing were the order of the day. Huge stocks of clothing arrived. We needed new rig-outs. We got them. Steel helmets were by then being issued. We were among the first to get them. We found, too, a vast improvement in our rations. Bully beef and biscuits were only memories. We took over the sector to the right of 'Plugstreet' Wood in early February. It had always been regarded as one of the 'cushy' sectors. But with the coming of Churchill it was soon evident that the live-and-let-live atmosphere that had always prevailed here was coming to an end. He visited the front line daily and nightly. He went on lone reconnaissance missions into 'no-man's-land'. We often got the order: 'Pass it along – no firing. The C.O.'s out in front.'

In the fire bays, he scrutinized every man and sometimes questioned them. I remember one night when Sergeant McGee, an old Regular, was rousing men in the dugouts to form a fatigue party. He was doing this by his usual method – pulling up the sacking flap, kicking the soles of protruding feet, then addressing the inhabitants in a mixture of profanity and Hindustani. He was unaware of Churchill's presence until a narrow beam of light from the C.O.'s torch pinpointed his three stripes.

'Ah, a sergeant.'

'Sir.' McGee's heels clicked.

'Are you an old soldier, sergeant?'

'Yes, sir. Eight years' service.'

'Well, sergeant, if you live to see the end of this war – which I hope you do – you are going to be a *very* old soldier.'

ROBERT FOX

217.

When Robert Graves (1895–1985) was transferred from the 2nd Welsh Regiment to the Royal Welch Fusiliers, he found himself subjected to petty humiliations such as many regular messes traditionally inflicted upon newcomers. A fellow sufferer briefed him.

'They treat us like dirt; in a way it will be worse for you than for me because you're a full lieutenant. They'll resent that with your short service. There's one lieutenant here of six years' service and second-lieutenants who have been out here since the autumn. They have already had two Special Reserve captains foisted on them; they're planning to get rid of them somehow. In the mess, if you open your mouth or make the slightest noise the senior officers jump down your throat. Only officers of the rank of captain are allowed to drink whisky or turn on the gramophone. We've got to jolly well keep still and look like furniture. It's just like peace time. Mess bills are very high; the mess was in debt at Quetta last year and we are economizing now to pay that back. We get practically nothing for our money but ordinary rations and the whisky we aren't allowed to drink.

'We've even got a polo-ground here. There was a polo match between the First and Second Battalions the other day. The First Battalion had had all their decent ponies pinched that time when they were sent up at Ypres and the cooks and transport men had to come up into the line to prevent a breakthrough. So this battalion won easily.

'Can you ride? No? Well, subalterns who can't ride have to attend riding-school every afternoon while we're in billets. They give us hell, too. Two of us have been at it for four months and haven't passed off yet. They keep us trotting round the field, with crossed stirrups most of the time, and they give us pack-saddles instead of riding-saddles. Yesterday they called us up suddenly without giving us time to change into breeches. That reminds me, you notice everybody's wearing shorts? It's a regimental order. The battalion thinks it's still in India. They treat the French civilians just like "niggers", kick them about, talk army Hindustani at them.

'It makes me laugh sometimes. Well, what with a greasy pack-saddle, bare knees, crossed stirrups, and a wild new transport pony that the transport men had pinched from the French, I had a pretty thin time. The colonel, the adjutant, the senior major and the transport officer stood at the four corners of the ring and slogged at the ponies as they came round. I came off twice and got wild with anger, and nearly decided to ride the senior major down. The funny thing is that they don't realize that they are treating us badly – it's such an honour to be serving with the regiment. So the best thing is to pretend you don't care what they do or say.' I protested: 'But all this is childish. Is there a war on here or isn't there?'

'The battalion doesn't recognize it socially,' he answered. 'Still, in trenches I'd rather be with this battalion than in any other that I have met. The senior officers do know their job, whatever else one says about them, and the N.C.O.s are absolutely trustworthy.' The battalion was peculiar in having a battalion mess instead of company

messes. The Surrey-man said grimly: 'It's supposed to be more socia-
ble.' This was another peace-time survival. We went together into the
big chateau near the church. About fifteen officers of various ranks
were sitting reading the week's illustrated papers or (the seniors at
least) talking quietly. At the door I said: 'Good morning, gentlemen,'
the new officer's customary greeting to the mess. There was no answer.
Everybody looked at me curiously. The silence that my entry had
caused was soon broken by the gramophone, which began singing
happily:

> *We've been married just one year,*
> *And Oh, we've got the sweetest,*
> *And Oh, we've got the neatest,*
> *And Oh, we've got the cutest*
> *Little oil stove.*

I found a chair in the background and picked up *The Field*. The door
burst open suddenly and a senior officer with a red face and angry eye
burst in. 'Who the blazes put that record on?' he shouted to the room.
'One of the bloody warts I expect. Take it off somebody. It makes me
sick. Let's have some real music. Put on the *Angelus*.' Two subalterns
(a subaltern had to answer to the name of 'wart') sprang up, stopped
the gramophone, and put on *When the Angelus is ringing*. The young
captain who had put on *We've been married* shrugged his shoulders and
went on reading, the other faces in the room were blank.

'Who was that?' I whispered to the Surrey-man.

He frowned. 'That's "Buzz Off",' he said.

Before the record was finished the door opened and in came the
colonel; 'Buzz Off' reappeared with him. Everybody jumped up and
said in unison: 'Good morning, sir.' Before giving the customary greeting and asking us to sit down he turned spitefully to the gramophone:
'Who on earth puts this wretched *Angelus* on every time I come into
the mess? For heaven's sake play something cheery for a change.' And
with his own hands he took off the *Angelus*, wound up the gramophone and put on *We've been married just one year*. At that moment a
gong rang for lunch and he abandoned it.

We filed into the next room, a ballroom with mirrors and a decorated ceiling. We sat down at a long, polished table. The seniors sat at
the top, the juniors competed for seats as far away from them as possible. I was unlucky enough to get a seat facing the commanding officer,
the adjutant and 'Buzz Off'. There was not a word spoken except for
an occasional whisper for the salt or for the beer – very thin French
stuff. Robertson, who had not been warned, asked the mess waiter for
whisky. 'Sorry, sir,' said the mess waiter, 'it's against orders for the
young officers.' Robertson was a man of forty-two, a solicitor with a
large practice, and had stood for Parliament.

I saw 'Buzz Off' glaring at us and busied myself with my meat and
potatoes. He nudged the adjutant. 'Who are those two funny ones
down there, Charley?' he asked. 'New this morning from the militia.
Answer to the names of Robertson and Graves.' 'Which is which?'
asked the colonel.

'I'm Robertson, sir.'

'I wasn't asking you.'

Robertson winced, but said nothing. Then 'Buzz Off' noticed
something.

'T'other wart's wearing a wind-up tunic.' Then he bent forward and
asked me loudly, 'You there, wart. Why the hell are you wearing your
stars on your shoulder instead of your sleeve?'

My mouth was full and I was embarrassed. Everybody was looking
at me. I swallowed the lump of meat whole and said: 'It was a regimental order in the Welsh Regiment. I understood that it was the same
everywhere in France.'

The colonel turned puzzled to the adjutant: 'What on earth's the
man talking about the Welsh Regiment for?' And then to me: 'As soon
as you have finished your lunch you will visit the master-tailor. Report
at the orderly room when you're properly dressed.' There was a severe
struggle in me between resentment and regimental loyalty. Resentment for the moment had the better of it. I said under my breath: 'You
damned snobs. I'll survive you all. There'll come a time when there
won't be one of you left to remember battalion mess at Laventie.' This
time came, exactly a year later.

218.

A routine 'hate' – a German barrage upon British trenches.

Towards the end of May [1916], a dozen recruits joined the company, young reinforcements, boyish and slight. Early one morning the enemy began to shell the trench with whizz-bangs; it was a sudden angry storm, too fierce and too localized to last long. I had just passed the fire-bay in which Delivett was frying a rasher of bacon, with five of these lads watching him and waiting their turn to cook. I stopped in the next bay to reassure the others. Suddenly a pale and frightened youth came round the corner, halting indecisively when he saw me, turning again, but finally going back reluctantly to his fire-bay in despair of finding any escape from his trap. Between the crashes of the bursting shells a high-pitched sing-song soared up. 'You'll 'ev 'em all over,' … Crash … 'All the milky ones.' … Crash … 'All the milky coco-nuts …' '… You'll 'ev 'em all over … All the milky ones.' … Crash … 'Therree shies a penny … All the milky coco-nuts … You'll 'ev 'em all over' … Crash – and then silence, for the morning hate had ended as suddenly as it began. I walked to find Delivett still frying bacon, and the five youths smiling nervously, crouched below the fire-step. I sent them away on some improvised errand and faced Delivett. 'That's a fine thing you did then, Delivett,' I said. He looked up, mess-tin lid in his hand, saying nothing, but the lines round his mouth moved a little. 'You saved those lads from panic – they were frightened out of their wits,' I added.

'Yes, sir, they was real scared,' he replied.

'Delivett, you've spent a lot of time on Hampstead Heath.'

'Yes sir … I ran a coco-nut shy there once …'

With these words a man and an environment fused into a unity, satisfying and complete in itself; here at last was a credible occupation for this quiet stranger. 'I'm going to tell the Colonel all about this,' I said. Delivett thought hard for several seconds, and put his bacon back on the fire.

'Well, sir,' he said diffidently, 'if it's all the same to you, I'd much rather you made me Sanitary man.'

'Do you mean that you'd really like to go round with a bucket of chloride of lime, picking up tins and …'

'Yes, sir, I'd like that job.'

'You shall have it here and now. You are made Sanitary man for valour in the field, this very moment.'

In half an hour Delivett was walking round with a bucket, his head a little higher in the air, spitting a little more deliberately than before, as his new dignity demanded. He had found a vocation.

WYN GRIFFITH

219.

Another trench tale.

This is what happened the other day. Two young miners disliked their sergeant, who had a down on them and gave them all the most dirty and dangerous jobs. When they were in billets he crimed them for things they hadn't done. So they decided to kill him. Later they reported at Battalion Orderly Room and asked to see the adjutant. This was irregular, because a private is not allowed to speak to an officer without an N.C.O. to act as go-between. The adjutant happened to see them and said: 'Well, what is it you want?' Smartly slapping the small-of-the-butt of their sloped rifles they said: 'We've come to report, sir, that we are very sorry but we've shot our company sergeant-major.'

The adjutant said: 'Good heavens, how did that happen?' They answered: 'It was an accident, sir.' 'What do you mean? Did you mistake him for a German?' 'No, sir, we mistook him for our platoon sergeant.' So they were both shot by a firing squad against the wall of a convent at Béthune. Their last words were the battalion rallying-cry: 'Stick it, the Welsh!' (They say that a certain Captain Haggard first used it in the battle of Ypres when he was mortally wounded.) The French military governor was present at the execution and made a little speech saying how gloriously British soldiers can die.

ROBERT GRAVES

220.

The epitome of tragic contrast between the spirit and reality of the Battle of the Somme was achieved by the footballers of zero hour on 1 July 1916. The episode has been attributed to several units at different times, but the best documented was sponsored by 21-year-old Captain Wilfred Nevill of the 8th East Surreys.

Nevill was a young officer who liked to stand on the fire-step each evening and shout insults at the Germans. His men were to be in the first wave of the assault near Montauban and he was concerned as to how they would behave, for they had never taken part in an attack before. While he was on leave, Nevill bought four footballs, one for each of his platoons. Back in the trenches, he offered a prize to the first platoon to kick its football up to the German trenches on the day of the attack. One platoon painted the following inscription on its ball:

The Great European Cup
The Final
East Surreys v. Bavarians
Kick-off at Zero

Nevill himself kicked off. 'As the gunfire died away [wrote a survivor], I saw an infantryman climb onto the parapet into No Man's Land, beckoning others to follow. As he did so he kicked off a football; a good kick, the ball rose and travelled towards the German line. That seemed to be the signal to advance.' The winning footballers of the 8th East Surreys were unable to collect the prize money from their commander. Captain Nevill was dead.

MARTIN MIDDLEBROOK

221.

Another moment on the Somme, experienced by the 9th Royal Sussex, 7 July 1916.

As I was travelling light and the men were loaded with all sorts of junk, I got to the enemy line alone. It was blown all to hell, but the dug-outs were obvious. There was not a soul in the trench and I realized I had got there before the Germans had come out of their burrows. I sat down facing the dug-out doors and got all the Germans as they came up. They had no idea I was there even, the ground was so blown up; they never knew what hit them. Broughall, a plucky little Canadian, was about the next of our people to arrive; he was very excited at my bag, but disgusted at their 'hats' as they all had steel helmets or flat caps and he had promised himself the best spiked helmet (*pickelhaube*) in France.

While we were cleaning up the front line we put up a big German wearing a very smart spiked helmet. Broughall, unaware that there were dozens of other helmets in the dug-outs, at once gave chase to the wearer. His platoon rallied to the cry of 'Get that bloody hat' and followed him. The quarry ran up a communication trench and was finally pulled down in the German fourth line, where Broughall and his platoon settled him and held their own against repeated counter-attacks. While we were rounding up the prisoners I came upon one of the Fusiliers being embraced round the knees by a trembling Hun who had a very nice wrist-watch. After hearing the man's plea for mercy the Fusilier said, 'That's all right, mate, I accept your apology, but let's have that ticker.'

CAPTAIN HENRY SADLER

222.

Elsewhere in the line British soldiers were less merciful, as an officer recorded after an attack.

Blake's face was slack and haggard, but not from weariness. He greeted me moodily, and then sat silent, abstracted in some distant perplexity. 'What's the matter, Terence?' I asked. 'Oh, I don't know. Nothing – at least. Look here, we took a lot of prisoners in those trenches yesterday morning. Just as we got into line, an officer came out of a dugout. He'd got one hand above his head, and a pair of field glasses in the other. He held his glasses out to S—, you know the ex-sailor with the Messina

earthquake medal – and said "Here you are, Sergeant, I surrender."
S— said "Thank you, sir," and took the glasses with his left hand. At
the same moment, he tucked the butt of his rifle under his arm and
shot the officer straight through the head. What ought I to do?'

'I don't see that you can do anything,' I answered slowly. 'What can
you do? Besides, I don't see that S— is really to blame. He must have
been half-mad with excitement when he got into the trench. I don't
suppose he even thought what he was doing. If you start a man killing,
you can't turn him off again like an engine.'

'It wasn't only him: another did exactly the same thing.'

'Anyhow, it's too late to do anything now. I suppose you ought to
have shot both on the spot. The best thing is to forget it.'

GUY CHAPMAN

223.

A casualty of the Somme.

There was a man with one side of his face blown away. The skin had
grown over it, but he was still bandaged and I was told to syringe his
face and to put a screen round him while I did it. I chatted to him
while I was taking the dressing off and I must have smiled. He said,
'How can you smile when you look at me, Nurse?' It must have meant
a lot to him, because later he wrote in my autograph book:

> 'Remember, dear Nurse, to keep that sweet smile,
> It helps us lame dogs over many a stile.'

GRACE BIGNOLD

224.

Face-saving by top brass.

When Rumania declared war in August 1916, it is said that one of the
first army orders, after mobilization, was that only officers above the
rank of major were allowed to use make-up.

RONALD LEWIN

225.

*A British battery commander, my own great-uncle Major Lewis
Hastings MC (1881–1966), who in 1939–45 was Military
Commentator of the BBC and made his first parachute jump at the
age of sixty-two, discovered unexpected virtue in one of his gunners.*

Briggs brought off one feat which endeared him to the battery. He
couldn't read or write, but he was about the best and most intelligent
scrounger we had – an extremely useful accomplishment in those hard
times. The late winter of 1916–17 was extremely bitter, and the ground
that we were holding had frozen iron hard. The issue of fuel was down
to a minimum, and conditions were arctic. Men shivered in the shallow dugouts round the guns and were too cold to sleep. One morning
I had the Maltese cart sent up to the Battery, called for Briggs, and told
him to go off into the blue anywhere and do his damnedest to come
back with any sort of fuel.

There wasn't much hope, for all the ruined frostbound villages had
been stripped to the bone. However, Briggs set off. Some distance in
the rear he came to shell-torn Colincamps. There was a tall post in the
pitted market square with a placard on it: DANGER! DON'T LOITER!
Briggs naturally tied his horse and cart to this post and started to
reconnoitre. Most of the little houses were rubble, but the walls of the
church still stood. Briggs clambered over broken stones and entered.
The roof was gone and so were the rafters, the pews, and the benches,
and apparently everything else combustible. Dangling the battery axe
thoughtfully in one hand, Briggs wandered round. He saw at
man-height from the ground a series of niches, each occupied by one
highly-coloured member of the select company of the Saints. Their
primitive colours of scarlet and blue and gold were still untarnished
amid this desolation. Were they marble? Stone? Plaster? Concrete?
Briggs approached the nearest, and tapped Saint Peter on the shin. He
tapped again hard. A light as of pious awe illuminated his Yorkshire
face.

Some hours later the Maltese cart arrived at the Battery position.
Piled high with chopped and painted timber, it was seized upon by the

shivering but happy gunners, and soon a dozen varieties of fug were making things homely and comfortable. Before morning dawned the good news went down the wagon lines, the sergeant-major collected the rest of the heavenly choir in a G.S. wagon, and for many days after the troops cooked their morning bacon and stoked their midnight fires with the chopped anatomies of Saints, Virgins, and Apostles. It gave an added zest to know that half the British Army must have seen these images and passed them by, and that it had been left to our Briggs with his little battery axe to discover that the saints were made of wood.

LEWIS HASTINGS

226.

A youthful experience of Lt. William Slim (1891–1970). He was a Bristol ironmonger's son who taught in a primary school and worked as a clerk before becoming an army officer in 1914 by way of Birmingham University Officers' Training Corps, though he was never a student. Slim later became Britain's best general of World War II.

The iron deck of a barge under a single awning in the fantastic temperature of a Mesopotamian summer is about as near hell as one can get this side of the Styx – and our boat was firmly embedded in the palms. I decided to explore. I found a couple of large tents inside a barbed-wire enclosure that was stacked with crates, boxes, sacks and supplies of all kinds. I passed an Indian sentry at the gate and made for one of the tents. Inside, seated at a packing-case fitted roughly as a desk, was a lieutenant-colonel of the Supply and Transport Corps. He was a tall, cadaverous, yellow-faced man with a bristling moustache. He looked very fierce and military – officers who dealt with bully-beef and biscuit in the back areas so often did – and he gave short shrift to my timid suggestion that his dump might possibly provide something in the way of additional awnings or tents for us. No, his Supply Depot contained nothing but supplies. Then, perhaps, a little something extra in the way of rations …? I was informed that his supplies were not for issue to any casual subaltern who cared to ask for them, and, if my detachment had not got everything that was necessary for its comfort, it was because either:

– I was incompetent.

– The staff at the Reinforcement Camp was incompetent.

– Or a combination of (a) and (b).

I gathered he rather favoured the first alternative. He ended with the final warning: 'And don't let your fellows come hanging round here. The British soldier is the biggest thief in Asia and his officers encourage him.' It is not a very profitable pastime for subalterns to quarrel with lieutenant-colonels, so I swallowed all this as best I could. We spent two more infernal nights on those moored barges. We had, however, one pleasant surprise. Our rations, which up to then had been limited to the regulation bully, biscuit, dried vegetables – horrible things – tea and sugar, were suddenly supplemented by a liberal issue of tinned fruit. As I squatted on my valise, making a leisurely choice between pineapple and peaches, I thought of the kind heart that S. and T. colonel must hide beneath his fierce exterior. Next morning when we all breakfasted off first-class bacon, followed by admirable Australian quince jam, while tinned milk flowed in streams and every man seemed to have a handful of cigarettes, I meditated on how one could be misled by first impressions.

I will not deny that certain suspicions did flit across my mind. There was a tinge of apprehension on the mahogany face of my quartermaster-sergeant when I suggested it would be a graceful act of courtesy if he would accompany me to thank the good colonel for his generosity. Well, well; perhaps the colonel was one of those splendid fellows who rejoiced in doing good by stealth and thanks might be embarrassing – most embarrassing. On the last afternoon of our stay another subaltern and I were standing in the stern of a barge, clad only in our topis, heaving buckets of tepid water over one another, when an agitated quartermaster-sergeant interrupted our desperate attempt to avert heat-stroke.

'They've caught 'im, sir!' he panted, as if announcing the fall of a second Kut.

'Caught who?'

'Chuck, sir!'

I groped for a pair of shorts.

'Who's caught him and why?' I demanded.

'The colonel at the dump, sir. Says Chuck's been pinchin' 'is comforts, sir. There's a warrant officer and a gang of natives come to search the barges, sir.'

'Search the barges?'

'Yes, sir, to see if any of the stuff's 'idden.'

One look at my sergeant's face told me what to expect if the search took place.

'How long do you want?' I asked.

''Arf an hour, sir,' he answered hopefully.

With as much dignity as I could muster I walked to the gangway and confronted the warrant officer, who informed me with the strained politeness of a hot and angry man that his colonel had sent him to search the barges. With the utmost indignation I spurned the idea that any unauthorized supplies could be concealed on my barges. Did he think my men were thieves? He made it quite clear that he did. I shifted my ground. What authority had he? No written authority! I could not think of permitting a search without written authority until I had seen the colonel. We would go back to the colonel.

I dressed, and we went to the supply depot. It was in an uproar. Indian *babus* and British N.C.O.s were feverishly checking stores in all directions, while from the office tent came roars of rage as each fresh discrepancy was reported. I entered in some trepidation to be greeted by a bellow. 'Do you know how much those Birmingham burglars of yours have looted from my hospital comforts? Look at this!'

He thrust a list under my nose, item after item: condensed milk, tinned fruit, cigarettes, jam.

'But – but how do you know my men have taken all this, sir?' I gasped.

'Caught 'em! Caught 'em in the act! Bring that hulking great lout who said he was in charge!'

Chuck, seemingly quite unmoved, and if anything slightly amused by the uproar, was marched in between two British sergeants.

'That's the feller!' exploded the colonel, stabbing a denunciatory pencil at Chuck. 'Caught him myself, marching out as bold as brass with a fatigue party of your robbers and a case of lump sugar – the only lump sugar in Mesopotamia! Lifting it under my very nose! Said he'd picked it up by mistake with the other rations, blast his impudence!'

Chuck stood there stolidly, his jaws moving slowly as he chewed gum – more hospital comforts, I feared. His eyes roamed over the tent, but as they passed mine they threw me a glance of bored resignation.

'He'll be court-martialled,' continued the colonel, 'and' – he glared at me – 'you'll be lucky if he's the only one that's court-martialled! Now I'm going to search those barges of yours.'

Chuck grinned ruefully as I passed and I caught a whisper of, 'It's a fair cop, all right.' I was afraid it was. We left him chewing philosophically while he and his escort awaited the arrival of the provost-marshal's police. The colonel, to give him his due, searched those barges thoroughly. He even had the hatches off and delved among the sacks of *atta* that formed the cargo. He and his men grew hotter, dustier, and more furious, but not an empty condensed milk tin, not the label of a preserved pineapple could they find. He turned out the men's kits, he rummaged in the cooks' galley; he even searched the sick-bay we had rigged up in the bows for the sick.

It was empty, except for one man, who lay stretched out flat on his blankets, under a mosquito-net. The colonel glared through the net at the wretched man who with closed eyes was breathing heavily. 'Suspected cholera!' the quartermaster-sergeant whispered hoarsely. The sick man groaned and clasped his stomach. I thought his complexion was rather good for a cholera case, but then I am no clinical expert. Nor was the colonel. He called off his men, and, breathing threats, left us. That evening we sailed. After much chuffing and chugging, to the accompaniment of a great deal of yelling in good Glasgow Scots and bad British Hindustani, a squat little tug had hauled our steamer out of the palm grove. We were lashed, a barge on each side, and staggered off up-river. But we left Chuck behind us, and I was thinking, rather sadly and not without some prickings of conscience, that we should miss him, especially at mealtimes, when the quartermaster-sergeant interrupted my gentle melancholy.

'Will you 'ave peaches or pineapple for dinner, sir?'

'Good lord, Quartermaster-Sergeant, I thought you'd chucked it all overboard?' I gasped.

'So we did, sir,' he grinned. 'But we tied a rope to it with a bit of wood for a float, and when the colonel 'ad gone we pulled it up again. Some of the labels've come off the tins, but that's all.'

'What about the cigarettes; they weren't in tins?'

'Oh, that chap in the sick-bay, 'e lay on 'em!'

VISCOUNT SLIM

227.

By 1917, any spirit of generosity towards the foe was long gone, as illustrated by the experience of a nurse on the Western Front.

I had spent a few months in Stuttgart before the war and had had a very happy time and made some friends, but I really hated the Germans then and I was still suffering the agony of losing my brother, whom I dearly loved, at Ypres. My feelings towards them were less than Christian, but I spoke German pretty fluently, so ironically I was one of the ones who used to be allocated to the prisoners. There were four or five ambulances full of Germans, and the last one was just due to go off.

I jumped in to have a quick glance around and a very pale-faced boy looked up at me and he said, 'Pain, pain.' So I spoke to him in German. I said, 'Where is the pain?' He said, 'In the leg, the knee.' I pulled back the blanket and there, to my horror – a spouting artery! I knew what to do, though I had never had to do it before: 'Search for the femoral pressure point, a hand's span from the hip to the groin, and press strongly down with both thumbs.' I used to practise it in my bath every night. I dragged his clothes aside, hoped that I'd found it, got my thumbs down and pressed for all I was worth, and at the top of my voice I shouted, 'Haemorrhage, Haemorrhage,' which was what we had to do.

I kept holding on until I heard the voice of the one trained nurse. She said, 'Good, dear. Good, dear, that's right. Now you can let go and I'll take over.' The orderly was standing behind her with a tourniquet, which they always had at the ready. On the next convoy I looked out for the driver of the ambulance and asked him what had happened to the boy. He said, 'You were lucky. You stopped that in time. He would

have been dead before we were out of the station. He was just spared.'
And I was glad. I really was glad. It melted a little bit of my hatred and
I felt quite differently towards Germans after that. Of course, I was
hopelessly teased by the other girls. They used to say to me, 'You've
saved the Kaiser another soldier.' And they kept it up for days, asking
me, 'Have you got your telegram from the Kaiser yet? You know,
thanking you for saving some of his army.'

CLAIRE TISDALL

228.

*The British leader of an Arab guerrilla army in Ottoman-ruled
Arabia, an officer of classical education, is moved by their marching
songs to recall the ballads of the Roman legions.*

There came a warning patter from the drums and the poet of the right
wing burst in strident song, a single invented couplet, of Feisal and the
pleasure he would afford us at Wejh. The right wing listened to the
verse intently, took it up and sang it together once, twice and three
times, with pride and self-satisfaction and derision. However, before
they could brandish it a fourth time the poet of the left wing broke out
in extempore reply, in the same metre, in answering rhyme, and
capping the sentiment. The left wing cheered it in a roar of triumph,
the drums tapped again, the standard-bearers threw out their great
crimson banners, and the whole guard, right, left and centre, broke
together into the rousing regimental chorus,

> *I've lost Britain, and I've lost Gaul,*
> *I've lost Rome, and, worst of all,*
> *I've lost Lalage –*

only it was Nejd they had lost, and the women of the Maabda, and
their future lay from Jidda towards Suez. Yet it was a good song, with
a rhythmical beat which the camels loved, so that they put down their
heads, stretched their necks out far and with lengthened pace shuffled
forward musingly while it lasted.

T. E. LAWRENCE

229.

A British officer in the Balkans discovers that his enemies are gentlemen. He wrote this in 1917.

Our first impressions, formed in Serbia in 1915, were that the Bulgar was little better than an uncivilized savage, who lived for a lust of blood. We heard terrible stories of the tortures inflicted on French soldiers who fell into their hands, and I have with my own eyes seen a Bulgar thrust his bayonet through an unarmed British soldier, who was offering to surrender. The more we have seen of the Bulgar soldier, however, the more we have come into the way of thinking that he is not such a bad sort of fellow after all, and that he will play the game as long as his opponent plays the game too. I am perfectly sure that, given the choice he would much prefer to fight with us than against us. The Germans have done their best to instil a feeling of hatred for us, but apparently, with not much success. It is quite the Teutonic way to tell the less 'kultured' Bulgar that horrible treatment awaits him at the hands of the British should he fall into our hands. One deserter who came in assured us that he had been told he would be eaten alive.

'That is why you came across?' questioned our Intelligence officer, cynically.

'I didn't believe it,' replied the deserter.

In the Struma fighting the Bulgar has revealed a sporting quality with which few people would credit him. The most striking instance of this was given after a fiercely delivered counter-attack had temporarily given the enemy a slight footing at the far end of the village. Both sides were engaged in the invigorating pastime of pouring 'rapid' into each other at a distance of 100 yards, when three of our men, observing that three wounded comrades were lying in the open between the Bulgars and ourselves, dashed over the top to bring them in.

The Bulgar fire where this very gallant deed was being performed immediately ceased, the three wounded men were safely brought in, and their rescuers were untouched. I remember being on outpost duty at Topalova, in front of which a troop of Yeomanry were keeping the

Bulgar patrols in check. The cavalry had dismounted when an unexpected Bulgar H.E. fell among the horses, and one frightened animal dashed away. A trooper promptly jumped on to his horse and galloped after the runaway. So exciting went the chase that in a few seconds our infantry were following it from their parapets, standing up in full view of any wily sniper who might be waiting for an exposed head.

Then we realized that the garrison of Prosenik had followed our example and was breathlessly following the race. It was a strange spectacle – Bulgar and British standing up in full view of each other, watching a runaway horse. At length the trooper headed the animal back towards our own lines, and returned with his captive without a single shot having been fired at him. 'The Bulgar is a humorous devil', is a remark one often hears passed, and he possesses a deal greater sense of humour than his friend, the Bosche. There was rather a rage for some weeks to post up little messages for the edification of enterprising patrols on either side. I forget what the particular message was, but on the day following its posting, pinned to the tree on which our message had been fixed, was the reply, and a P.S. which read, 'For goodness' sake, Englishmen, write in English next time. Your French is awful.'

When told in one message from us that any Bulgar who would like to look us up would be welcomed and given plenty of bread, a reply was sent to the effect that any Britisher who thought of doing likewise would be warmly welcomed and that they had enough bread to feed all who came across. The Bulgars invariably commenced their message with the prefix 'Noble Englishmen', and often reproached us for having invaded their 'peaceful Macedonian soil'. A subtle but amusing reference to Ireland was frequently included.

One day a scratch football match was in progress behind our line, and well within view and range of the enemy guns, and it certainly was somewhat surprising that not a single shell came over to interrupt our game. In consequence of this forbearance a conscientious O.C. Company, seeking an opportunity to fill a sleepy hour by improving his company, paraded in as unexposed a spot as he could find and commenced arm drill. A shell quickly dissipated the idea that this could be indulged in with impunity (to the delight of the company), and the following day a patrol discovered a message worded, as near as I can recollect – 'We like to see you playing football, and we shall not shell you, and we are if we are going to watch you doing company drill.'

DONOVAN YOUNG

230.

A Guards officer enforces discipline in London.

Late one morning, I was sent for to the Regimental Orderly Room. Hastening to the temple, I was commanded to proceed at once, with the regulation escort of a lieutenant and ensign, to St Pancras Station, and there arrest, as he stepped out of the train, a 'young officer'. The charge against him was unspecified, but, I was given to understand, of a serious nature. I must use force, if necessary, to nab him. And let me remind the reader that the term 'young officer' was technical, youth consisting in short regimental service rather than lack of years and that the accused might be old enough to be my father.

And so, indeed, he proved. The poor old chap seemed very surprised when, placing his hat on his grey hair, he stepped out of the carriage and found two other officers and myself waiting for him, with, if I remember correctly military procedure, our swords drawn. He enquired, with an engaging air of puzzlement and timidity, what we could have against him. I replied that he would shortly learn, in the Regimental Orderly Room. We marched there in fine style, and were received with pomp. 'Mr Crouchend,' the Lieutenant-Colonel observed, looking over the top of a pair of beautifully made spectacles with a terrifying mildness, 'there is a serious charge against you!'

'Sir!' Mr Crouchend replied dutifully, in the sacrificial monosyllable that is the correct reply to a superior officer in the Brigade of Guards.

'You gave a false address when on leave last week,' the Lieutenant-Colonel continued in the voice of an oracle. 'You wrote in the book 42 Clarges Street: we have evidence that you were staying at 12 Half Moon Street.'

With the cry of a wild animal that has been snared, poor Mr Crouchend broke away from tradition and wailed,

'But it's the same building, sir! It's Fleming's Hotel!'

Silence of a rather portentous kind followed this disclosure, and the junior officers of the escort stared in front of them with peculiarly unseeing eyes. Already the mind's ear could detect the thunder of rebuke that would, when we had departed, roll through the room: for the idol would speak. At last he roused himself from the coma into which he seemed to be descending, and called, 'March that officer out at once! I will go into the matter later.'

OSBERT SITWELL

231.

The British Army's March 1918 retreat was attended by a sniper of the Black Watch, Eric Linklater (1889–1974). A Welsh-born Scottish poet, he later became a well-known humorous novelist, author of Private Angelo.

The first two or three days, the bloodiest and most feverishly distraught of that magnificent rearguard action, were finished before myself was added to it; and though I was involved in some disagreeable episodes, my roundest recollections are of consuming, with great relish, a bottle of looted champagne and a jar of the best plum jam I have ever tasted; of sleeping very comfortably between two convenient graves in a shattered churchyard; and of listening, during a whole day of battle, to an Irishman discussing his next war. Fragments and wandering details of many regiments were fighting together, and by chance I found myself in company with two men of the South Irish Horse – a plump and rather surly young trooper, and an elderly leather-skinned band-sergeant, a talkative fellow with a single mind. His opinion of the war in which we were then so closely engaged was simple and instructive: it was useful training for the approaching conflict with England. Even without their presence it would have been an interesting day, for we were retreating across open country, and behaving exactly as we had been taught: two platoons going back while others gave covering fire, and the latter retiring under the shelter of the rifles and Lewis guns of those who had preceded them. To the Irish sergeant, it was a schoolroom exercise. 'Now stop shoving about,' he would cry to the trooper, as we settled into a fold of the ground. 'Haven't I been telling you all morning to run like hell when you mean to run, and lie like a bloody rock when you're down? It's open country we'll be in when we're fighting the English, and now's your time to learn how to behave, and not make a target of yourself by thrusting your great round bottom in the air like a porpoise under the Old Head of Kinsale. What are your sights at?'

'Four hundred,' shouted the plump young trooper, puffing and blowing.

'God Almighty! Give me patience! They were four hundred, weren't they, when you were down in the ditch beyond? And now you're a hundred yards nearer the sea, God help you, and by the same extent farther from Jerry, you fool, so put them up, you devil, and remember the wind that's blowing.'

I lay on his other side, and turning his head he spoke earnestly to me. 'If there's one thing I hate to see more than another,' he said, 'it's wasting good ammunition, and the waste that goes on here would frighten a millionaire. It breaks my heart to see the boys throwing off a bandolier because it's too heavy to carry, and to think what we could do with it in Ireland.'

'So you're going to fight the English?'

'We are that.'

'When?'

'As soon as we get finished with this blasted pigsty of a war, though I shouldn't be speaking against it, for it's good practice for the like of that lad there, who didn't know the touch of a trigger from the toe of my boot when he came out. But he's learning. He's learning every day, and he'll be a good boy yet.'

We retired again when a couple of light field-guns opened fire on us; and the Irish sergeant and I lay side by side in a stony gully. 'You ought to come in with us yourself,' he said. 'Scotland and Ireland together: we'd knock the bloody English to hell.'

'I haven't any quarrel with England,' I said.

'Ach, everybody has a quarrel with England, if they'd only the bloody sense to see it! It's England that has spoiled the whole world with its scheming and money-making and the righteousness, God save us, that says England knows the way of it, and all the rest are niggers or talking nonsense like children. Let the Scotch and Irish join together, and give England what Cromwell gave Drogheda, the bastard. – And what in hell are you thinking of, Michael James? Can you not see that bloody machine-gun there? And for the love of God put your sights up. Now all together: three rounds rapid – fire!'

ERIC LINKLATER

232.

Charles Yale Harrison (1898–1954) was born in Philadelphia, raised in Canada and served briefly with the Royal Montreal Regiment in France in 1917–18. This fictionalized account of the looting of Arras in his 1930 bestseller Generals Die In Bed *caused controversy on publication, because of its claims about the misbehaviour of Canadian troops.*

It seems as though a pestilence had swept over this part of the country. We do not see any signs of fighting, not even a solitary shell-hole. Soon we are in cobble-paved streets. We see shops. No shopkeepers. There are hotels, churches, stores, wine-shops. It is broad daylight now, but there is not a single soul in sight other than the marching troops. Our heavy footsteps echo down the empty streets. On both sides are stores – grocery stores, tobacco shops, clothing stores, wine-shops. In the windows we see displays of food and cigarettes temptingly displayed – tins of lobster, glass jars of caviare, tinsel-capped magnums of champagne. I look through a glass window and read: 'Veuve Cliquot' – the bottle looks important and inviting. In another window I read: 'Smoke De Reszke cigarettes.'

We ask our captain – a fidgety, middle-aged man by the name of Penny – why the town is deserted. He explains that the Germans dropped a few shells a few days ago, and the inhabitants, thinking that Heinie was about to enter, fled leaving the city as we now see it. We rest on the kerb, looking hungrily at the food and cigarettes behind the thin glass partitions. Little knots of soldiers gather and talk among themselves. As I stand talking a man in the company ahead of us idly kicks a cobble-stone loose. He picks it up and crashes it through a

wide, gleaming shop window. The crash and the sound of the splinter-
ing, falling glass stills the hum of conversation. The soldier steps
through the window and comes out with a basketful of cigarettes. He
tosses packages to his comrades.

Another crash! More men stream through the gaping windows. The
street is a mass of scurrying soldiers. Discipline has disappeared. I step
through an open, splintered window and soon come out laden with
tins of peas, lobster, caviare, bottles of wine. Broadbent and I visit
many shops. In each are crowds of soldiers ransacking shelves,
cupboards, cellars. Some of them are chewing food as they pillage.

When we have filled our bags with food, drink and cigarettes, we
make off to look for a place to rest. We climb through a window of a
pretentious-looking dwelling. It is deserted. We prowl through the
house. In the dining-room the table is set for the next meal. We dump
our sacks down in the centre of the room and begin to prepare the
food. In a little while we are tackling lobster salad, small French peas,
bread and butter, and washing it down with great gulps of Sauterne.
We do not speak, but simply devour the food with wolfish greed.

At last we are sated. We search in the sacks and find tins of choice
Turkish cigarettes. We light up, putting our dirty feet on the table and
smoke in luxury. We hunt through the house and find the owner's
room. Water is boiled and soon we are shaved and powdered with the
late owner's razor and talcum. We throw ourselves on the valanced
beds and fall asleep. We are wakened by the sound of crashing noises
downstairs. We descend. A party is going on in the drawing-room.
Some of our men have found the house. They are drunk. Some sprawl
on the old-fashioned brocaded gilt furniture. Some dance with each
other. One of the recruits, a machine-gunner, draws his revolver from
his holster and takes pot-shots at a row of china plates which line a
shelf over the mantelpiece. His companions upbraid him: 'Hey, cut
out that bloody shooting; you're filling the damned room with smoke.'

Broadbent and I go out into the street. It is nearly dark. Men stagger
about burdened with bags of loot. They are tipsy. The officers are
nowhere to be seen. Up towards the line the sky is beginning to be lit
with the early evening's gun flashes. Over to the south side of the town
a red glow colours the sky. Some of our men must have set fire to some

houses. As we look we see flames and a shower of sparks leap into the air.

We look at each other in amazement.

'Do you know that this is looting a town?' Broadbent says.

'… God, who would've thought that plain gravel-crushers like us would ever get rich pickin's like this …'

'… the soldier's dream come true, all right, all right …'

'… hey, the frogs is supposed to be our allies …'

'What, with vin rouge at five francs a bottle?'

'Well, why the hell didn't they bring the grub up …'

CHARLES YALE HARRISON

233.

11 November 1918, last day of World War I.

Very early on the last morning Shadbolt was watching the men dragging the heavy howitzers into a little clearing in the wood. The day was grey and overcast and the raindrops from a recent shower were dripping sadly off the trees. Above them a few pigeons, disturbed by the movements and cries of the men, circled and wheeled. A despatch rider rode up and handed him a message form. 'Hostilities will cease at 11a.m. today. A.A.A. No firing will take place after this hour.' He sat down on the stump of a tree. This, then, was the end. Visions of the early days, their hopes and ambitions, swam before his eyes. He saw again his prehistoric howitzer in the orchard at Festubert, and Alington's long legs moving towards him through the trees. He was back with the Australians in their dug-out below Pozieres. He saw the long slope of the hill at Heninel, covered with guns, ammunition dumps, tents and dug-outs. Ypres, the Salient, Trois Tours, St Julien – the names made unforgettable pictures in his mind. Happy days at Beugny and Beaussart, they were gone and the bad ones with them. Hugh was gone, and Tyler and little Rawson; Sergeant Powell, that brave old man; Elliot and James and Johnson – the names of his dead gunners strung themselves before him. What good had it all been? To serve what purpose had they all died? For the moment he could find no answer. His brain was too numb with memories.

'Mr Straker.'

'Sir.'

'You can fall the men out for breakfast. The war is over.'

'Very good, sir.'

Overhead the pigeons circled and wheeled.

FRANKLIN LUSHINGTON

234.

Another snapshot of the same momentous day.

The grim business of war itself went on as usual, right up to 11 a.m., and, at one or two points along the line, even beyond. Thus a captain commanding an English cavalry squadron which took the Belgian village of Erquelinnes wrote that morning: 'At 11.15 it was found necessary to end the days of a Hun machine-gunner on our front who would keep on shooting. The armistice was already in force, but there was no alternative. Perhaps his watch was wrong but he was probably the last German killed in the war – a most unlucky individual!' Elsewhere on the British front an officer commanding a battery of six-inch howitzers was killed at one minute past eleven – at which his second-in-command ordered the entire battery to go on firing for another hour against the silent German lines. But generally, any firing still going on ended at the last second of the tenth hour, sometimes with droll little ceremonies – as on the British front near Mons, where another and more fortunate German machine-gunner blazed off his last belt of ammunition during the last minute of the war and then, as the hour struck, stood up on his parapet, removed his steel helmet, bowed politely to what was now the ex-enemy opposite, and disappeared. The British division on whose front that little incident took place had lost, during that one final week of the war, two officers killed and twenty-six wounded, and among the other ranks one hundred and seventeen killed, six hundred and ninety-three wounded and sixty-one missing. Small wonder that its historian recorded 'no cheering and very little outward excitement'.

GORDON BROOK-SHEPHERD

235.

Much to the chagrin of Montgomery and other 1939–45 British generals, World War I proved the last conflict for which Parliament continued a nineteenth-century tradition of making cash grants to victorious commanders, supplanting the heroic corruption that enabled the 1st Duke of Marlborough to build Blenheim Palace. Below is the list of officers who benefited from passage of the Commons vote of 6 August 1919.

Motion made, 'That a sum, not exceeding £585,000 [over £300 million at 2021 prices] be granted to His Majesty, to be issued to those officers who commanded and directed His forces by sea, on land, and in the air, in recognition of their eminent services during the late War, namely:

Navy:
    Admiral of the Fleet, Sir David Beatty – £100,000
    Admiral of the Fleet, Viscount Jellicoe – £50,000
    Admiral Sir Charles Madden – £10,000
    Admiral Sir Doveton Sturdee – £10,000
    Rear-Admiral Sir John de Robeck – £10,000
    Vice-Admiral Sir Roger Keyes – £10,000
    Commodore Sir Reginald Tyrwhitt – £10,000

Army:
    Field-Marshal Sir Douglas Haig – £100,000
    Field-Marshal Viscount French – £50,000
    Field-Marshal Sir Edmund Allenby – £50,000
    Field-Marshal Sir Herbert Plumer – £30,000
    Field-Marshal Sir Henry Wilson – £10,000
    General Sir Henry Rawlinson – £30,000
    General The Hon. Sir Julian Byng – £30,000
    General Sir Henry Horne – £30,000
    General Sir William Robertson – £10,000
    General Sir William Birdwood – £10,000

Lieut.-Colonel Sir Maurice Hankey [Secretary of the Imperial War cabinet] – £25,000

Air:

Air Vice-Marshal Sir Hugh Trenchard – £10,000

[In proposing the motion approving the grants, Prime Minister David Lloyd George began his speech:]

'It is an honoured tradition of this country that it rewards liberally those who have rendered it conspicuous and distinguished services in time of peril, and that, I venture to say, is a sound tradition for a country. Ingratitude chills the ardour of service, and no State has long thriven which does not display its gratitude to those who have served it well in its time of peril. It was really one of the marked features of the distinction between Rome and Carthage, and the lesson is not without its value. I am proud of the fact that the pensions we have voted in this House – the scale of pensions – to the men who have served us well in the field and in the air and on the sea, is incomparably the most generous in Europe. I think the amount aggregates £98,000,000 a year, which is half the whole of the national expenditure before the War. There is no scale in Europe which compares with it, and I am proud of it. But we do wisely, we shall do wisely, if we accept this Motion, in maintaining the tradition that exceptional rewards should be given to those who have borne exceptional responsibilities with exceptional success. That is the proposition which I would invite the [House] to accept.

'The Duke of Wellington had voted to him two pensions, in the aggregate £4,000 a year, for three lives, and after the Battle of Waterloo a sum of £500,000 was voted to him. That was for him alone. Lord Wolseley had £25,000 voted to him for the Ashanti campaign, and £30,000 for the Egyptian campaign. Those were moved by Mr Gladstone, who was, undoubtedly, one of the most rigid and stern of all the great economists. Lord Roberts had £12,500 voted to him for the Afghan campaign and £100,000 for the Boer War. Lord Kitchener had £30,000 voted to him for the Soudan campaign, and £50,000 for the

Boer War. Those two were moved by another famous and very rigid economist, Sir Michael Hicks-Beach. That was the view taken in the past of the kind of reward that ought to be accorded to men who had rendered services of this kind. In magnitude, in fatefulness for this country, there is no comparison between those wars and the War which has just come to a conclusion. Therefore, the sums which we are inviting the Committee to agree to err in comparison, not on the side of over-liberality, but quite the reverse.'

HANSARD

236.

A decade after the Armistice, a former British officer returned to an old battlefield. Henry Williamson (1895–1977) became famous as author of Salar the Salmon *and* Tarka the Otter, *but like J. F. C. Fuller his reputation suffered from an association with fascism, and indeed Nazism.*

The village was flooded when last I was here; the Germans had gone the night previously, and no guns were firing, for the enemy had walked out of range. I remember the Bengal Lancers filing through the ruins: bearded, turban'd, dark-skinned soldiers, with the pennons on their upheld lances scarcely fluttering in the windless air. A battalion of the Yorkshire Regiment was 'on fatigue', laying baulks of timber over the liquid mud of the broken road. As I passed on my horse I saw a pallid hand sticking out between two balks, and a sodden grey uniform cuff; a young soldier laughingly put the handle of a broken spade between the stiff fingers, saying, 'Now then, Jerry, get on wi' it; no bluudy skrimshankin' 'ere.'

I entered an estaminet near the station and asked for bread and cheese and wine. It was an untidy place like the village, a place of shapeless shacks and sheds made of rusty sheets of wartime corrugated sheet-iron among buildings partly rebuilt. There were several young men in the room watching two playing billiards. An idiot child was running about the room, and seeing me, but without human recognition, it came up and took my stick out of my hand. Its father, a man of about thirty-five, in slovenly clothes, shouted something as he raised himself from a leaning position over the billiard table, and then resumed his preparation for a stroke. A female voice replied shrilly and rapidly from the unseen kitchen, and a moment later a woman ran out, seized the child by its wrist, wrenched the stick out of its fingers, returned it to me without a glance, and dragged the child through a door into the kitchen.

I waited five minutes, ten minutes, but no one took any notice of me; the game continued with much jabbering of voices, which is another way of saying that I understood only one word in fifty of the language of the country in which I was a foreigner. At the end of a quarter of an hour madame came back, less untidy and less worried, and agreed to cook me an omelette.

I ate in silence when it arrived; and I had just finished it when the game ended and the *patron* came to speak with me. I told him in my weak French that I was a returned soldier; I gave him a laboured account of the village as it had appeared in the winter of 1916–17. He said that his wife had been there until November, 1916, when the English advanced up the valley from Beaumont Hamel; *le pauvre petit* – he indicated the idiot child – had been a baby of two years then, and had been struck in the head by a piece of English shell – *c'est la guerre* … The other men stopped talking, and listened; I explained that I was meditating a novel, or novels, of the War, the story of an insignificant and obscure family which had helped, in its small way, to prepare and make the Great War.

Their eyes lit up; they exclaimed with enthusiasm at that name. That was reality, *la verité*! Only the week before a German soldier, looking for the grave of his brother, had come to the village, and it so chanced that he had read Henri Barbusse, and had declared that it was true for the German as well as for the French! He was a comrade, that Boche … no, Boche was a bad word, part of the old world: *pas vrai*! He was a man like themselves, but in the War his uniform happened to be a different colour. He was a brother!

It was amazing, the animation on the faces of those men. Their eyes were lit by inner fire; they smiled eagerly, their gestures and attitudes were vital and happy. What had brought this miracle – to use a term of the old world? A stranger had come, a German; a stranger had come, an Englishman; after the mention of a name there was no reserve, no suspicion, no distrust; all shared a common humanity. Something not supernatural, but supernational. '*Bonne chance, camarade! Bonne chance!*' I settled the very small bill, adding a few francs extra, and left gaily in the rain, and turned to see them watching me from the doorway. All of us waved together.

HENRY WILLIAMSON

237.

An act of misguided chivalry by the US Army's Col. George S. Patton (1885–1945), in character with the legend he later established.

On a summer night in 1922, while driving his roadster from a horse show to his hotel in Garden City, he spotted three rough-looking characters with a damsel in apparent distress. They seemed to be pushing the girl into the back of a truck. Patton stopped his car, jumped out and forced the men at gunpoint to release the young woman. Then it developed that the girl was the fiancée of one of the men, who merely were helping her to climb into the truck. The incident was reminiscent of Don Quixote's encounter with the six merchants of Toledo on the road to Murcia and his spirited defence of Dulcinea's unquestioned virtue. When later Patton laughingly related the story of his gratuitous intervention to a spellbound lady of Long Island society, he was asked, 'How come, Georgie, that you go armed to a civilian horse show?' 'I believe in being prepared,' he told her. 'I always carry a pistol, even when I'm dressed in white tie and tails.'

LADISLAS FARAGO

238.

A timeless sample of service bureaucracy, as recorded by the Military Assistant to the Chief of the Imperial General Staff.

I hope I shall not fall foul of the Official Secrets Act if I quote verbatim the following branch memorandum. On March 10, 1926, the C.I.G.S. asked me for a pair of dividers, costing, I suppose, eighteen pence. How did I obtain them? Here is the answer:

1 'Q.M.G.9.

'Would you be good enough to obtain a pair of dividers for the use of
C.I.G.S.?

    J. F. C. Fuller, Colonel G.S., M.A. to C.I.G.S.'

    10th March, 1926.

2 'X Compass, drawing, shifting leg, double jointed, II. Q.M.G.F. (b).

'We propose to convey approval to the permanent issue of a Compass
as at X above. Have you any remarks from a financial point of view?

    Q.M.G.9 (a).

    J. Gardner,

    11th March, 1926.

    For D.A.D.E.O.S.'

3 'Q.M.G.9 (a).

'No financial objection. Will you let us have this B.M. again when
the Compass has been issued?

    Q.M.G.F. (b).

    G. Lillywhite.'

    15th March, 1926.

4 'A.D.O.S.P.

'Will you please arrange issue of a Compass, drawing, shifting leg,
double-jointed II. to – M.A. to C.I.G.S. Room 217, War Office?
Issue will be permanent.

    Q.M.G.9 (a).

    J. Gardner,

    16th March, 1926.

    For D.A.D.E.O.S.'

5 'Q.M.G.9 (a).

'Issue has been arranged.

    'I.O.P.4.D/4186 dt. 18.3.26.

    H. E.,

    18th March, 1926.

    For A.D.O.S. Provision.'

6 'M.A. to C.I.G.S.'
'Please let me know when the compass is received.
    Q.M.G.9. (a).
    J. Gardner,
    19th March, 1926.
    For D.A.D.E.O.S.'

7 'Q.M.G.9 (a).
    'Compass received, thank you.
    J. F. C. Fuller,
    24th March, 1926.
    Colonel G.S., M.A. to C.I.G.S.'

J. F. C. FULLER

239.

The parade sergeant-major addressed the King's Squad of the Royal Marines, training to participate in the annual Royal Tournament:

'On the 26th June, 1926, you will be marching into the arena at Olympia. Seated in the Royal Box will be 'er Majesty, Queen Mary. 'Er 'usband, King George V, our Colonel-in-Chief, will not be present as 'e is sick-a-bed at Buckingham Palace. You will go through your stuff, and when you 'ave done, 'er Majesty will get into 'er carriage and drive back to Buckingham Palace … She will say: "I've bin to Olympia, George", and 'e will say: "Oh, 'ave you, Mary, and 'ow was those young Marines of mine?" and she will say: "Well, excuse me, George, but they was bloody awful" – AND SO YOU ARE.'

ROBERT BRUCE LOCKHART

240.

David Niven (1910–83), who became a Hollywood star, spent five years of his youth as a regular soldier, graduating from Sandhurst in 1929.

When I sat for the final exams I discovered with pleasure mixed with surprise that they came quite easily to me and as I had also accumulated a very nice bonus of marks for being an under-officer, my entry into the Argylls seemed purely a formality. Everything in the garden was beautiful – a fatal situation for me. Just before the end of term, all cadets who were graduating were given a War Office form to fill in:

'Name in order of preference three regiments into which you desire to be commissioned.'

I wrote as follows:

The Argyll and Sutherland Highlanders
The Black Watch

and then – for some reason which I never fully understood, possibly because it was the only one of the six Highland Regiments that wore trews instead of the kilt – I wrote

Anything but the Highland Light Infantry.

Somebody at the War Office was funnier than I was and I was promptly commissioned into the Highland Light Infantry.

DAVID NIVEN

241.

Late in life, the British Army's official historian of World War I recounted the fortunes of his fellow-students on the Staff College course of 1896. After cataloguing generals and battle casualties, the roll became much less admirable.

Placed on the retired list for quelling a riot by machine-gun fire in
  India [Brigadier-General Reginald Dyer, architect of the 1919
  Amritsar massacre of 379 unarmed Sikh demonstrators] – 1
Retired on coming into money – 2
Shot his mother-in-law and her lawyer and committed suicide – 1
Last heard of keeping a brothel in Smyrna – his father married a
  Levantine during the Crimean War – 1

SIR JAMES EDMONDS

242.

John Royal, a fellow officer of David Niven, found his commission abruptly terminated after an unfortunate incident in India.

Soon after his arrival, the officers were invited to a ball given by the local maharajah and John became sleepy, so after dinner he lay down behind some potted palms and stole forty winks. He was awakened by a captain of a cavalry regiment who stirred him, none too gently, with his foot.

'Stand up,' said the Captain. John stood.

'You are drunk,' said the Captain.

'You are right,' said John and flattened him with a left hook.

He then composed himself once more behind the potted palms. Pretty soon he was awakened again, this time by a full colonel.

'Stand up,' said the colonel. John stood.

'You are drunk,' said the colonel, and collected a right cross.

John was court-martialled and insisted on conducting his own defence. He had been dropped on his head as a baby, he said, and this had the unfortunate effect of making him lash out at the first person he saw when he was woken from a deep sleep.

The prosecuting officer smiled faintly. 'Perhaps you would tell the court what happens to your batman when he wakes you up in the morning?'

'Nothing,' said John, unmoved. 'I have issued him with a fencing mask.'

DAVID NIVEN

243.

Perils of frontier life, described in a 1930 memoir immortally entitled The Lives of a Bengal Lancer. *The name (though emphatically not the plot) was borrowed for a 1935 movie starring Gary Cooper, of which completion was delayed because footage shot in Indian locations proved to have been ruined by the sky-high local temperatures.*

I do not know how far discipline of the sex life is a good thing. But I know that a normal sex life is more necessary in a hot than a cold country. The hysteria which seems to hang in the air of India is aggravated by severe continence of any kind; at the end of Ramzan, for instance, my fasting squadron used to become as lively as a basket of rattlesnakes. Many good brains in India have been bound like the feet of a mandarin's wife, so that they can only hobble ever after; and such cramping of the imagination may lose us the Empire. In this grey London weather, I cannot believe that I am not exaggerating. I cannot believe that it was too hot to bear a sheet on my skin, that I ingested six glasses of milk and soda for breakfast, had a malaria temperature twice a week for months on end, that my brain grew addled, and my liver enlarged, and my temper liable to rise like the fires of Stromboli. Yet so it was.

Men's brains and bodies, like other machines, work differently at different temperatures; and India would be a happier country if we could always remember that, especially in Whitehall. One night, when the temperature had risen apoplectically (for a ceiling of thunderclouds had closed in on us) and I lay gasping on the roof of my quarters, a revolver shot rang out from a neighbouring bungalow. A moment before I had been drinking tepid soda water, and thinking of England, and cursing this stifling night through which the angel of sleep would not come. But now Providence had sent something better – raiders?

Voices cries '*Halaka ghula di*!' (''Ware thief!'). Khushal arrived with the first weapon to his hand, a lance.

I went out in my pyjamas to explore. [An] elderly Major had been celebrating his approaching departure with more than enough champagne. On reaching his bed he had lain down quietly; in a stupor, no doubt. Then his shattered nerves began to conjure up visions, and by the glimmer of the night-light which he always kept burning beside him, he saw a skinny outline at the foot of the bed. When he moved, it moved. Seizing a revolver in his trembling hand, he fired; then he roared with pain, for he had shot not a face, but his own foot. Next morning, he was hurried down to Kohat, with an orderly to put ice on his mangled toes and on his poor, deluded head.

FRANCIS YEATS-BROWN

244.

John Masters (1914–83) became a famous writer with a sequence of novels about British India – in old age, he discovered that he was descended from the Indian mistress of one of his nineteenth-century forebears. Until 1949, he served as a career Gurkha officer. Here, he describes a 1934 experience in the subcontinent, as a subaltern newly attached to the Duke of Cornwall's Light Infantry.

My first day in the awe-inspiring and gloomy splendour of an officers' mess was in Bareilly, and I was still nineteen. I sat down to tea at the polished table and cautiously admired a large fruit cake. After licking my lips for five minutes, I screwed up my courage: and asked the captain next to me whether I might be allowed to have some. He turned, looked at me with an indescribable expression of scorn and astonishment, and said coldly, 'Yes. That's a mess cake. You're not at your prep school now.' Christmas came. I had ceased to be nineteen, and it was my turn to be orderly officer. On this day my duties included following the colonel, second-in-command, and adjutant round every mess hall in the battalion – seven of them – and, in each mess, drinking toasts from half-tumblers of whisky thinly diluted with fizzy lemonade. The British soldiers, far from home and families, tried to forget their exile in a riot of snowballing, singing, and drinking. The sergeants waited on them at table.

After a couple of hours our bedraggled and hardly conscious convoy reached the officers' mess and sat down, without hunger, before an enormous Christmas dinner. The lieutenant at the head of the table suddenly remembered that this day was traditionally a topsy-turvy one for rank and discipline. He picked a leg of turkey off his plate and flung it accurately at the officer sitting at the other end, who happened to be a senior major. The colonel smiled, collected the majors with his eye and left us. As the field officers crept silently out the air grew thick with flying potatoes, pudding, turkey, gravy, and oranges.

I watched in amazement. If we had behaved like this at the Royal Military College we would have been rusticated and told we were quite unfit to be officers. Apparently when one actually became an

officer the rules were different. Amazement gave place to loneliness
and a despair born of all those toasts, for I was not used to drinking. I
have tried so hard to be an officer, I thought tearfully. I want so much
to be treated as one of the family. But no one is throwing anything at
me. They haven't forgiven me about the cake yet.

Then – oh, ecstasy! – hard fingers were rubbing brandy butter into
my hair and stuffing Christmas pudding into my ears. I was forgiven,
accepted! I flung myself with abandon into the riot, and the steaming
rum punch flew faster round the table and the snow flew thicker
outside the windows. When all our fiendish energy had at last been
spent, and we were preparing to go and clean ourselves, someone heard
a faint muttering from the floor. We knelt and looked under the wide
table. We saw two doctors, both medical majors, lying comfortably on
their sides, their heads pillowed on cushions. A bottle of brandy stood
between them and they were arguing in an involved way about
horse-breeding for they were, of course, like most army doctors, Irish-
men. We sent another bottle of brandy down to them, with the
compliments of the regiment, and trooped out into the snow.

JOHN MASTERS

245.

*One of the most celebrated, albeit romanticized, incidents in the
Spanish Civil War took place in the summer of 1936, during the
defence of Toledo by a 1,300-strong Nationalist – Francoist – garrison
under Colonel José Moscardó.*

From Madrid, the Republican minister of education, the minister of
war, and General Riquelme had been furiously telephoning the
58-year-old infantry colonel, Moscardó, commander of the nationalist
garrison still holding out in the Alcázar, in an attempt to persuade him
to surrender. Finally, on 23 July, Cándido Cabello, a republican barris-
ter in Toledo, telephoned Moscardó to say that if Moscardó did not
surrender the Alcázar within ten minutes, he would shoot Luis
Moscardó, the Colonel's 24-year-old son, whom he had captured that
morning. 'So that you can see that's true, he will speak to you,' added
Cabello.

'What is happening, my boy?' asked the Colonel. 'Nothing,' answered the son, 'they say they will shoot me if the Alcázar does not surrender.' 'If it be true,' replied Moscardó, 'commend your soul to God, shout *Viva España*, and die like a hero. Goodbye my son, a last kiss.' 'Goodbye father,' answered Luis, 'a very big kiss.' Cabello came back on to the telephone, and Moscardó announced that the period of grace was unnecessary. 'The Alcázar will never surrender,' he remarked, replacing the receiver. Luis Moscardó was not, however, shot there and then, but was executed with other prisoners in front of the Tránsito synagogue on 23 August, in reprisal for an air raid.

HUGH THOMAS

246.

The legend of the communist International Brigades, which fought for the Republican cause against Franco in Spain, glosses over some cruel realities.

One of the most tragic episodes involved the men of the [American] Lincoln Battalion, who had arrived in the middle of February 1937 fresh in their 'doughboy' uniforms. They were put under the command of an English charlatan who pretended to have been an officer in the 11th Hussars. He ordered them into attack after attack, losing 120 men out of 500. The Americans mutinied, nearly lynched the Walter Mitty character who had been imposed on them, and refused to go back into the line until they could elect their own commander.

Damage to morale had been increased by events behind the lines. Communist preferment and proselytizing at the front had reached such levels that former communist supporters among the regular officers were horrified. Prieto was shaken when he heard that non-communist wounded were often refused medical aid. Battalion commanders who rejected invitations to join the Party found weapon replacements, rations or even their men's pay cut off. Those who succumbed were given priority over non-communists. They were promoted and their reputations were boosted in despatches and press accounts.

The American mutiny in the early spring had been successful, though that was seen as an aberration which had been put right. Some

Italians from the Garibaldi Battalion deserted to join the liberal and anarchist Giustizia e Libertá column. During the Segovia offensive XIV International Brigade had refused to continue useless frontal attacks on La Granja and foreigners in a penal battalion mutinied when ordered to shoot deserters.

The anger at futile slaughter was accompanied by a growing unease at the existence of 're-education' camps, run by Soviet officers and guarded by Spanish communists armed with the latest automatic rifles. At these camps labour was organized on a Stakhanovite basis, with food distribution linked to achieving or exceeding work norms. The prisoners were mostly those who wanted to return home for various reasons and had been refused. (It was not known until later that several Brigaders in this category were locked up in mental hospitals, a typically Soviet measure.) One of the most sordid camps was at Júcar, where disillusioned British, American and Scandinavian volunteers were held. Some British members were saved from execution by the intervention of the Foreign Office, but those from other nationalities were imprisoned.

The persistent trouble in the International Brigades also stemmed from the fact that volunteers, to whom no length of service had ever been mentioned, assumed that they were free to leave after a certain time. Their passports had been taken away on enlistment and many of them were sent to Moscow by diplomatic bag for use by NKVD agents abroad. Brigade leaders who became alarmed by the stories of unrest filtering home imposed increasingly stringent measures of discipline. Letters were censored and anyone who criticized the competence of the Party leadership faced prison camps, or even firing squads. Leave was often cancelled and some volunteers who, without authorization, took a few of the days owing to them were shot for desertion when they returned to their unit. The feeling of being trapped by an organization with which they had lost sympathy made a few volunteers even cross the lines to the nationalists. Others tried such unoriginal devices as putting a bullet through their own foot when cleaning a rifle.

247.

George Orwell (1903–50), real name Eric Blair, one of the twentieth
century's great truth-tellers, served in 1937 as a British volunteer with
the Spanish Republican army. He was wounded in the throat,
unsurprisingly, since at 6ft 2in he was taller than most Spanish
trenches. Here, he reflects upon the tragi-comic incompetence which
characterized the conduct of the war, recorded in his 1938 memoir
Homage to Catalonia.

The difficult passwords which the army was using at this time were a
minor source of danger. They were those tiresome double passwords
in which one word has to be answered by another. Usually they were
of an elevating and revolutionary nature, such as *Cultura – progreso*, or
Seremos – invencibles, and it was often impossible to get illiterate
sentries to remember these highfalutin' words. One night, I remember,
the password was *Cataluña – eroica*, and a moon-faced peasant lad
named Jaime Domenech approached me, greatly puzzled, and asked
me to explain.

'*Eroica* – what does *eroica* mean?'

I told him that it meant the same as *valiente*. A little while later he
was stumbling up the trench in the darkness, and the sentry challenged
him:

'*Alto! Cataluña!*'

'*Valiente!*' yelled Jaime, certain that he was saying the right thing.
Bang!

However, the sentry missed him. In this war everyone always did
miss everyone else, when it was humanly possible.

GEORGE ORWELL

248.

Peter Kemp (1913–93) was a lifelong adventurer who served with
Franco's Nationalist Foreign Legion throughout the civil war. In 1938
his duties included the execution of a fellow-countryman, of which his
account is understandably apologetic.

I found [the company commander] talking with some legionaries who had brought him a deserter from [the Republican] International Brigades – an Irishman from Belfast; he had given himself up to one of our patrols. The man explained that he had been a seaman on a British ship trading to Valencia, where he had got very drunk one night, and been picked up by the police. The next thing he knew, he was impressed into the International Brigades. He had bided his time until he reached the front, when he had taken the first opportunity to desert. Translating his account I urged that the man was a deserter, not a prisoner. De Mora was sympathetic. 'You seem to have a good case,' he said. 'Unfortunately my orders from Colonel Peñaredonda are to shoot all foreigners. If you can get his consent I'll be delighted to let the man off. You'll find the Colonel over there. Take the prisoner with you, in case there are any questions.' It was an exhausting walk of nearly a mile with the midday sun blazing on our backs. 'Does it get any hotter in this country?' the deserter asked as we panted up the steep sides of a ravine, the sweat pouring down our faces and backs. 'You haven't seen the half of it yet. Wait another three months,' I answered, wondering grimly whether I should be able to win him even another three hours of life.

I found Colonel Peñaredonda sitting cross-legged with a plate of fried eggs on his knee. He greeted me amiably enough as I stepped forward and saluted; I had taken care to leave my prisoner well out of earshot. I repeated his story, adding my own plea at the end. 'I have the fellow here, sir,' I concluded, 'in case you wish to ask him any questions.' The Colonel did not look up from his plate: 'No, Peter,' he said casually, his mouth full of egg, 'I don't want to ask him anything. Just take him away and shoot him.' My heart seemed to stop beating. Motioning the prisoner and escort to follow, I started down the hill; I would not walk with them, for I knew that he would question me. I decided not to tell him until the last possible moment. I was so numb with misery and anger that I didn't notice where I was going until I found myself in front of de Mora once more. When I told him the news he bit his lip: 'Then I'm afraid there's nothing we can do,' he said gently. 'You had better carry out the execution yourself. Someone has got to do it. Try to get it over quickly.'

It was almost more than I could bear to face the prisoner, where he stood between my two runners. As I approached they dropped back a few paces, leaving us alone; they were good men and understood what I was feeling. I forced myself to look at him. I am sure he knew what I was going to say. 'I've got to shoot you.' A barely audible 'Oh my God!' escaped him. Briefly I told how I tried to save him. I asked if he wanted a priest, or a few minutes by himself, and if there were any messages he wanted me to deliver. 'Nothing,' he whispered, 'please make it quick.' 'That I can promise you. Turn around and start walking straight ahead.' He held out his hand and looked me in the eyes, saying only: 'Thank you.' 'God bless you!' I murmured. As he turned his back and walked away I said to my two runners: 'I beg you to aim true.' They nodded and raised their rifles. I looked away. The two shots exploded simultaneously. 'On our honour, sir,' the senior of the two said to me, 'he could not have felt a thing.'

PETER KEMP

249.

John Verney (1913–93) later served with distinction in special forces, became an author and artist who inherited a baronetcy. He learnt the business of soldiering on pre-war exercises with the North Somerset Yeomanry.

The squadron huddled together on horseback under cover of a wood. Tense with the excitement of the chase, Victor Bone sat erect on his horse, his heavy cavalry moustache stirred gently by the morning breeze. He sniffed the air as if to sense from it the whereabouts of the missing enemy. Then, with a few curt orders, he sent a patrol forward to reconnoitre. I wondered whether the time would ever come when I would attain to a comparable grasp of these complex matters. The patrol assembled itself without seeming haste; and without seeming haste trotted off ahead. Obviously we had some time to wait before fighting could recommence. I took out a packet of cigarettes and offered one to my idol. 'Have a smoke, sir?' His prominent, rather red-rimmed blue eyes stared at me in outraged astonishment. 'We never smoke during battle,' he said angrily.

JOHN VERNEY

250.

John Frost (1912–93) was an ambitious young officer who in September 1944 commanded a battalion of the Parachute Regiment in the battle at Arnhem bridge. Here, he recalls 3 September 1939.

I can still clearly remember the exact text of one message: 'War has broken out with Germany only.' The first person I told was a Captain who leapt for joy, saying: 'Marvellous! Marvellous! I was terrified that old Chamberbottom would settle up once again.' However, most other people heard the news rather solemnly. My foremost thought was that it might mean promotion.

JOHN FROST

251.

'Officer under instruction' in 1939, from a post-war memoir bathetically entitled A Very Quiet War.

I arrived at the racecourse grandstand, which was then serving the Infantry Training Centre as its officers' mess, at about eight o'clock in the evening. The Adjutant, summoned from his dinner by the mess corporal, was young, a regular soldier, and properly proud of his regiment and its traditions. 'Those trousers,' he muttered, averting his eyes. 'You can't,' he kept on repeating, 'come into the mess in those Godawful trousers.' I had been travelling all day. I was tired, hungry and rather frightened. Visions of a drink, followed by dinner, began to fade. The Adjutant asked if I had brought a civilian suit. I told him I had. 'In that case,' he said with evident relief, 'you must go and change into it at once. At once.' He organized transport to take me to my billet, an opulent villa, full of pink lampshades. The driver waited while I changed out of the faulty uniform which I had put on so proudly. I was driven back to the mess, where a late supper had been laid on. The Adjutant darted in and out like a flustered mother-bird. He did his best to dispel an unfortunate first impression, but I was painfully aware that whenever he set eyes on me during my sojourn at the ITC the recollection of those khaki slacks came between him and a proper appreciation of my military potentialities.

He had, poor chap, plenty of other troubles in connection with the newly joined wartime officers, of whom I was only one among many. Three days after we had reported for duty he announced that a special parade had been laid on for us in the dining-room. When we assembled we observed that one place, and one place only, had been laid on the long mahogany dining-table which had been transported to the grandstand from the regimental mess. There was a lavish display of knives, forks, spoons and wine-glasses. Had we, the Adjutant inquired, brought our notebooks? Some of us had; others, imagining perhaps that we were to receive encouragement in the shape of a buckshee and supplementary meal, had not.

While the ill-equipped retired to make good their deficiencies, the rest of us regarded the dining-room table with puzzled concern. When we were all in a state of readiness the object of the exercise was explained. The Colonel, the Adjutant told us, had been concerned and shocked on the previous evening – a guest night – to observe that some of the newly-joined officers had been in doubt about the correct implements and glasses to employ for the successive courses. If we would be kind enough to pay attention and take notes, he would give us a practical demonstration. Without batting an eyelid this impeccable young man then sat down at the table and an equally solemn mess waiter served him first with token soup, then with token fish, then with token meat, then with a token pudding and finally with a token savoury. The wine waiter went through the motions of pouring out sherry, burgundy, port and brandy. Somewhere I still possess the valuable notes I made.

RALPH ARNOLD

252.

On the beaches of Dunkirk in the dying days of May 1940. The writer, South African-born David Divine (1905–87), crewed a small boat during the evacuation; he crossed the Channel three times, was severely wounded and decorated. He later became a well-known naval war correspondent and author.

The picture will always remain sharp-etched in my memory – the lines of men wearily and sleepily staggering across the beach from the dunes to the shallows, falling into little boats; great columns of men thrust out into the water among bomb and shell splashes. The foremost ranks were shoulder deep, moving forward under the command of young subalterns, themselves with their heads just above the little waves that rode in to the sand. As the front ranks were dragged aboard the boats, the rear ranks moved up, from ankle deep to knee deep, from knee deep to waist deep, until they, too, came to shoulder depth and their turn. Some of the big boats pushed in until they were almost aground, taking appalling risks with the falling tide. The men thankfully scrambled up the sides on rope nets, or climbed the hundreds of ladders, made God knows where out of new, raw wood and hurried aboard the ships in England.

The little boats that ferried from the beach to the big ships in deep water listed drunkenly with the weight of men. The big ships slowly took on lists of their own with the enormous numbers crowded aboard. And always down the dunes and across the beach came new hordes of men, new columns, new lines. On the beach was the skeleton of a destroyer, bombed and burnt. At the water's edge were ambulances, abandoned when their last load had been discharged.

There was always the red background, the red of Dunkirk burning. There was no water to check the fires and there were no men to be spared to fight them. Red, too, were the shell bursts, the flash of guns, the fountains of tracer bullets. The din was infernal. The batteries shelled ceaselessly and brilliantly. To the whistle of shells overhead was added the scream of falling bombs. Even the sky was full of noise – anti-aircraft shells, machine-gun fire, the snarl of falling planes, the angry hornet noise of dive bombers. One could not speak normally at any time against the roar of it and the noise of our own engines. We all developed 'Dunkirk throat', a sore hoarseness that was the hallmark of those who had been there.

Yet through all the noise I will always remember the voices of the young subalterns as they sent their men aboard, and I will remember, too, the astonishing discipline of the men. They had fought through three weeks of retreat, always falling back, often without orders, often without support. Transport had failed. They had gone sleepless. They

In the twentieth century citizen armies and mass communications enforced a belated rendezvous with reality about the ghastliness of war. (*Above*) Otto Dix portrayed the Flanders battlefield in WWI as a desolate moonscape. (*Below*) By June 1944, civilian audiences were allowed to glimpse some of the horrors concealed from those of Jane Austen's era, though these casualties of D-Day appear less shocking than were many badly wounded men.

(*Clockwise*) Leo Tolstoy;
Lakshmibai; Edward Louis
Spears; Marcel Proust; Fyodor
Dostoevsky; Constance Edwina
Lewis, Duchess of Westminster.

(*Clockwise*) Dorothie Feilding; Robert Graves;
Adrian Carton de Wiart; Winston Churchill;
Siegfried Sassoon; Theodore Roosevelt;
Millicent Leveson-Gower,
Duchess of Sutherland.

(*Above*) The woman pays: here is a twentieth-century reprise of Rubens in the photo of a terrified Vietnamese woman cradling her baby amid a firefight. (*Below*) The woman enters the front line: US Army cook Shoshana Johnson was wounded and taken prisoner during the 2003 invasion of Iraq. Here, she is seen after her liberation following some terrified days in Iraqi captivity.

(*Clockwise*) George Orwell,
Evelyn Waugh, Eric Linklater,
Erwin Rommel, Enoch Powell.

(*Clockwise*) Auberon Waugh, Karl Marlantes, Hiroo Onoda, David Niven, Patrick Leigh Fermor, Michael Howard.

(*Clockwise*) Arkady Babchenko,
Andrew Milburn, Johnson Beharry,
Jane Blair, Kayla Williams.

New visions of combat: (*above*) the SAS, seen here during their 1980 storming of the terrorist-occupied Iranian embassy in London, have achieved iconic status as an acceptable face of warriorhood – though their ruthlessness is as studiously concealed from the public as their features. (*Left*) British troops in Afghanistan vanish into the fog of war. Some observers believe that the long age of the soldier as mortal man is approaching an end as risk-averse societies seek to enforce their purposes through remote killing, robots and drones. Marshal Ney would lament the passing of glory, but in truth the military art has always been suffused with suffering and sacrifice.

had been without food and water. Yet they kept ranks as they came down the beaches, and they obeyed commands. While they were still filing back to the beach and the dawn was breaking with uncomfortable brilliance, we found one of our stragglers – a navy whaler. We told her people to come aboard, but they said that there was a motor-boat aground and they would have to fetch off her crew. They went in, and we waited. It was my longest wait, ever. For various reasons they were terribly slow. When they found the captain of the motor-boat, they stood and argued with him and he wouldn't come off anyway. Damned plucky chap. He and his men lay quiet until the tide floated them later in the day. Then they made a dash for it, and got away.

We waited for them until the sun was up before we got clear of the mole. By then, the fighting was heavy in-shore, on the outskirts of the town, and actually in some of the streets. Going home, the dive-bombers came over us five times, but somehow left us alone, though three times they took up an attacking position. A little down the coast, towards Gravelines, we picked up a boatload of Frenchmen rowing off. We took them aboard. They were very much bothered as to where our 'ship' was, and said quite flatly that it was impossible to go to England in a thing like ours. Too, too horribly dangerous!

One of the rare touches of comedy at Dunkirk was the fear of the sea among French *poilus* from inland towns. They were desperately afraid to forfeit solid land for the unknown perils of a little boat. When, on the last nights of the evacuation, the little boats got to the mole many refused to jump in, despite the hell of exploding shells and bombs behind them.

DAVID DIVINE

253.

On 1 June 1940, during the final phase of the evacuation, the monocled, one-eyed cavalry officer Lt.-Col. Alfred Wintle (1897–1966) – he had been badly wounded in 1917 – determined that only direct action by himself, working with old friends of the French air force, might enable some portion of its strength to be flown to Britain. What followed was an epic of absurdity, described in his autobiography The Last Englishman.

I realized that I hadn't a hope in hell of getting official sanction. Immediate action was called for. If I failed to pull off the stunt I had in mind
I would either be shot by the enemy or court-martialled by our side.
But I considered it a worthwhile risk. I rang the Commandant at
Heston Airfield, giving every password and code sign I knew, and
claiming to be a senior staff officer at the Air Ministry. 'Please prepare
an aircraft with all speed to take a Colonel A. D. Wintle to Bordeaux
immediately. He will arrive shortly in a French Mission car. Absolute
secrecy must be observed.'

It appeared to go down quite well. I was satisfied and drove out to the
airport in the Mission car I had commandeered through my influential
French friend. Unfortunately, as I reached Heston the Commandant
was on the 'phone to the Air Ministry, crowing that the plane was ready.
The response, of course, was: 'What bloody plane …?' The flight was
cancelled. I thought quickly and attacked. I called the Commandant a
blithering idiot for not maintaining secrecy and stormed out. I then
drove straight to the Air Ministry and stormed into the office of Air
Commodore Boyle, the Director of Intelligence, whose name and
manner I had assumed when making the 'phone call ordering the plane.

I pleaded with him to allow me to fly to France that night. I begged
him, as a patriot, to take this chance of salvaging France's air strength.
He answered sharply that I would be court-martialled and that I could
meanwhile, if I wished, make my suggestion through the normal channels. I saw red and cursed him fluently. 'While you sit there,' I
concluded, 'blood is flowing in France, not ink. And I am deadly
serious.' I then drew my revolver and waved the muzzle under his nose
like a warning finger. 'You and your kind ought to be shot,' I snorted.
And I named a few top people who deserved it as much as he did. I
then broke my revolver, spilled out the bullets to show I had not been
bluffing and left him shaking with fright.

I returned to my unit in the North that night. Next morning I was
arrested. This seemed a trivial happening while the fate of Europe was
in the balance. But I was given an escort and a warrant, to be placed
under close arrest in the Tower of London while charges were prepared.
The rest of this story is more like a comedy play than an episode in a
major war. My escort, a timid little re-employed officer of the Lanca-

shire Fusiliers, spent the whole journey from the camp to Liverpool fumbling through his pockets to see that the documents were safe. A dozen times he opened and shut my revolver holster, which he was carrying, to make sure the gun was still there. Yet, for all this worrying, when we got to the station the travel warrant was missing. We had about ten minutes left to catch the train. So I said to him: 'For heaven's sake stop fumbling. Wait here by the baggage while I go and get another warrant.'

He obeyed meekly and stood watching with an unhappy expression. I found the RTO's office and asked for a warrant. As no other officer was present, I signed it myself. This must surely make me the first prisoner who has ever signed his own travel warrant to the Tower of London. Nothing else happened on the journey to disturb my escort's peace of mind. At the end of it he saw me lodged in a room in the forbidding stone tower with heavily barred windows. A Scots Guardsman, with fixed bayonet, took up a post at the door. Officers at first cut me dead, thinking I was some kind of traitor; but when the news of my doings leaked out they could not do enough for me. I was as well cared for as if I had been at the Ritz. I would have a stroll in the moat after breakfast for exercise. Then sharp at eleven each morning, Guardsman McKie, detailed as my servant, would arrive from the officers' mess with a large whisky and ginger ale. He would find me already spick and span, for, though I have a great regard for the Guards, they have not the gift to look after a cavalry officer's equipment. The morning would pass pleasantly. By noon visitors would begin to arrive.

One or two always stayed to lunch. They usually brought something with them. I remember a particularly succulent duck in aspic – it gave me indigestion – and a fine box of cigars brought by my family doctor. Tea-time was elastic and informal. Visitors dropped in at intervals, usually bringing bottles. I dined sharp at eight and entertained only such guests as had been invited beforehand. After a few days of settling in, I was surprised to find that being a prisoner in the Tower of London had its points. If there hadn't been a war on I might have tried to get a life sentence.

Only one visitor was unwelcome. He was an officer of the Judge-Advocate's department. I didn't in the least mind his questions, but he took the liberty of calling me 'old boy' – a slovenly form of address which I don't like from friends and can't bear from strangers. This gentleman told me he wanted to be 'perfectly frank' – a figure of speech which always puts me on my guard. After a lot of beating about the bush, it turned out that the War Office wanted to kiss and make up. Some of the brasshats were getting cold feet. 'After all, old boy,' said my 'frank' visitor, in honeyed tones, 'we don't want a Court Martial, do we?'

'You may not,' I answered tartly, 'but I do. I came here through the "usual channels" and that is the way I propose to leave. Now, sir, I will not detain you if you have pressing business elsewhere. Good day.' Looking like a little boy who has just been ticked off at school, he departed. He even forgot to say 'old boy'.

The day of my court-martial came. I was escorted to the Duke of York's Headquarters, Chelsea. Then I read the charges and had great difficulty in stopping myself from laughing aloud. This, I could see, was going to be great fun. Most of the original charges (some dozens of them) had been dropped. Only three remained, and these were: 'Stating that certain of His Majesty's Ministers ought to be shot – thereby committing common assault ... Endeavouring to evade active service by feigning blindness in one eye ... Threatening a superior officer with a revolver.'

I was delighted to see that whoever was responsible for bungling the trial had made a first-class job of it. I did not deny the first charge. I carefully explained that in the Army there is a slogan 'no names, no packdrill'. 'It is a corporal's slogan,' I went on, 'and as I have never been a corporal I do not uphold it. I propose to name names. And if I forget anybody it will only be because of my inadequate memory and the fact that my list is so long.' I then carefully and loudly named all those of His Majesty's Ministers whom I still thought should be shot, but I was not allowed to finish. When I got to Kingsley Wood, about seventh on my mental list, the prosecuting advocate interrupted. In an embarrassed tone of voice he rose to say that he did not propose to proceed with this charge. I was almost disappointed.

On to the second charge. Here, I proved conclusively that I had volunteered for active service several times. I had not faked blindness or anything else. Instead, to try to get into action I had pretended not to be blind in an eye which was blind. I called Sir Edmund Ironside, CIGS, as a witness and he gave me an excellent reference as an eccentric and a fighter for England. If anything, this charge was dropped rather more quickly than the first. Then came the third and last. It was argued at length. Great stacks of books were brought out. Learned counsel nodded and whispered together. I felt like a criminal specimen under a legal microscope.

Then my 'friend in court', J. D. Casswell, QC, revealed that the man I was alleged to have threatened was not my superior officer at all, nor was he anyone else's. He was a perfectly ordinary civil servant, who had merely been given permission to wear the King's uniform while the war was on. The prosecution looked as though they'd had something for lunch that had disagreed with them badly. But, of course, they had to have a face-saver. This they achieved by the swift reduction of the third charge to one of ordinary civil assault. So, on these minor grounds, I was found guilty and sentenced to a severe reprimand.

The brasshats' faces were saved. I had a sense of victory. Technically, I was a naughty boy. Actually, I felt grand. Back at the Tower the Scots Guards greeted me like a conquering hero. Their CO almost ran across the square to hand back my sword-belt and revolver and insisted on making me his guest for the evening. I was invited to continue using the room which had been my cell, the difference being that I now had the key of the door. Now these kindnesses, I am sure, were not demonstrations of affection for myself. They were an expression of the frustration that all true soldiers feel at seeing their country's flag dragged in the mire by a bunch of dithering incompetents.

ALFRED WINTLE

254.

The British Expeditionary Force was commanded during the last stages of the evacuation from Dunkirk by its senior Corps commander, Lt.-Gen. Sir Harold Alexander (1891–1969). As the great drama drew to a close, Alexander set off personally to ensure that every possible man had been taken off the beaches.

As soon as it was dark on June 2nd, the remnants of the BEF began to embark at the mole. The arrangements worked without a hitch. All the men were aboard by 11.40 p.m. When the destroyers sailed for Dover, Alexander and half a dozen others boarded a motor-boat, ordering a single destroyer to await them. There was no shelter from the incessant gunfire. They zig-zagged out of the harbour, then turned east parallel to the beaches for about two miles, as close inshore as the draught of the boat would allow. Twice they grounded on sandbars. The sea was covered with a film of oil, in which were floating the corpses of many soldiers. Alexander took a megaphone and shouted over and over again, in English and French, 'Is anyone there? Is anyone there?' There was no reply. They returned to the harbour, shouted the same question round the quays, and then boarded the waiting destroyer. Unharmed, they reached Dover as dawn was breaking. Alexander went immediately to see Anthony Eden at the War Office. 'After he had given me an account of what had passed,' Eden wrote, 'I congratulated him, and he replied, with engaging modesty, "We were not pressed, you know."'

NIGEL NICOLSON

255.

The Dunkirk spirit, exemplified by a captain of engineers
in the Orkneys.

I was summoned, with other local commanders, to a conference at headquarters in Stromness. Our general, sometime a Horse-Gunner in India, was vigorous, swiftly moving, with a thin, bony, claret-coloured face, lively blue eyes, a meagre sward of ginger hair, and a patch of ginger bristles on a long upper lip. His temper could flare fierce as a blow-lamp – there was much occasion for it – and it consumed in its fire many of the obstructions that our ignorance created. We had been summoned to discuss what could best be done to meet an attempted landing on our shores. We listened to explanation of the general plan for defence, and then, each in turn, we were asked what local and particular instructions we had given to the soldiers in our own commands. I, responsible for a stony segment of the island of Flotta in which it was impossible to dig trenches, replied meekly, 'Don't shoot till you see the whites of their eyes.'

'Don't be a damned fool!' exclaimed our fiery general. 'This is no time for joking, this is a serious occasion.'

'I have, sir, two Lewis guns, each with two drums of ammunition, and only ten rounds to a rifle. I thought it would be advisable to conserve ammunition.'

ERIC LINKLATER

256.

James Lees-Milne (1908–97), a junior officer who later became celebrated as a diarist and prime mover of the National Trust, faced a personal moment of crisis in the summer of 1940.

France had fallen. Great Britain stood alone. Hitler's invasion was expected at any moment. Day and night an officer was kept on duty awaiting from some higher intelligence the warning code signal, 'Oliver Cromwell'. When this ominous name came down the telephone the officer knew that the invasion was on the way. He must instantly ring through to the Colonel and arouse the battalion. At 3 o'clock one morning it was my turn to be on duty. Rather drowsily I was reading *Barchester Towers*. The telephone rang. 'This is Higher Command QE2X speaking,' came from a rather sissy voice a long way off. 'I say, old boy, sorry to tell you – Oliver Cromwell.' 'What?' I screamed, my heart in my boots. 'Are you sure? Are you absolutely sure?' I had no reason for questioning the man's words beyond the utter horror of the announcement. 'Well, I may have got it wrong,' the voice said affectedly. 'Then for dear Christ's sake,' I pleaded, 'do get it right.' There was a pause, during which I had my finger on the special telephone to the Colonel's bedroom, as it were on the pulse of England. 'Sorry, old chap,' the voice came back again. 'It's only Wat Tyler. I get so confused with these historical blokes.'

JAMES LEES-MILNE

257.

*The difficulties of service chiefs compelled to indulge the Prime
Minister's enthusiasms were vividly described by Lt.-Gen. Sir James
Marshall-Cornwall, in a reminiscence he entitled 'Mad Hatter's
Dinner Party'.*

On Friday 26 July 1940, I was rung up by General Sir John Dill, the
CIGS, to say that I was commanded to spend Saturday night with the
Prime Minister at Chequers. Thunderstruck, I asked Jack Dill what it
was all about. He replied that he hadn't the faintest notion, but that
he had been invited also. I reached Chequers about six o'clock. Dinner
was indeed a memorable meal. I was placed on the PM's right, and on
my right was Professor Frederick Lindemann, Churchill's scientific
adviser. The others around the oval table were Mrs Churchill, Duncan
Sandys and his wife, 'Pug' Ismay, Jack Dill, Lord Beaverbrook, and one
of the PM's private secretaries.

Churchill was bubbling over with enthusiasm and infectious gaiety.
I marvelled how he could appear so carefree with the enormous load
of anxieties on his shoulders, and I wish that I could remember some
of the splendid sentences that rolled off his tongue. As soon as the
champagne was served he started to interrogate me about the condition of my Corps. I told him that I had found all ranks obsessed with
defensive tactical ideas. I had issued orders that only offensive training
exercises were to be practised, and that the III Corps motto was
'Hitting, not Sitting', which prefaced every operation order.

This went down tremendously well with the PM, who chuckled and
chortled: 'Splendid! That's the spirit I want to see.' He continued, 'I
assume then that your Corps is now ready to take the field?' 'Very far
from it, Sir,' I replied. 'Our re-equipment is not nearly complete, and
when it is we shall require another month or two of intensive training.'

Churchill looked at me incredulously and drew a sheaf of papers
from the pocket of his dinner-jacket. 'Which are your two Divisions?'
he demanded. 'The 53rd (Welsh) and the 2nd (London),' I replied. He
pushed a podgy finger on the graph tables in front of him and said,
'There you are; 100 per cent complete in personnel, rifles and mortars;

50 per cent in field artillery, anti-tank rifles and machine-guns.' 'I beg your pardon, Sir,' I replied. 'That state may refer to the weapons which the ordnance depots are preparing to issue to my units, but they have not yet reached the troops in anything like those quantities.' The PM's brow contracted; almost speechless with rage, he hurled the graphs across the dinner-table to Dill, saying, 'CIGS, have those papers checked and returned to me tomorrow.'

An awkward silence followed; a diversion seemed called for. The PM leant across me and addressed my neighbour on the other side: 'Prof! What have you got to tell me today?' The other civilians present were wearing dinner-jackets, but Professor Lindemann was attired in a morning-coat and striped trousers. He now slowly pushed his right hand into his tail-pocket and, like a conjuror, drew forth a Mills hand-grenade. An uneasy look appeared on the faces of his fellow-guests and the PM shouted: 'What's that you've got, Prof, what's that?' 'This, Prime Minister, is the inefficient Mills bomb, issued to the British infantry. It is made of twelve different components which have to be machined in separate processes. Now I have designed an improved grenade, which has fewer machined parts and contains a 50 per cent greater bursting charge.' 'Splendid! Splendid! That's what I like to hear. CIGS! Have the Mills bomb scrapped at once and replaced by the Lindemann grenade.'

The unfortunate Dill was completely taken aback; he tried to explain that contracts had been placed in England and America for millions of Mills bombs, and that it would be impracticable to alter the design now, but the PM would not listen. To change the subject he pointed a finger at Beaverbrook across the table. 'Max! What have you been up to?' Beaverbrook replied, 'Prime Minister! Give me five minutes and you will have the latest figures.' He rose and went to a telephone box at the far end of the room; after a very few minutes he returned with a Puckish grin on his face. 'Prime Minister,' he said, 'in the last 48 hours we have increased our production of Hurricanes by 50 per cent.'

The brandy and coffee had now circulated and the PM lit his cigar. 'I want the Generals to come with me,' he said, and stumped off to an adjoining room, followed by Dill, Ismay and myself. On a table was a large-scale map of the Red Sea. The PM placed his finger on the Italian port of Massawa. 'Now, Marshall-Cornwall,' he said, 'we have command of the sea and the air; it is essential for us to capture that port; how would you do it?' I saw Dill and Ismay watching me anxiously and felt that I was being drawn into some trap. I looked hard at the map for a minute and then answered, 'Well, Sir, I have never been to Massawa. It is a defended port, protected by coast defence and anti-aircraft batteries. It must be a good 500 miles from Aden, and therefore beyond cover of our fighters. The harbour has a very narrow entrance channel, protected by coral reefs, and is certain to be mined, making an opposed landing impracticable. I should prefer to wait until General Wavell's offensive against Eritrea develops; he will capture it more easily from the land side.'

The PM gave me a withering look, rolled up the map and muttered peevishly, 'You soldiers are all alike; you have no imagination.' We went to bed. On our way back to London Jack Dill said to me, 'I'm thankful, Jimmy, that you took the line you did last night. If you had shown the least enthusiasm for the project, I should have been given orders to embark your Corps for the Red Sea next week.'

JAMES MARSHALL-CORNWALL

258.

The boundary between genius and insanity is as ill-defined in military high command as in other fields of human endeavour, but Orde Wingate (1903–44), a Churchillian favourite, certainly crossed it. The guerrilla leader is here sketched by another individualist, Wilfred Thesiger, who served with Wingate – then a major – in 1940–1 Abyssinia.

While expounding his seemingly impossible plans, he strode about his office, his disproportionately large head thrust forward above his ungainly body; in his pale blue eyes, set close together in a bony, angular face, was more than a hint of fanaticism. As he shambled from one office to another, in his creased, ill-fitting uniform and out-of-date Wolseley helmet, carrying an alarm clock instead of wearing a watch, and a fly-whisk instead of a cane, I could sense the irritation and resentment he left in his wake. His behaviour certainly exasperated General Platt, whom I once heard remark, 'the curse of this war is [T. E.] Lawrence in the last'. Wingate was often deliberately rude and aggressive. Once, breakfasting at a hotel, he gratuitously accused two young officers at a nearby table of cowardice, for taking staff jobs. He would make appointments with officers, then receive them lying naked on his bed; one described to me how Wingate had been brushing his body-hair with a toothbrush as he gave instructions. [In the field] Wingate was accompanied by a civilian Jew called Akavia who had worked with him in Palestine and was now virtually his chief-of-staff. I once watched the two of them roast a chicken over a fire. They had no pots or pans, no plates or knives or forks, and when it was cooked Wingate, squatting on the ground, tore it to pieces with his teeth, getting his hands and face smothered with grease. Having finished he said: 'Here, Akavia, catch!' and threw the remnants. He appeared to live in a lonely world of his own making, driven by some perverse urge to alienate his fellows. In Khartoum he once asked me if I was happy. I answered that in general I was. He reflected, and said: 'I'm not happy, but I don't think any great man ever is.' [In 1943–4 Wingate became creator and commander of the Chindit irregular columns in Burma, but fortunately perished in an air crash before Churchill could fulfil a threat to make him army commander-in-chief.]

WILFRED THESIGER

259.

Admiral Sir Walter Cowan, KCB, DSO (1871–1956), commanded a Nile gunboat during the 1898–9 Sudan campaign, served as ADC in the Boer War to both Lords Kitchener and Roberts, fought the Bolsheviks in the Baltic in 1919, and in 1941 at the age of seventy-four was attached at his own request to Eighth Army in the Western Desert, where he was captured. Cowan was affronted to find himself released by the Italians as 'of no further use' to the British war effort. Evelyn Waugh met him when both men were serving with the Commandos.

[He was] a very old, minute hero who came out as a kind of mascot. Most of his experience had been with the Army and he showed great intolerance of signalmen, who he could not bear anywhere near him. He neither smoked nor drank wine, and ate all his food on the same plate, porridge, fruit, meat, eggs in a single mess. He was exquisitely polite, almost spinster-like in conversation; churchgoing with a belief in British-Israel. He sat behind me in the boat at Bardia bearing the weight of fifteen men. I could feel him fluttering like a bird in the hand. Later he said 'Young Waugh is uncommon heavy.' He read nothing but Surtees. He was popularly believed to spend his leisure in sniping at Italian prisoners with a catapult. He certainly loathed them almost as violently as he loathed signallers. I was once talking to him when a group marched past us with distinguishing patches in the seats of the trousers. He had been asking solicitously, like an aunt, after my health. He suddenly broke off and said with extravagant venom, 'That's the place to mark the sods.' He went with the *Aphis* party on their abortive raids and greatly enjoyed the bombing. 'They ought to have got us,' he said very regretfully.

EVELYN WAUGH

260.

A dilemma for Michael Howard (1922–2019), in 1941 a Christ Church undergraduate and later one of Britain's most distinguished historians.

I played in the Oxford orchestra, but was very surprised indeed to be visited one day by a musician who was talent-spotting for the RAF orchestra. Would I be interested in joining the RAF, he asked, and training as an oboist? The idea was not unappealing. I went to consult my history tutor, Keith Feiling. He listened puffing his pipe, looking like a wise old owl. 'Y-yes, Howard,' he said when I had finished. 'I-I can see that it doesn't look a b-b-bad idea. But afterwards, when they ask you w-w-what you actually did during the war, and you said that you spent it p-p-playing the oboe …' [Howard joined the Coldstream Guards.]

MICHAEL HOWARD

261.

John Masters encountered the Yeomanry during the 1941 Iraq campaign, and describes a staff conference he attended with these amateur cavalrymen, who were peacetime rural neighbours and fellow-foxhunters.

The General: 'Well, I think we should send a patrol up the Euphrates for fifty miles or so, to make sure no one's lying up in the desert out there.'

ONE OF HIS YEOMANRY COLONELS: 'Good idea, George.'
THE GENERAL: 'From your regiment, I thought, Harry. About a troop, with a couple of guns, eh?'
THE COLONEL: 'Oh yes, George … I think I'll send Charles.'
THE GENERAL (HORROR IN HIS FACE): 'Charles? Charles? Do you think he'll go?' We learned later that though 2nd Lieutenant Charles was distinctly vexed at being sent on such a piddling mission, he did finally agree to go, since George and Harry seemed to set so much store on it. They were delightful people. My own favourite story about them concerns an early inspection, by the general, of one of the regiments. The Yeomanry colonel, going down the line introducing his officers, stopped before one captain, and said, 'This is Captain … Captain …' He shook his head, snapped his fingers and cried genially, 'Memory like a sieve! I'll be forgetting the names of me hounds next.'

JOHN MASTERS

262.

Evelyn Waugh writes to his wife Laura, 31 May 1942.

No. 3 Commando was very anxious to be chums with Lord Glasgow [1874–1963, a notorious fascist who lived at Kelburn Castle, Ayrshire], so they offered to blow up an old tree stump for him and he was very grateful and he said don't spoil the plantation of young trees near it because that is the apple of my eye and they said no of course not we can blow a tree down so that it falls on a sixpence and Lord Glasgow said goodness you are clever and he asked them all to luncheon for the great explosion. So Col. Durnford-Slater DSO said to his subaltern, have you put enough explosive in the tree. Yes, sir, 75 lb. Is that enough? Yes sir I worked it out by mathematics it is exactly right. Well better put a bit more. Very good sir. And when Col. D Slater DSO had had his port he sent for the subaltern and said subaltern better put a bit more explosive in that tree. I don't want to disappoint Lord Glasgow. Very good sir.

Then they all went out to see the explosion and Col. DS DSO said you will see that tree fall flat at just that angle where it will hurt no young trees and Lord Glasgow said goodness you are clever. So soon they lit the fuse and waited for the explosion and presently the tree, instead of falling quietly sideways, rose 50 feet into the air taking with it ½ acre of soil and the whole of the young plantation. And the subaltern said Sir, I made a mistake, it should have been 7½ lb not 75. Lord Glasgow was so upset he walked in dead silence back to his castle and when they came to the turn of the drive in sight of his castle what should they find but that every pane of glass in the building was broken. So Lord Glasgow gave a little cry and ran to hide his emotion in the lavatory and there when he pulled the plug the entire ceiling, loosened by the explosion, fell on his head.

This is quite true.

EVELYN WAUGH

263.

Britain's most successful general of the desert war, then titled Sir Bernard Montgomery, was not celebrated for his hospitality.

Montgomery's allegedly austere personal habits gave rise to a number of disobliging comments and anecdotes. His appalled reaction to the clouds of cigar smoke with which Churchill had filled his tent in the desert was the subject of much amusement, as was Churchill's reply to a questioner in the House of Commons who complained that Montgomery had invited Wilhelm von Thoma, the defeated German general, to dinner in his desert caravan. 'Poor von Thoma,' said Churchill gravely. 'I, too, have dined with Montgomery.'

LORD CHALFONT

264.

The great Russian war correspondent Vasily Grossman (1905–64), who went on to write the greatest Russian novel of the war, Life And Fate, *was awed by the courage and fortitude he saw displayed throughout the conflict by Russian women, but dismayed by the systemic exploitation of uniformed women by Red Army officers.*

Grossman wrote: 'The PPZh is our great sin'. The PPZh was the slang term for a 'campaign wife', because the full term, *pokhodno-polevaya zhena*, was similar to PPSh, the standard Red Army sub-machine gun. Campaign wives were young nurses and women soldiers from a head-quarters – such as signallers and clerks – who usually wore a beret on the back of the head rather than the fore-and-aft *pilotka* cap. They found themselves virtually forced to become the concubines of senior officers. Grossman scribbled some bitter notes on the subject, perhaps for use in a story later:

Women – PPZh. Note about Nachakho, chief of administrative supplies department. She cried for a week, and then went to him.

'Who's that?'

'The general's PPZh.'

'And the commissar hasn't got one.'

Before the attack. Three o'clock in the morning.

'Where's the general?' [someone asks]

'Sleeping with his whore,' the sentry murmurs.

And these girls had once wanted to be 'Tanya' or Zoya Kosmode-myanskaya [a legendary partisan who used Tanya as a *nom de guerre*].

'Whose PPZh is she?'

'A member of the Military Council's.'

Yet all around them tens of thousands of girls in military uniforms are working hard and with dignity.

VASILY GROSSMAN

265.

*The 1942–3 Soviet defence of Stalingrad was sustained by savage
disciplinary sanctions against displays of weakness. Some 13,500 men
were executed for alleged cowardice or desertion. Here Grossman notes
laconically a grotesque episode, in which a man was convicted for
'betrayal of the Motherland'.*

'An extraordinary event. Sentence. Execution. They undressed him and
buried him. At night, he came back to his unit, in his blood-stained
underwear. They shot him again.' There was a similar episode in the
45th Rifle Division, when the execution squad from the NKVD
Special Department attached to the division failed to kill the
condemned man, perhaps because their aim was affected by alcohol.
This soldier, like so many others, had been condemned to death for a
self-inflicted wound. After shooting him, the execution squad buried
him in a nearby shell-hole, but the condemned man dug himself out
and returned to his company, only to be executed a second time.
Usually, however, the prisoner was forced to undress before being shot
so that his uniform could be issued to somebody else without too
many discouraging bullet-holes.

VASILY GROSSMAN

266.

A portrait of Eighth Army's commander by his ADC.

Monty's habits in the desert and thereafter were simple and regular. He
would be called by his soldier servant, Corporal English, at 6.30 every
morning with a cup of tea and would not come out of his caravan till
8 to walk across to the mess tent for breakfast. You could set your
watch by his visit to the W.C. He would retire to bed at 9.30 in the
evening no matter who was visiting the headquarters. Even when
George VI came he would say: 'If you will excuse me, Sir, we have the
battle to win and I must go to bed. These lads will sit up all night
drinking and I trust they will look after you.' He went to bed as usual
at 9.30 on the eve of the Normandy landing, saying 'Come and tell

me the news at 6 a.m.' He was awake when I went in and just said quietly, 'What's the news?'

Living conditions were sometimes quite spartan. In southern Italy we were once in a very primitive house. As I walked down the passage I noticed Monty trying to have a bath where there was no running water. He was calling out to bring him some cold water. Noticing a jerrycan in the passage which I presumed to be full of water, I opened the door to come to the rescue. He was standing up in the bath lifting his feet in and out of the very hot water. He took the can and poured what turned out to be Marsala wine over a vulnerable area above the knees. He was not at that moment as amused as I was, but afterwards frequently enjoyed telling the story of how I was unable to tell the difference between wine and water.

Dinner in the evening was the only period of relaxation. In the desert there were often not more than three or four of us, though there would be a greater number in Europe. Nearly every night he would provoke a conversation, often of a trivial, bantering nature, which would run something like this:

MONTY: 'Johnny, have you read my latest pamphlet on military training?'
REPLY: 'No, matter of fact I haven't.'
MONTY: 'Well, you will never make a soldier if you don't get down to that.'
REPLY: 'But I don't want to be a soldier.'
MONTY: 'Alright, what do you want to do?'
REPLY: 'I don't really know, but perhaps I might go into the City.'
MONTY: 'Oh, you want to make money, do you? That won't do you any good.'

One evening in Belgium he asked us for our definition of a gentleman. He clearly thought we weren't doing very well, as indeed we weren't, so he said: 'Well, we'll ask Winston when he comes out next week.' On Winston's first night we had not been sitting down long before Monty duly remarked: 'I have been asking these fellows for a definition of a gentleman and they aren't very good at it – what's yours?' Winston thought for a moment and said: 'I know one when I see one,' and then added, 'I suppose one might say – someone who is only rude intentionally.'

Monty had a good sense of humour and could tell a story really well, often against himself. He enjoyed recounting a story which dated from pre-war days when he was commanding a battalion of the Royal Warwickshire Regiment in Egypt. One of his young officers was, in Monty's view, doing himself no good by being out too often with the girls. 'So I gave him an order not to have another woman without my permission, though if I thought it necessary I would give it.' Some weeks later he was dining in Cairo with the Ambassador, Lord Killearn. During dinner the butler announced that there was a telephone call for Colonel Montgomery. The Ambassador said 'Ask who it is and what he wants.' The butler returned and gravely announced: 'It is Lieutenant X and he wants to ask the Colonel if he can have a woman.' Permission was granted.

JOHNNY HENDERSON

267.

A miniature of Gen. George S. 'Blood and Guts' Patton by a British officer – who confusingly pseudonymized himself as Meego and was addressed as Larry – attached to his staff in 1942–3, before and during the TORCH landings in North Africa. Col. Robert Henriques (1905–67) called at the general's house before they sailed together for the Mediterranean.

In the hall he met Mrs Patton, a gracious and minute lady of wonderful charm and spirit. She was carrying in both hands a lump of primitive carving in lava rock. Somewhat battered, it seemed to be a carved head with the hair curled up in a peak. 'Meet Charlie!' she said to Meego. 'He's very, very old, and very, very distinguished. They gave him to the General in Hawaii. He's a warrior. You can tell he's a warrior from his peak.'

'Do they have warriors in Hawaii?' Meego asked.

'They used to have, one time,' Mrs Patton said. 'This is a warrior idol. If you're a warrior yourself, he brings you good luck. But if you're not a warrior, it's quite the reverse. Would you care to have him, Larry?'

'Doesn't the General want him?'

'I don't know what you've done to the General today,' Mrs Patton said. 'He's in a terrible wax.'

'The General', Mrs Patton continued, 'came right home and threw Charlie right in the pond. I just fished him out.' She thrust the idolized lump into Meego's hands. It was quite a weight. 'I warned you about being a warrior,' she said. They were walking towards the drawing-room. 'One other thing, Larry ... If you take him, you must take him with you.'

'Take him where?' Meego asked.

'Take him in the little boats,' she said. 'I must be promised that. He has to go to Casablanca with the General to bring him luck.' Meego was shocked at the indiscretion and must have shown it. 'Doesn't your wife know where you're going when you go off?' Mrs Patton asked. 'The General didn't exactly tell me,' Mrs Patton added, 'I must have just kind of guessed. I'm kind of psychic like that.'

[On 6 November 1942, Patton landed in North Africa.] Meego met him with Charlie in his hands. Patton said: 'Goddamit, Larry, didn't I tell you to stay in your goddam ship?' Meego said: 'Mrs Patton made me promise to bring Charlie ashore with the assault.' Patton said: 'So she did!' Then he asked what had happened and was happening. Meego told him and explained also that the chaos, which Patton was reviewing at the time, was due mainly to the dereliction of the 'Shore parties'.

Patton stood there, with his two pearl-headed pistols, surveying the long beach. The scene was silent, sunlit and tranquil. An Arab was leading a donkey laden with sacks along the stranded craft. He was filling the sacks with such oddments and trifles of American equipment as he thought were of worth. Nobody else was doing any work. They had even stopped digging, for they had no longer any pretext of danger to dig against. This picture of total idleness and untroubled chaos provoked Patton to say – and he said it then, and he said it to Meego and to nobody else – he said: 'Jesus, I wish I were a corporal!'

He saw the itinerant Arab with his donkey come upon a rifle which some American soldier had let fall rather than carry. As the Arab was stowing it away in his sack, Patton sent a shot across his bows from one of his pearl-handled pistols. The result was satisfactory: the rifle let go, and the donkey sent scurrying in panic, with the Arab after him.

ROBERT HENRIQUES

268.

Field-Marshal Earl Wavell (1883–1950), Viceroy of India in 1942, was described by his best biographer, Ronald Lewin, as having been suited to greatness in any field of endeavour save that of high command in war. Wavell diminished himself by soliciting his final rank from Churchill, who had sacked him as Commander-in-Chief Middle East. Literary passions formed an important part of his legend: in 1944 he edited Other Men's Flowers, *an anthology of poetry which has never since been out of print. His Military Assistant recorded a domestic incident in the worst days of the war.*

When the Japanese menace to India was at its height, an order had gone forth from GHQ in Delhi that all officers should have pistol practice; so everyone, myself included, suddenly became very pistol-minded. One evening Wavell sent for me to his office. His desk was littered with books, and all the drawers were open as if he had been looking for something. 'Peter,' he said, 'I can't find my Browning. You did not borrow it, I suppose?' I spent the next hour frantically searching for a Browning pistol, though it was the Collected Works the C-in-C had meant.

PETER COATS

269.

The British people cherish an intensely sentimental view of the army's Gurkha regiments – Nepalese mercenaries who have 'taken the Queen's shilling' for more than two centuries. This is the account by one Gurkha, Jemadar Sing Basnet, of an exploit in Tunisia in April 1943, when he led a night patrol to the capture of German-held high ground, demonstrating a courage for which he and his comrades were famous, as well as a directness that is today unfashionable.

I was challenged in a foreign language. I felt it was not the British language or I would have recognised it. To make sure I crept up and found myself looking into the face of a German – I recognized him by his helmet. He was fumbling with his weapon, so I cut off his head with my kukri. Another appeared from a slit trench and I cut him down also. I was able to do the same to two others, but one made a great deal of noise, which raised the alarm. I had a cut at a fifth, but I am afraid I only wounded him. I was now involved in a struggle with a number of Germans, and eventually, after my hands had become cut and slippery with blood, they managed to wrest my kukri from me. One German beat me over the head with it, inflicting a number of wounds. He was not very skilful, however, sometimes striking me with the sharp edge, but oftener with the blunt. They managed to beat me to the ground, where I lay pretending to be dead. I could not see anything, for my eyes were full of blood. I wiped the blood out of my eyes and quite near I saw a German machine-gun. By now it was getting light and as I lay thinking of a plan to reach the gun, my platoon advanced and started to hurl grenades. I thought that if I did not move I really would be dead. I managed to get to my feet and ran towards my platoon. They recognized my voice and let me come in. My hands being cut about and bloody, and having lost my kukri, I had to ask one of my platoon to take my pistol out of my holster and put it in my hand. I then took command again.

SING BASNET

270.

A tale of one of many clever civilian eccentrics who became successful staff officers in World War II, though devoid of conventional military skills. This was Professor J. Enoch Powell (1912–88), of whom it was said that 'he has one of the most brilliant minds in England – until he makes it up'.

Breakfast was not a success. The fire smouldered dejectedly until the Professor teased it with a gill of petrol, and then it sprang up in a fury and singed his moustache; when he assaulted the sausages the tin counter-attacked and cut his finger; the water refused to boil, and while he was not looking tipped itself over into the fire. 'Oh, the malice – the cursed, diabolical malice of inanimate objects!' muttered the Professor ferociously between clenched teeth. 'Here, let me help,' I said. 'You keep away,' he snarled. 'If they want to be bloody-minded, I'll show them, by God I will,' booting the empty sausage-tin into a cactus bush. I knew from experience that it was no good interfering. I picked the tin of cigarettes out of the ration box, sat down on the tailboard and watched him begin to rekindle the fire.

It was half-past six on a June morning in 1943. 'Pinafore', a thirty-cwt truck, was parked in a disused railway cutting sixty miles east of Algiers. The Professor really had occupied a distinguished chair of learning until the outbreak of war, when he had enlisted as an infantryman. Now he was my senior officer, and if he was a singular and in some ways unorthodox lieutenant-colonel, he certainly looked more like a soldier than an absent-minded scholar. He was still in his early thirties, stockily built, with a pale face and brown hair *en brosse*. His eyes were greenish, very penetrating and rather sinister; they indicated something of their owner's intellectual brilliance and force of character.

The Professor's life had been crammed with scholarly achievement; but he had never been a Boy Scout, and during the war he had been too busy for sixteen hours a day creating confusion in the ranks of the enemy to have time to learn how to look after himself in the open. He had never cooked his own food. Another consideration of some importance was that he could not drive. He possessed a certificate, carefully tucked away in his pocket-book, stating that he was authorized to drive any Government vehicle; but I gathered that this had been presented to him when he had been a cadet by an over-optimistic Sergeant Instructor before his first and only driving lesson.

The Professor recounted all these shortcomings to me in detail, and it may well be wondered why I consented to make the journey with him at all, let alone agree to teach him to drive on the way. I felt sure that the Professor was minimizing his abilities. Moreover, although I did not really profess to be a hardened campaigner or knowledgeable mechanic, I rather enjoyed the prospect of looking after the two of us and 'Pinafore'. The Professor insisted on preparing breakfast that first morning, and it made a bad start. The sausages were cold and flabby, tea-leaves floated on top of a grey, tepid liquid which I tactfully consumed with feigned relish. But the Professor was not deceived and went about shaking his head muttering 'Bloody inefficient! bloody inefficient!' too angry to eat. If he had a failing it was an overbearing intolerance of stupidity and inefficiency. People less acute and less energetic than himself, that is practically every other human being with whom he came in contact, were very liable to incite his wrath. I myself had of course already been the cause of many outbursts, and I had witnessed more than one explosive scene when the offenders had been very senior officers indeed, and had not taken kindly to his blunt exposure of their brainlessness.

But this was something new. I had never before seen the Professor really angry with himself. I had often almost exasperated him, but he had been very patient and had taught me a great deal. I wondered just how far our roles were now to be reversed. By half-past seven we were ready to start, and the Professor decided to take his first spell at the wheel. I do not remember feeling particularly apprehensive as we jerked and jolted off towards the main road. I had taught other people to drive and was quite confident in my ability to teach him. The main difficulty, according to the Professor, lay in the steering. His diagnosis was at least partly correct, as I discovered when we had to turn back on to the main road. Instead of slowing down he suddenly accelerated, at the same time swaying about in his seat as though wrestling for possession of the wheel. We turned neither to right nor left, but shot straight on towards a stone wall on the far side. We stopped with a lurch a few inches short, and I found that I had subconsciously pulled the handbrake hard on.

'You see what I mean?' asked the Professor, quite unperturbed.

'Yes, I see,' I replied, wiping the sweat off my hands. 'Now, to turn her it's no good just shifting your weight about in the seat; you must take a grip of the steering-wheel and turn it like this.'

'I'll manage it,' muttered the Professor with the most ferocious look of resolution. Next time he certainly did manage it. We turned a good deal more than the necessary right angle and narrowly missed the ditch on the wrong side of the road. 'Done it!' beamed the Professor as we swerved back into the centre of the road. The road to Bougie forks short of Tizi Ouzou, and at the fork was a large notice stating that the coast road was 'closed to W.D. transport'. The Professor chose it accordingly. We soon realized the reason for the notice. The road was narrow and began to mount and wind in tortuous convolutions through the hills. 'Good practice for steering,' he said, crashing his gears as he negotiated a steep hairpin from which we looked down into the plain, hundreds of feet below. As a matter of fact his steering improved remarkably rapidly, and I had just told him how pleased I was with his progress when quite suddenly and unexpectedly 'Pinafore' slewed hard to port and was only prevented from slipping over the edge of a precipice by hitting the end of a stone bridge. 'How did that happen?' asked the Professor innocently as we inspected the damage. We were relieved to find that we had only stove in an iron bracket and crumpled the oil and water cans held in place by it. They were now oozing their contents into the dust. The Professor seemed to take the accident very much as a matter of course.

We soon found our flow of random conversation running dry, and might sit for two or three hours in dead silence. This was boring and could be dangerous; in the heat of the driving-cab, lulled by the desolate monotony of the road and the steady note of the engine, it was all too easy for the driver to fall asleep. Some weeks previously I had implied in the course of conversation with the Professor that Xerxes had been the opponent of Alexander. The Professor had started as though jabbed deeply with a needle and had glared at me as if I had caused him some personal injury. I now proposed that the long hours of driving should be employed in his remedying my ignorance. The Professor consented on condition that I in turn should improve his knowledge. My difficulty was to choose a subject on which I was better informed than he was. The Professor's range of knowledge was mortifyingly wide, as I discovered when I made a blundering attempt with French painting. But with his usual obstinacy he made it plain that unless I talked he would not. I cast about desperately, until the narrow scope of my own experience, and the discovery that the Professor had never ridden horseback, resulted in agreement to my talking discursively on horses and foxhunting.

I was treated to brilliant impromptu lectures on Greek and Roman history, art and literature. We anticipated the chronological sequence of events so that the Punic Wars could be described as we passed through the territory of Carthage. Cato died as we traversed the battle-field of Utica, and as we bowled along the Via Balbia I received a truly peripatetic introduction to the philosophy of Aristotle. It seemed ludi-crous at first to intersperse the story of the Odyssey with episodes from 'Mr Sponge's Sporting Tour', or to follow the Professor's discourse on Roman military strategy with a description of what I knew about the art of working a pack of foxhounds. But my diffidence at trying to deliver a 'sporting lector' was soon dispelled by the Professor's unex-pectedly eager interest. We each became so engrossed in the other's subject that in the evenings after supper the Professor would illustrate verse forms and teach me the Greek alphabet, while I would draw for him bits of harness, the points of a horse, and other intricacies.

Of the various characters I introduced into my talks Squire 'Mad Jack' Mytton was one of his favourites. I wondered why this should be so, and it occurred to me that perhaps the Professor recognized in Mytton's mentality a kindred streak of what the euphemists call eccen-tricity. An anecdote which particularly amused him was that of Mytton driving an acquaintance in a gig. The acquaintance nervously suggested that they should slacken speed lest they should overturn, an unpleas-ant experience which he had so far been spared. 'What!' cried Mytton, 'never upset in a gig?' and drove straight into the ditch.

We set off on the road to Tunis with the Professor at the wheel. I remarked rather acidly that the speedometer needle was wavering at over forty m.p.h. 'Just repeat the ingredients of the bran mash again – I must get all this quite clear,' replied the Professor with his foot hard down on the accelerator. At that moment I saw two hundred yards ahead a barrier across the road and a military policeman waving us down a diversion. Forty yards from the barrier the Professor had still not slackened speed. I yelled. The policeman began to run. The Professor suddenly wrenched the wheel over and we careered down the side road into a three-foot ditch. There was a moment's silence while we sprawled at an angle of forty-five degrees and collected our wits. Then the Professor began to roar with laughter; 'Ha-ha-ha!' he gasped, 'never been upset in a gig?'

MICHAEL STRACHAN

271.

The Japanese treatment of their prisoners in World War II remains a blot upon their nation's history. Here one of the victims, Russell Braddon (1921–95), an Australian who became a best-selling author, describes a moment in his own experience, from his memoir The Naked Island.

Outside Kanu, in the small stream that trickled down from the mountain above, lay a naked man. When asked why he lay there, he pointed to his legs. Tiny fish nibbled at the rotten flesh round the edges of his ulcers. Then he pointed inside the camp. Other ulcer-sufferers, reluctant to submit to this nibbling process, wore the only dressings available – a strip of canvas torn from a tent soaked in Eusol. Their ulcers ran the whole length of their shin bones in channels of putrescence. Looking back at the man in the stream it was impossible to decide, even though he was insane, which treatment was the best.

We reached the next camp, and found it practically deserted – almost all the original inhabitants were dead. There were proud signs of the struggle for survival those men had put up. Carefully constructed latrines, spotless surrounds, an overhead pipeline made from bamboos which brought cholera-free water from its source two or three miles away and hundreds of feet up at the top of the mountain. This pipeline led to a shower-centre with a bamboo floor and separate cubicles (pathetic symbol of man's desire for even a little privacy) and to a cookhouse that was all clean wood and carefully-swept packed earth. All of these were refinements installed after gruelling sixteen-hour shifts of work at the expense of sleep.

But none of this had been enough – flies carried the cholera germs and mosquitoes the malarial parasites. Starvation and slavery did the rest. Now, as we set out back to our own camp, there were only a few skeleton-like travesties of humanity left and the big fire where they burnt their dead. These fires flared at every camp where cholera struck. They lighted the way out to work in the dark before dawn: they guided the men back through the dark wetness of the jungle long after dusk. And, always, lying round them in stick-like-bundles, were the bodies that awaited cremation – bodies at which the returning men peered closely as they came in, to see if any of their mates lay among them. And every now and then, as they filed past, came that muttered: 'Half his luck.'

At one camp the task of attending to the pyre and of consigning the bodies to the flames was given to an Australian who, being without brains or emotions or finer susceptibilities, was more than happy in his work. He stripped the dead of their gold tooth caps; he stole fearlessly from the guards who dared not touch him lest he contaminate them: he cooked what he stole – for one only stole food, or something that could be bartered for food or tobacco – on the fire where he burnt the bodies. It was his practice before dealing with the fresh batch of bodies that arrived each morning to boil himself a 'cuppa char'. He liked watching the working parties fall-in because he stayed at home by his fire where, even in the monsoonal rains, he could keep warm and do his cooking. Upon one particular morning he sipped his tea out of the jam tin that served as a mug and watched the parade. As he watched he rolled some tobacco in a strip of the tissue that clings to the inside of a bamboo: then, his fag completed, he picked up a body and tossed it easily from yards off (for it was only light) on to the fire. He enjoyed the revulsion this caused. He did it every morning just before the workers marched out. Grinning at them as they glowered angrily, he then shambled to the fringe of the fire to light his cigarette.

As he leant forward to pick up a faggot the body he had just tossed into the flames, its sinews contracted, suddenly sat bolt upright and grunted, and in its hand thrust out a flaming brand on to the cigarette in the moron's mouth. With a scream of terror the man who had burnt hundreds of bodies with callous indifference fell backwards, his hands over his eyes. When the workers reached him he was jabbering and mad. They took him to the hut that housed the sick, an attap roof draped over a patch of mud in which – all over one another – lay hundreds of men. For days he lay there silent, knowing nothing. Then one night he suddenly remembered and screamed, screamed piercingly and long so that, even though it was forbidden, the medical orderly lit a resin flare and rushed down to where he lay to see why he screamed.

RUSSELL BRADDON

272.

*An example of unchivalrous conduct to animals in the field, recorded
by David Smiley (1916–2009), who served as an SOE officer with
the Albanian partisans before becoming a luminary of the Secret
Service. Hon. Alan Hare (1919–95), the villain of this piece,
afterwards became chairman of the* Financial Times *and of Château
Latour.*

I bought a riding mule for myself costing five sovereigns, and I chris-
tened her Fanny (we always referred to our mentor in Cairo as Fanny
Hasluck). Fanny had a very sweet nature and I became devoted to her;
she carried me everywhere for six months of my stay in Albania. On
leaving the country I handed her over to Alan Hare of the Life Guards.
While on leave in England I sent a signal asking news of Fanny. It was
a very severe winter, and our mission at the time was not only on the
run from the Germans, but was very short of food. Even so, Hare
could not have been a true cavalryman, as his reply was short and to
the point: 'Have eaten Fanny.'

DAVID SMILEY

273.

*In Sicily in the late summer of 1943, having established a reputation
as a driving and inspirational commander, George Patton committed
a series of acts which resulted in his dismissal from field command,
and would have terminated his career but for the support of
Eisenhower, who secured his reinstatement in time for Patton to lead
the US Third Army in North-West Europe.*

Patton was told that an increasing number of men had gone to the
hospital with nothing but combat fatigue, a form of neurosis for which
he had neither understanding nor sympathy. The departure of these
men from the tough battle for Troina was felt seriously by the hard-
pressed regiments. On this 3rd August Patton spied signs to the 15th
Evacuation Hospital. He told Sergeant George Mims, his driver, 'Take
me to that Evac,' to see for himself how crowded it was with

combat-neurosis cases. The grand round cheered Patton, because the men appeared to be legitimate casualties. He was about to leave the tent when his eyes fell on a boy in his mid-twenties who was squatting on a box with no bandage on him to indicate that he had been wounded. He was Private Charles Herman Kuhl of Mishawaka, Indiana, a bright-faced, good-looking young soldier eyeing the General with what Patton thought was a truculent look.

'I just get sick inside myself,' Patton later told his friend, Henry J. Taylor, a millionaire businessman who doubled as a war correspondent in Sicily, 'when I see a fellow torn apart, and some of the wounded were in terrible, ghastly shape. Then I came to this man and asked him what was the matter.

'The soldier replied: "I guess I can't take it."

'Looking at the others in the tent, so many of them badly beaten up, I simply flew off the handle.'

What happened next was described the day after by Kuhl himself in a letter to his father. 'General Patton slapped my face yesterday,' he wrote, 'and kicked me in the pants and cussed me. This probably won't get through, but I don't know. Just forget about it in your letter.' When Patton was through, he turned to Colonel Wasden.

'Don't admit this sonovabitch,' he yelled. 'I don't want yellow-bellied bastards like him hiding their lousy cowardice around here, stinking up this place of honour.' Still shouting at the top of his high-pitched voice, he ordered: 'Check up on this man, Colonel. And I don't give a damn whether he can take it or not! You send him back to his unit at once – you hear me, you gutless bastard,' he was now shrieking at Kuhl again, 'you're going back to the front, at once!' Kuhl was picked up by a group of corpsmen attracted to the scene by the noise. They took him to a ward, where he was found to have a temperature of 102.2°. It also developed that he had been suffering from chronic diarrhoea. A blood test then showed that he had malaria. Neither the medical staff, nor Kuhl nor his folks, followed up the incident, and that seemed to close the case.

On 10th August, he saw signs of the 93rd Evacuation Hospital. He walked unannounced to the admission tent, where Colonel Donald E. Currier caught up with him. Patton greeted Currier amicably. The surgeon was from Boston and was a friend of his family. What followed was described most graphically by Major Etter in the report he prepared for 'Surgeon, II Corps':

On Monday afternoon August 10, 1943, at approximately 1330, General Patton entered the Receiving Ward of the 93rd Evacuation Hospital and started interviewing and visiting the patients who were there. There were some ten or fifteen casualties in the tent at the time. The first five or six that he talked to were battle casualties. He asked each what his trouble was, commended them for their excellent fighting; told them they were doing a good job, and wished them a speedy recovery.

He came to one patient who, upon inquiry, stated that he was sick with high fever. The General dismissed him without comment. The next patient was sitting huddled up and shivering. When asked what his trouble was, the man replied, 'It's my nerves,' and began to sob. The General then screamed at him, 'What did you say?' He replied, 'It's my nerves. I can't stand the shelling any more.' He was still sobbing.

The General then yelled at him, 'Your nerves Hell, you are just a Goddamn coward, you yellow son of a bitch.' He then slapped the man and said, 'Shut up that goddamned crying. I won't have these brave men here who have been shot seeing a yellow bastard sitting here crying.' He then struck at the man again, knocking his helmet liner off and into the next tent. He then turned to the Receiving Officer and yelled, 'Don't you admit this yellow bastard, there's nothing the matter with him. I won't have the hospitals cluttered up with these sons of bitches who haven't the guts to fight.'

He turned to the man again, who was managing to 'sit at attention' though shaking all over, and said, 'You're going back to the front lines and you may get shot and killed, but you're going to fight. If you don't, I'll stand you up against a wall and have a firing squad kill you on purpose. In fact,' he said, reaching for his pistol, 'I ought to shoot you myself, you Goddamned whimpering coward.' As he went out of the ward he was still yelling back at the Receiving Officer to 'send that yellow son of a bitch to the front lines'. Colonel Currier was beside himself. He rushed back to his office and put a call to Colonel Arnest,

Surgeon of II Corps. 'Dick,' he told Arnest, 'this is Currier, 93rd Evac. You better come over here as fast as you can.'

LADISLAS FARAGO

274.

The perils of speaking truth to power.

In the summer of 1943, in Sicily fifty thousand Germans held half a million Allied soldiers at bay for five weeks. The invaders repeatedly failed to translate capture of ground into destruction of enemy forces. The German high command was baffled when most of the defenders and much of their heavy equipment and weapons were allowed to make good their escape across the Straits of Messina. Having expected a landing in Calabria to cut off their retreat, some of Hitler's officers cherished a fantasy, that for political reasons the allies had acquiesced in their withdrawal.

Lt. Col. Lionel Wigram [1907–44] was one of the most brilliant officers in the British Army. The son of a Sheffield tycoon, before the war he earned £30,000 a year as a Mayfair solicitor, while finding time also to serve as a Territorial Army officer. In England in 1941–42 he was among the originators of Battle Schools, to accustom troops to the realities of facing fire. After commanding a battalion in Sicily, he wrote a coruscating report on lessons that should be learned from what he considered a poor performance by the British Army. He criticized frontal attacks, over-dependence on artillery, refusal to exploit infiltration. He urged that every unit should be relieved of twenty-odd men, who in action invariably ran away. He concluded: 'the Germans have undoubtedly in one way scored a decided success in SICILY. They have been able to evacuate their forces almost intact, having suffered very few casualties. We all feel rather irritated as a result.'

This recklessly frank assessment reached the desk of the victorious Montgomery. His vanity pricked, he sacked Wigram from battalion command, denouncing him as 'not a proper chap'. Wigram was devastated by his demotion, from which at variance with all his experience as a lawyer, he had no recourse to appeal. He was killed in February 1944, aged thirty-seven, leading a night attack by Italian partisans on a German-held village near Civitella. He was described in an obituary as 'one of Britain's greatest amateur soldiers', however shabbily he was treated by the general convinced that he was Britain's greatest professional one.

DENIS FORMAN AND MAX HASTINGS

275.

A Coldstream subaltern's first experience of battle, at Salerno in September 1943.

Our objective was a small hill – the 'Pimple'. Securing it would straighten our own lines and give us a view over the valley below. Little opposition was to be expected, said Alan [the company commander] in an alarmingly deadpan voice. I would be lying if I pretended that I looked forward to it with anything except rather sick apprehension. In practice the whole affair proved wonderfully illustrative of Clausewitz's 'friction of war'. First we had to find our way to the starting-point, which meant leaving our scattered slit-trenches after dark, under spasmodic mortar-fire, shaking out into single file, and moving in the correct order over steep mountain paths. That took far longer than expected. The files lost one another, the platoons somehow got into the wrong order.

We eventually arrived at the start-line long after H-Hour, almost too late to catch up with the artillery barrage. Two out of my three sections had disappeared altogether, and turned up only as the attack began. From the dark hill facing us, streams of tracer bullets were already zipping over the low wall behind which we eventually formed up. Once again I had an absurd sense of déjà vu. This was just another B movie, and I was playing the David Niven role as the gallant platoon commander. All right, I thought, if I was cast as David Niven I had better behave like David Niven; so I hissed 'Right – over with me!'

Everything thereafter became so confused that it is hard to make any coherent narrative out of it. We stumbled down the slope, dodging under the fixed lines of enemy fire, and began to climb the opposite hill. The first obstacle was a terrace-wall about six feet high, up which the faithful Johanson pushed me. There were flashes and ear-splitting detonations as the Germans lobbed down grenades. Fire was coming from a dark patch of trees in front, into which we plunged, firing blindly. There were only four or five men still with me and I roared abusively, summoning the others. We pushed on, shouting like madmen and shooting at the dim figures we saw scuttling away ahead. By the time we reached the summit, the hill seemed clear.

My training clicked in: as the rest of the platoon came up, I disposed them in good positions of all-round defence, our fire-power considerably increased by the capture of half-a-dozen Spandaus and a good quantity of grenades. About four of my men had been wounded, one of them, alas, my precious Johanson. More had somehow, by accident or intent, 'got lost' – this happens quite often in battle – but turned up in time to share the triumph. Alan came chugging up the hill understandably delighted, and spoke kind words. I had done all right.

MICHAEL HOWARD

276.

Howard experienced a much less happy outcome some months later, on a night patrol in no man's land, accompanied by a single Guardsman.

This was fear – the sudden stop of the rhythm of breath and heartbeat, followed by agonized butterflies in the breast instead of lungs. I stopped. The voices stopped. Then came the challenge '*Halt! Wer da?*' We got down, and all was still. After a little we began cautiously to crawl away; then cautiously stood up and began to walk. We had gone only a few steps before I felt a stinging blow in the back of my legs and heard again that villainous little explosion – this time just behind me. 'Are you all right, Terry?' I whispered. 'No, sir – it's got my foot.' At once the German post opened fire. Fascinated, I watched the orange spurt of the machine-gun 15 yards ahead apparently firing straight at us. Pressed close to the ground, I heard the bullets swish overhead, firing as usual too high. Poor Terry began to scream in fear and pain, but the only answer was a flurry of grenades.

This is the end, I thought. No good firing back; they are well dug in. I am in the open and in the middle of a minefield. I can't get Terry away – he is almost twice my size. Seriously I thought of surrendering, but that would have been stupid. This is the hardest part to write. Deliberately, and fully aware of what I was doing, I left Terry and crawled away. The Germans were only yards away. I told myself; they would find him at daybreak and bring him in. I shouted at them that there was a British soldier here, *schwerverwundert*, but the only answer was a flurry of grenades. Meanwhile, rifles had joined the machine-gun fire. I found that I had myself been lightly wounded in the legs by Terry's mine, and could only move with difficulty. Terrified of more mines, I crawled on all fours, feeling ahead among the tufted grass as I went. The mist was thick, and I had now lost all sense of direction. Just as I thought it safe enough to stand upright, the firing began again, this time from half a dozen directions. Evidently the German company commander had called down defensive fire, and now machine-guns were sweeping the ground where I lay. I thought of that warm, well-lit room at battalion headquarters with its blazing fire. It seemed the summit of all earthly desire. Pressed into a tiny hollow as the machine-guns rattled round me, I wondered whether I would ever see it again.

Eventually the firing stopped, and I crawled cautiously on. Forcing my way through briars and brambles, I found the right track and stumbled back as quickly as I could. My mind was not so much a turmoil as a series of quite distinct layers of feeling: a layer of relief, a layer of shame, a layer of anxiety and a layer almost of amusement at the absurd figure I was cutting – dirty, bramble-torn, distraught and peppered in the hams. I had learned a great deal – too much – about myself; not least that I did not begin to deserve a Military Cross [which he had been awarded at Salerno, almost a year earlier]. It is easy enough to be brave when the spotlight is on you and there is an appreciative audience. It is when you are alone that the real test comes. Everyone at battalion headquarters was immensely kind. I offered rather unconvincingly to take a party back to find Terry, an offer which Colonel Billy Steele very sensibly refused. I was sent back for another brief spell in hospital. And Terry? Poor lad, he did not survive. Whether he bled to death before the Germans found him, or died in their care, I do not know. Years later I sought out his grave, and sat beside it for a long time, wondering what else I could have done. I still wonder.

MICHAEL HOWARD

277.

High on an Italian hillside south-west of Florence in 1943, travelling
with the BBC correspondent Wynford Vaughan Thomas (1908–87),
Eric Linklater, who was serving as an official chronicler of the
campaign, encountered an astounding oasis of culture.

It was a union of little castle and large villa, an ancient tower rising
among cypresses on a small hillside above the plain high walls of a
sixteenth-century building. It was, we discovered, the Castello di
Montegufoni, the property of Sir Osbert Sitwell. In recent years incon-
gruities have been as common as violent death, but, unless the mind
has been numbed by too much exposure, the latter still dismays and
the former continues to excite a curious pleasure. If it was engaging to
find the Mahratta Light Infantry in residence in a Tuscan castle, it was
delightful to learn that it belonged to Sir Osbert: the Indian soldiers
looked like new images, domiciled as urbanely as their many divers
predecessors, in the Sitwells' eclectic hospitality; and we may have
been fractionally prepared for the greater, the superb, the enchanting
surprise that awaited us within the walls.

Idly we looked into a courtyard, and within a minor entrance to the
house discovered to our surprise three or four pictures propped against
a wall. Elderly dark paint on wooden panels, and some tarnished gold:
a Virgin, the Child and the Virgin, a painted Crucifix. The yellow faces
were drawn with the severe and melancholy stare of the earliest Flor-
entine painting. We sat on our heels to examine them more closely.
One of us said with astonishment, 'But they're very good!'

'They must be copies,' said another. I answered like an auctioneer,
with the conviction of faith rather than of knowledge: 'Genuine Italian
primitives!'

We went into a room where many more pictures were stacked
against the walls, some in wooden cases, some in brown paper, and
others naked in their frames. Two or three of those that were exposed
to view aroused in us the dishonest pretence of recognition so common
in visitors to an art gallery. 'Why,' we said, 'surely that's by So-and-so.
Not Lippo Lippi, of course, but – oh, what's his name?' We were not

the only occupants of the room, however. Half a dozen soldiers were rummaging in a large desk which seemed to have been roughly opened. It would be altogether too harsh to accuse them of looting, they were only looking for small souvenirs. But they may have been careless, for there was broken glass on the floor and books which had been tidily stacked were scattered about in some confusion. I looked at some neatly tied bundles of yellowish paper, and saw that they were legal documents of Sir George Sitwell's time. I found a copy of *Before the Bombardment* inscribed by Osbert to his mother, and some invitation cards which announced that Lady Ida Sitwell would be At Home on such-and-such a date, when there would be Dancing.

Presuming on the slightest sort of acquaintanceship – I had met Osbert once or twice – I said to the soldiers, very mildly, 'I don't think you should take anything from here. I know the owner of the house.' A genial well-fed sergeant at once replied, 'Oh well, sir, that makes all the difference, doesn't it? If we'd known that, we wouldn't have touched a thing. Not a thing,' he repeated, as though shocked by the very idea. Vaughan Thomas, his rosy face tense with excitement, reappeared. 'The whole house is full of pictures,' he exclaimed, 'and some of the cases are labelled. They've come from the Uffizi and the Pitti Palace!'

Hastily I followed him into the next room, where a score of wooden cases stood against the walls, and then to the room beyond. There a very large picture lay upon trestles. It was spattered with little squares of semi-transparent paper, stuck for protection over imperilled areas where the paint was cracking or threatening to flake. On the near side there were cherubs, or angel-young, with delicate full lips, firm chins and candid eyes wide open over well-defined cheekbones. Against a pale blue sky the Virgin floated in splendour. Two reverent, benign and bearded figures held a crown above her head.

We failed to recognize it. We knew now that we were in the presence of greatness, and a bewildered excitement was rising in our minds. Recognition could not yet speak plainly, but baffled by the vast improbability merely stammered. Stupidly we exclaimed, 'But that must be …'

'Of course, and yet …'

'Do you think it is?'

By this time we had gathered a few spectators. Some refugees had been sleeping in the castello – their dark bedding lay on the floor – and now, cheerfully perceiving our excitement, they were making sounds of lively approval, and a couple of men began noisily to open the shutters that darkened the last of the suite of rooms. This was a great chamber that might have served for a banquet or a ball, and as we went in the light swept superbly over a scene of battle: over the magnificent rotundities of heroic war-horses, knight tumbling knight with point of lance – and beside it, immensely tall, an austere and tragic Madonna in dark raiment upon gold.

Vaughan Thomas shouted, 'Uccello!'

I, in the same instant, cried, 'Giotto!'

For a moment we stood there, quite still, held in the double grip of amazement and delight. Giotto's Madonna and Uccello's Battle of San Romano, leaning negligently against the wall, were now like exiled royalty on the common level. They had been reduced by the circumstance of war from their own place and proper height; and they were a little dusty. We went nearer, and the refugees came round us and proudly exclaimed, '*E vero, è vero! Uccello! Giotto! Molto bello, molto antico!*'

Now Vaughan Thomas is a Welshman, more volatile than I, quicker off the mark, swifter in movement, and while I remained in a pleasant stupefaction before the gaunt Virgin and the broad-bottomed cavalry, he was off in search of other treasures. A helpful Italian took down the shutters from the far end, and let in more light. Then I heard a sudden clamour of voices, a yell of shrill delight, and Vaughan Thomas shouting 'Botticelli!' as if he were a fox-hunter view-halloing on a hill. I ran to see what they had found, and came to a halt before the Primavera. I do not believe that stout Cortez, when he first saw the Pacific, stood silent on a peak in Darien. I believe he shouted in wordless joy, and his men with waving arms made about him a chorus of babbling congratulation. We, before the Primavera, were certainly not mute, and the refugees – some had been sleeping side by side with Botticelli – were as loudly vocal as ourselves. They had a fine sense of occasion, and our own feeling that this was a moment in history was vigorously supported by the applause they gave to our exclamation and delight.

Commanding officers who have lately been engaged in battle and are roused from their entitled sleep are sometimes difficult; but fortunately for us Colonel Leeming of the Mahrattas was a good-humoured man. He listened politely, then with growing attention to what I told him. He knew there were some pictures in the house, but he had had no time to look at them, he said, and he had supposed they belonged to the family. The castello was the property of the Sitwells, who were artistic people, weren't they?

To describe the wealth of treasure that lay below him, I used all the superlatives I could put my tongue to – and still the Colonel listened, unprotesting. To the north we could hear the noise of war, and so much concern for a few yards of paint may have seemed excessive to him, whose care was men; but he was very patient. He admitted that he knew little about art, and wistfully added that if his wife were there she would be more impressed. She took a great interest in pictures, he said. He came down to look at the Primavera. He stood silent for some time, and still without comment walked slowly past the other pictures, into the adjoining rooms and back again, as though he were making his rounds of a Sunday morning after church parade. He was evidently pleased with what he saw, and now permitted himself – with a decent restraint – to be infected by our enthusiasm. He would do everything in his power to keep the pictures safe, he promised.

Several other officers had appeared, and to one of them he said: 'Have all these rooms put out of bounds, and get a guard mounted. You'll have to find somewhere else for the refugees to sleep; there's plenty of room in the place.' We explained to the [guardian of the castello] that his pictures were now under official protection – Mahratta bayonets would guard them night and day, we told him – and at once he grew boisterously happy, and danced about thanking everyone in turn. In the morning we returned to Montegufoni, and found dark sentries, grave of feature and dignified in their bearing, outside the doors. The Colonel's orders had been strictly obeyed, and the rooms had now the untenanted peace of a museum on a fine morning. We opened the shutters, and with more leisure made further discoveries. Many pictures that we had scarcely noticed in our first excitement now appeared like distinguished guests at a party, obscured by numbers to begin with, who, when at last you meet them, are so dignified or decorative that it seems impossible they could have remained unrecognized even though their backs were turned and a multitude surrounded them. Lippo Lippis came forward smiling, a Bronzino was heartily acknowledged, Andrea del Sartos met our eyes and were more coolly received.

Beside the Giotto Madonna stood a huge equestrian portrait of Philip IV of Spain – by Rubens or Velasquez? I do not know which – and peering round Philip's shoulder, absurdly coy, was the stern and antique countenance of another great Virgin. In a room on the other side of the courtyard, that we had not visited before, we found Duccio's Sienese Madonna, the Rucellai Madonna. Here also were many altarpieces, triptychs in lavish gold, and painted crucifixes of great rarity in long-darkened colours with mouths down-drawn in Byzantine pain.

Then, privily, I returned to the great room and Botticelli's Primavera. I was alone with his enchanting ladies, and standing tiptoe I was tall enough, I kissed the pregnant Venus, the Flowery Girl, and the loveliest of the Graces: her on the right. I was tempted to salute them all, but feared to be caught in vulgar promiscuity. Some day, I said, I shall see you again, aloft and remote on your proper wall in the Uffizi, and while with a decently hidden condescension I listen to the remarks of my fellow-tourists, I shall regard you with a certain intimacy: with a lonely, proud and wistful memory.

ERIC LINKLATER

278.

John Verney was captured in Sardinia while serving with the SAS. After escaping from an Italian PoW train he walked south down the Apennines, until he at last approached the Allied lines.

I reached the hilltop as dawn broke, but the morning mist was too thick to see anything. I longed to be challenged by a British sentry, but there were none about. Descending the hill, I met three Italians on donkeys. 'Are there any Germans round here?' I asked them in Italian. Suspiciously, they asked if I was German. Hearing I was English, they embraced me and told me to keep going for half a mile when I would reach their village, Montenero, and find English soldiers everywhere. *'Niente tedeschi, niente tedeschi!'* they laughed. The sun broke through the morning mist as I came in sight of the village. Approaching it, I deliberately dawdled and finished the last mouthful of Italian bully beef. 'Well, here you are,' I said aloud. 'I suppose you'll remember the next minutes all your life.' And I was quite right.

From a hundred yards away the much-battered buildings appeared deserted. Then I noticed a Bren-carrier behind a wall, a few trucks under camouflage netting in a yard. As I limped slowly into the main street, a solitary shell whistled overhead, exploding somewhere at the farther end. A group of British soldiers, the first I had seen, in long leather waistcoats and khaki cap-comforters, chatted unconcernedly in a doorway. I glanced shyly at them, but they took no notice. Just another bedraggled peasant haunting the ruins of his home. The conventional inhibited Englishman is ill-equipped by temperament for such occasions. I would have liked to dance, to shout, to make some kind of demonstration. A Frenchman, an Australian, would have done it naturally, but somehow I could not. So I walked slowly on, holding off the pleasure of the long-awaited moment, the exclamation of surprise, the greeting from a compatriot. English voices and laughter came from a house. I crossed the threshold and found, in what had been the peasants' kitchen but was now the usual military desolation of a billet, two half-dressed privates cooking breakfast. One of them saw me, standing there grinning at them. 'Christ, Nobby, look what the cat's brought in,' he said. I tried to be hearty but failed. 'I've just come through. I'm soaked. Can I warm up by your fire?' I heard my bored voice say politely.

Neither of the soldiers was much surprised by this sudden entry of what was, to all appearances, a dank and bearded Italian, who spoke fluent English with a B.B.C. accent. Sensibly enough, they were more interested in breakfast.

'What are you? Escaped P.O.W. or something?'

'Yes.' It didn't cause much of a sensation and they asked no more questions. They treated me, as they might have a stray dog, with a sort of cheery kindness and without fuss. In a few minutes I was sitting naked before their fire, drinking a cup of char and feeling the

warmth return to my numbed body. A corporal and others of the section came in for their breakfast. My back view may have mildly surprised them.

'Bloke's an escaped P.O.W.,' Nobby explained.

'Lucky sod. They'll send you home,' the corporal said, offering me a Players. Long-anticipated pleasures seldom come up to expectation. Neither the tea, the tobacco nor even the warmth were now quite as delicious as I had imagined they would be. The scene was unreal, unbelievable. Though the soldiers' gossip around me was vivid enough.

'Ginger, go and swipe some clothes for him off the C.Q.M.S.'s truck,' the corporal said to one of the men.

'Do you think the C.Q.M.S. can spare me something?'

'The C.Q.M.S. won't know,' Ginger winked as he left the room.

Certainly I was back with the British Army all right.

Later, in the miscellaneous garments swiped by Ginger (the C.Q.M.S. would, I am sure, have supplied them voluntarily – but that, of course, would have been more trouble), I visited R.H.Q. in another part of the village. The C.O. refused at first to credit my story that I had walked through his outposts unobserved, until I traced my route for him on the map. Evidently I had indeed walked slap through C Company's position. 'Can't think why they didn't shoot you,' the colonel said irritably. 'Remind me to have a word with the company commander about that,' he added to his adjutant. He seemed to treat my personal survival as a discredit to his blasted Battalion. He was about my age, had a pink face and a toothbrush moustache and the manners of a bilious scoutmaster. We disliked one another on sight. I ventured a facetious joke. I hadn't made one for years. 'It was only eight o'clock. Perhaps they were all still asleep.'

'That is not very probable.'

We had little in common and I was glad when he told the Adjutant to take me over. 'Be a good chap,' the latter whispered as we parted, 'and don't mention that you walked through our lines unopposed.'

'Of course I won't,' I smiled. Nor have I, till this day.

279.

Alsatian Guy Mouminoux (b. 1927) was author of a notable 1965 work, The Forgotten Soldier, *in which he described under the pen name Guy Sajer lightly fictionalised experiences as a private soldier in Hitler's elite Gross Deutschland division, which he joined at the age of sixteen. In October 1943, when he and other stragglers were remustered after surviving the desperate retreat across the Dneiper, they looked yearningly to a long-promised leave.
And were disappointed.*

There were two noncoms and a lieutenant in our group. We went into the building, which had its own generator and was brightly lit. Our state of extreme filth suddenly made us feel awkward. Military men of all ranks and military police were sitting facing us behind a row of long tables. An *obergefreite* came up to us, yelling as in the old days at training camp. He told us to get over to the tables to be screened. We should be ready to produce on demand the papers and equipment entrusted to us by the army. This reception only increased our sense of astonished unease.

'First, your documents,' an M.P. shouted across the table.

The lieutenant, who was directly ahead of me in the line, was being interrogated.

'Where is your unit, lieutenant?'

'Annihilated, *Herr Gendarme*. Missing or dead. We had a hard time.'

The M.P. said nothing to this, but went on leafing through the lieutenant's papers.

'Did you leave your men, or were they killed?'

The lieutenant hesitated for a moment. We were all watching in frozen silence.

'Is this a court-martial?' The lieutenant's voice was exasperated.

'You must answer my questions, *Herr Leutnant*. Where is your unit?'

The lieutenant clearly felt caught in a trap, as did we all. Very few of us could have answered that question with any precision. He tried to explain. But there is never any point in explaining to an M.P.: their powers of comprehension are always limited to the form they wish to fill. Further, it appeared that the lieutenant was missing a great many things. This fact obsessed his interrogator. It didn't matter that the man in front of him was effecting a miracle simply by staying on his feet, and had lost at least thirty pounds since entering the army. The M.P. only noted that the Zeiss fieldglasses, which are part of an officer's equipment, were missing. Also missing were a mapcase and the section telephone, for which the lieutenant was responsible.

In fact, the lieutenant, who had managed to save only his life, was missing far too many things. The army did not distribute its papers and equipment only to have them scattered and lost. A German soldier is expected to die rather than indulge in carelessness with army property. The careless lieutenant was assigned to a penal battalion, and three grades were stripped from his rank. At that, he could think himself lucky. The lieutenant's eyes were wild, and he seemed to be fighting for breath. He was a pitiful and terrifying sight. Two soldiers dragged him off to the right, toward a group of broken men, who'd been dealt with in the same way.

Then it was my turn. I felt stiff with fright. I pulled my rumpled documents from an inside pocket. The M.P. rifled through them, throwing me a reproving look. His bad temper seemed to soften somewhat at the sight of my apprehensive, mortified face, and he continued his inventory in silence.

Fortunately, I had been able to reintegrate with my unit and had saved the scrap of white cardboard which stated that I had left the infirmary to take part in an attack. My head was swimming, and I thought I was going to faint. Then the M.P. read off a list of articles which ordinary soldiers like myself were supposed to carry at all times.

The words rolled off his tongue, but I didn't catch them quickly enough, and didn't immediately produce the items still in my possession. The M.P. then treated me to a certain German word, which I was hearing for the first time. It appeared I was missing four items, including that fucking gas mask I had deliberately abandoned. My paybook was passed from hand to hand to be inspected and stamped. In my panic, I made an idiotic move. Hoping to gain favour, I produced nine unused cartridges from my cartridge belt. The M.P.'s eyes lit on these like the eyes of an alpinist who spots a good foothold.

'You were retreating?'

'*Ja, Herr Unteroffizier.*'

'Why didn't you try to defend yourself? Why didn't you fight?' he shouted.

'*Ja, Herr Unteroffizier.*'

'What do you mean – *ja?*'

'We were ordered to retreat, *Herr Unteroffizier.*'

'God damn it to hell!' he roared. 'What kind of an army runs without shooting?'

I held back my tears with difficulty. Finally the M.P.'s right fingers handed back my liberty. I had not been assigned to a penal battalion, but my emotion overwhelmed me anyway. As I picked up my pack, I sobbed convulsively, unable to stop. A fellow beside me was doing the same. The crowd of men still waiting stared at me in astonishment. Like a miserable tramp, I ran past the line of tables and left by a door opposite the one we'd entered by. I felt that I had disgraced myself. I rejoined my comrades, who were standing in the rain in the other part of the camp. They weren't resting on the soft beds we'd dreamed of before coming to this place, and the rain streaming down their shoulders and backs was another hope disappointed.

GUY SAJER

280.

By 1943, this Scottish Yeomanry unit was training for the landing in North-West Europe.

The final fling in Yorkshire was exercise 'Eagle', probably the most realistic ever to be held in the country. One Yeoman remarked that the sooner the invasion began the better, as these exercises were killing him. The realism was due to the dreadful weather, which would have been enough to wash out most operations; to the deliberately short rations; the absence of Canadian umpires; the issue of only one blanket per head – all ranks. It was just before leaving, dressed in full marching order, steel helmets but no greatcoats, which were forbidden, they were finishing a glum, silent breakfast at 4.55 a.m. Outside, it was raw and lashing with rain; inside, tempers were rough and morale at rock bottom. Just as all were gloomily savouring the last warmth they would enjoy for the next fortnight, Gunner H. E. Barnicoat entered. Looking exactly like a stage butler, he had been hall porter at the Cambridge Hotel, Farnborough. On one occasion he was handing a glass of sherry to the Divisional Commander who was visiting the Yeomanry for lunch, when the General remarked that his face was familiar and asked where he had seen him before, to which the bold Barnicoat replied, 'At the Cambridge Hotel, sir, where I helped to put you to bed after the Staff College dinner.' He was now batman to Major the Lord Montgomerie. Standing solemnly to attention he announced: 'Excuse me, my Lord. Your Lordship's tank is at the door.'

W. S. BROWNLIE

281.

A timeless story of a soldier's return home – in this case a Scottish officer and pre-war Daily Express *correspondent in Paris, who had escaped from a German PoW train and returned across France and Spain to England in January 1944. George Millar (1910–2005) was reunited with his wife for the first time in three years, an encounter he describes in the third person.*

'What do you intend to do now?' she asked.

'I want to go back into France, to fight with the Resistance there, but I won't do it if you …'

'There you go again, always the little Boy Scout,' she said in a tired voice. 'You are hopeless, really hopeless.'

He made great efforts to talk smoothly, to reassure her, to let her see that he was nervous and only wanted to rest. Then everything crashed in on him. He was attempting to explain to her something of what he had felt in prison; and then on the previous night; and lastly, when he had sat drinking Tio Pepe in her big room.

'The only thing that would end everything,' he said to her, 'would be if you had fallen in love with somebody else.'

'Perhaps I have.'

She had answered in a hard, pondering voice. He looked into her face. He felt like an oyster, torn from its shell and wobbling ludicrously there in front of the table. He began then to wish that he had not come home.

[In 1945, having been parachuted back into France as an officer of Special Operations Executive, Millar published *Maquis*, one of the first, best, and best-selling accounts of life among the Resistance. He remarried in the following year, and lived happily ever after.]

GEORGE MILLAR

282.

Resistance and special operations gave birth to countless acts of audacity, courage and derring-do, many of scant value to the war effort. Few caught post-war popular imagination as vividly as the 26 April 1944 Ill Met By Moonlight *kidnapping of Major-General Karl Kreipe, who commanded a German infantry division on Crete, by Major Patrick Leigh Fermor (1915–2011), afterwards a celebrated travel writer, and Captain William 'Billy' Stanley Moss (1921–65) of SOE, together with a trusted band of Cretan partisans.* * *Here, Moss takes up the story as the two officers, disguised as German soldiers and accompanied by ten Cretan partisans, reached the T-junction where in darkness they planned to stop the general's car.*

It was eight o'clock. We had met a few pedestrians on the way, none of whom seemed perturbed at seeing our German uniforms, and we had exchanged greetings with them with appropriately Teutonic gruffness. When we reached the road we went straight to our respective posts and took cover. It was now just a question of lying low until we saw the warning torch-flash from Mitso.

There were five false alarms during the first hour of our watch. Two Volkswagens, two lorries, and one motor-cycle combination trundled past at various times, and in each of them, seated primly upright like

---

* Core abduction team of ten partisans was: Ilias Athanasakis, Michail Akoumianakis, Efstratios Saviolakis, Dimitrios Tzatzadakis, Nikolaos Komis, Antonios Zoidakis, Georgios Tyrakis, Emmanouil Paterakis, Grigorios Chnarakis, Antonios Papaleonidas.

tailors' dummies, the steel-helmeted figures of German soldiers were silhouetted against the night sky. It was a strange feeling to be crouching so close to them – almost within arm's reach – while they drove past with no idea that nine pairs of eyes were so fixedly watching.

It was already one hour past the General's routine time for making his return journey when we began to wonder if he could possibly have gone home in one of the vehicles which had already passed by. It was cold, and the canvas of our German garb did not serve to keep out the wind. I remember Paddy's asking me the time. I looked at my watch and saw that the hands were pointing close to half-past nine. And at that moment Mitso's torch blinked.

'Here we go.'

We scrambled out of the ditch. Paddy switched on his red lamp and I held up a traffic signal, and together we stood in the centre of the junction. In a moment – far sooner than we had expected – the powerful headlamps of the General's car swept round the bend and we found ourselves floodlit. The chauffeur, on approaching the corner, slowed down.

Paddy shouted, '*Halt!*'

The car stopped. We walked forward rather slowly, and as we passed the beams of the headlamps we drew our ready-cocked pistols from behind our backs and let fall the life-preservers from our wrists.

As we came level with the doors of the car Paddy asked, '*Ist das das General's Wagen?*'

There came a muffled '*Ja, ja*' from inside.

Then everything happened very quickly. There was a rush from all sides. We tore open our respective doors, and our torches illuminated the interior of the car – the bewildered face of the General, the chauffeur's terrified eyes, the rear seats empty. With his right hand the chauffeur was reaching for his automatic, so I hit him across the head with my cosh. He fell forward, and George, who had come up behind me, heaved him out of the driving-seat and dumped him on the road. I jumped in behind the steering-wheel, and at the same moment saw Paddy and Manoli dragging the General out of the opposite door. The old man was struggling with fury, lashing out with his arms and legs. He obviously thought that he was going to

be killed, and started shouting every curse under the sun at the top
of his voice.

The engine of the car was still ticking over, the handbrake was on,
everything was perfect. To one side, in a pool of torchlight in the
centre of the road, Paddy and Manoli were trying to quieten the
General, who was still cursing and struggling. On the other side
George and Andoni were trying to pull the chauffeur to his feet, but
the man's head was pouring with blood, and I think he must have been
unconscious, because every time they lifted him up he simply collapsed
to the ground again.

This was the critical moment, for if any other traffic had come along
the road we should have been caught sadly unawares. But now Paddy,
Manoli, Nikko, and Stratis were carrying the General towards the car
and bundling him into the back seat. After him clambered George,
Manoli, and Stratis – one of the three holding a knife to the General's
throat to stop him shouting, the other two with their Marlin guns
poking out of either window. It must have been quite a squash.

Paddy jumped into the front seat beside me.

The General kept imploring, 'Where is my hat? Where is my hat?'

The hat, of course, was on Paddy's head.

Suddenly everyone started kissing and congratulating everybody else;
and Micky, having first embraced Paddy and me, started screaming at
the General with all the pent-up hatred he held for the Germans. We
had to push him away and tell him to shut up. Andoni, Grigori, Nikko,
and Wallace Beery were standing at the roadside, propping up the chauffeur between them, and now they waved us goodbye and turned away
and started off on their long trek to the rendezvous on Mount Ida.

We started.

The car was a beauty, a brand-new Opel, and we were delighted to
see that the petrol-gauge showed the tanks to be full. We had been
travelling for less than a minute when we saw a succession of lights
coming along the road towards us; and a moment later we found
ourselves driving past a motor convoy, and thanked our stars that it
had not come this way sooner. Most of the lorries were troop transports, all filled with soldiery, and this sight had the immediate effect
of quietening George, Manoli, and Stratis, who had hitherto been

shouting at one another. When the convoy had passed Paddy told the General that the two of us were British officers and that we would treat him as an honourable prisoner of war. He seemed mightily relieved to hear this and immediately started to ask a series of questions, often not even waiting for a reply. But for some reason his chief concern still appeared to be the whereabouts of his hat – first it was the hat, then his medal. Paddy told him that he would soon be given it back, and to this the General said, '*Danke, danke.*'

It was not long before we saw a red lamp flashing in the road before us, and we realized that we were approaching the first of the traffic-control posts through which we should have to pass. We were, of course, prepared for this eventuality, and our plan had contained alternative actions which we had hoped would suit any situation, because we knew that our route led us through the centre of Heraklion, and that in the course of our journey we should probably have to pass through about twenty control posts.

Until now everything had happened so quickly that we had felt no emotion other than elation at the primary success of our venture; but as we drew nearer and nearer to the swinging red lamp we experienced our first tense moment. A German sentry was standing in the middle of the road. As we approached him, slowing down the while, he moved to one side, presumably thinking that we were going to stop. However, as soon as we drew level with him – still going very slowly, so as to give him an opportunity of seeing the General's pennants on the wings of the car – I began to accelerate again, and on we went. For several seconds after we had passed the sentry we were all apprehension, fully expecting to hear a rifle-shot in our wake; but a moment later we had rounded a bend in the road and knew that the danger was temporarily past. Our chief concern now was whether or not the guard at the post behind us would telephone ahead to the next one, and it was with our fingers crossed that we approached the red lamp of the second control post a few minutes later. But we need not have had any fears, for the sentry behaved in exactly the same manner as the first had done, and we drove on feeling rather pleased with ourselves.

In point of fact, during the course of our evening's drive we passed twenty-two control posts. In most cases the above-mentioned formula sufficed to get us through, but on five occasions we came to roadblocks – raisable one-bar barriers – which brought us to a standstill. Each time, however, the General's pennants did the trick like magic, and the sentries would either give a smart salute or present arms as the gate was lifted and we passed through. Only once did we find ourselves in what might have developed into a nasty situation.

Paddy, sitting on my right and smoking a cigarette, looked quite imposing in the General's hat. The General asked him how long he would have to remain in his present undignified position, and in reply Paddy told him that if he were willing to give his parole that he would neither shout nor try to escape we should treat him, not as a prisoner, but, until we left the island, as one of ourselves. The General gave his parole immediately. We were rather surprised at this, because it seemed to us that anyone in his position might still entertain reasonable hopes of escape – a shout for help at any of the control posts might have saved him.

According to our plan, I should soon be having to spend twenty-four hours alone with Manoli and the General, so I thought it best to find out if we had any languages in common (for hitherto we had been speaking a sort of anglicized German). Paddy asked him if he spoke any English.

'*Nein*,' said the General.

'Russian?' I asked. 'Or Greek?'

'*Nein*.'

In unison: '*Parlez-vous francais?*'

'*Un petit peu.*'

To which we could not resist the Cowardesque reply, 'I never think that's quite enough.'

W. STANLEY MOSS

283.

Captain Evelyn Waugh told his friend Frank Pakenham that one merit of the war should be to convince artists, such as both were, that they were not men of action. So it proved with Waugh, for whom the difficulty of finding military employment became insuperable, as his diary records.

March 2nd 1944

Luncheon with General [G. I. 'Butcher'] Thomas, who accepted me as ADC in spite of my warnings against it. I thought him a simple soldier but heard later that he is a man of insatiable ambition and unscrupulous in his means of self-advancement. On Tuesday I went to his headquarters for a week's trial – today returned unaccepted. This is a great relief. The primary lack of sympathy seemed to come from my being slightly drunk in his mess on the first evening. I told him I could not change the habits of a lifetime for a whim of his.

May 11th 1944

Interview in the afternoon. I found the room at Hobart House full of the scouring of the Army, pathetic old men longing for a job, obvious young blackguards. We were seen in turn by a weary but quite civil lieutenant-colonel and a major. The colonel said 'We've two jobs for you. I don't know which will appeal to you the more. You can be a welfare officer in a transit camp in India.' I said that was not one I should choose. 'Or you can be assistant registrar at a hospital.' I said if I had to have one or the other I would have the latter. Then he said, 'By the way are you educated? Were you at a university?'

'Yes, Oxford.'

'Well, they are very much in need of an educated officer at the War Office, G3 Chemical Warfare.'

'My education was classical and historical.'

'Oh, that doesn't matter. All they want is education.'

EVELYN WAUGH

284.

Montgomery's Military Assistant, Lt.-Col. Kit Dawnay (1909–89), recalls an incident in London in May 1944.

About a month before the invasion Monty, whose personal vanity was markedly not diminished by his new status as the nation's conquering hero, decided that he must have his portrait painted. Since I knew Augustus John slightly I suggested him as the most appropriate artist. John agreed to do the portrait for a very modest fee. After the first sitting my bell rang loudly and I found myself confronted with an outraged Monty – 'Kit, who is this terrible man you have sent me to? His clothes are dirty, I think he was rather tight and there are dozens of women in the background. I shan't go again unless you come with me.' So I found myself attending all future sittings, where I was usually made to sit in the corner with my face to the wall. This far from cordial atmosphere was considerably lightened during a visit made to the studio one day by Bernard Shaw with whom Monty struck up an immediate rapport. What transpired is described in letters written by Shaw:

4, Whitehall Court,
London S.W.1

26th February 1944

Dear Augustus John,
This afternoon I had to talk all over the shop to amuse your sitter and keep his mind off the worries of the present actual fighting. And as I could see him with one eye and you with the other – two great men at a glance – I noted the extreme unlikeness between you. You, large, tall, blonde, were almost massive in contrast with that intensely compacted hank of steel wire, who looked as if you might have taken him out of your pocket.

A great portrait painter always puts himself as well as his sitter into his work; and since he cannot see himself as he paints (as I saw you) there is some danger that he may substitute himself for his subject in the finished work. Sure enough, your portrait of B.L.M. immediately reminded me of your portrait of yourself in the Leicester Gallery. It fills the canvas, suggesting a large tall man. It does not look at you, and Monty always does this with intense effect. He concentrates all space into a small spot like a burning glass; it has practically no space at all, you haven't left room for any.

Now for it. Take that old petrol rag that wiped out so many portraits of me (all masterpieces), and rub out this one until the canvas is blank. Then paint a small figure looking at you straight from above, as he looked at me from the dais. Paint him at full length

(some foreground in front of him) leaning forward with his knees
bent back gripping the edge of his campstool, and his expression one
of piercing scrutiny, the eyes unforgettable. The background; the vast
totality of desert Africa. Result; a picture worth £100,000. The
present sketch isn't honestly worth more than the price of your keep
while you were painting it. You really weren't interested in the man.

Don't bother to reply. Just take it or leave it as it strikes you.

What a nose! And what eyes!

Call the picture infinite horizons and one man.

Fancy a soldier being intelligent enough to want to be painted by
you and to talk to me!

Always yours, (Sgd) GBS

KIT DAWNAY

285.

*The debacle in Berlin on the evening of 20 July 1944, following the
failure of the generals' conspiracy to kill Hitler and seize power,
sounded the death knell of the old German army.*

About nine o'clock Field-Marshal Witzleben, muttering, 'This is a fine
mess,' climbed into his car and drove back to the Lynar estate at Seesen,
and almost at the same time it became known that the commander of
Wehrkreis III (von Thüngen) and the military Commandant of Berlin
(von Hase) had accepted the order of General Reinecke, a notorious
Nazi who had also been on the telephone to Hitler, to withdraw their
troops, surrender their authority and consider themselves under arrest.
Olbricht thereupon assembled in his room [at the Berlin Bendlerstrasse OKH headquarters] those officers who were privy to the
conspiracy and begged them to resist the assault which was now inevitably imminent and to fight it out to the end. This they agreed to do
and orders were given to put the building into some state of defence.
With Beck and Hoepner, von Quirnheim, von Stauffenberg and
Werner von Haeften, Olbricht then retired to Fromm's old room (now
Hoepner's) on the floor above to hold a last council of war.

Scarcely had they assembled than shots were heard on the stairs and
in the corridors outside, and a group of officers, headed by Colonel

Bodo von der Heyde, all armed with tommy-guns and grenades, forced their way into the room and at pistol point demanded that Fromm should be released and handed over to them. It was the counter-Putsch and the nemesis of misplaced mercy. It was an error of judgment to have allowed Fromm and von Kortzfleisch and their fellow-prisoners to have remained in open custody in the same building with the conspirators. If they had not been shot out of hand – a fate which most of them richly deserved and which they did not hesitate to mete out when their turn came – they should at least have been closely confined. Instead, they had been accorded the honours of war and food and wine had been provided for them; in the confusion they had eluded their guards and found arms. Olbricht showed fight and was overpowered. Von Stauffenberg was shot in the back as he was retreating into his own room next door. The others remained rooted where they stood. Then Fromm appeared.

This wretched man had been all things to all men for many years. An ardent Nazi when the fortunes of the Führer were in the ascendant, he had been privy to and compliant with the conspiracy which was being hatched in his own office, and, had the attempt upon Hitler's life succeeded, would have been among the first to hail the new régime in Germany. There were many who were aware of how much he knew and his conduct in the early afternoon had been anything but unequivocal. Now at the last moment he sought to rehabilitate himself in the eyes of the winning side by eliminating the chief conspirators, ostensibly as a proof of his undying loyalty to the Führer but actually to destroy the incriminating evidence against himself. It is of some satisfaction to know that, though he succeeded in carrying out this weasel plan, it profited him nothing.

Urged on by the knowledge that retribution was hard on his track if he did not act at once, Fromm proceeded with ruthless and indecent haste. He ordered the prisoners to be disarmed and constituted himself and his recently released fellow-prisoners a drum-head court-martial of summary procedure. Beck, who had sat as if stunned throughout this last swift passage of events, asked to keep his pistol as he wished to use it for 'private purposes'. 'You would not deprive an old comrade of this privilege,' he said to Fromm with quiet dignity.

So this was the end. He, Ludwig Beck, had seen it coming for a long time. In his deepest heart he had never believed in success for the Putsch but he had been convinced that it must be attempted as an act and gesture of expiation. 'There is no use. There is no deliverance,' he had said to a friend only a few weeks before. 'We must now drain little by little the bitter cup to the bitterest end.' And why? Because men of high character such as himself had once allowed themselves to be beguiled by the enticements and seductions of National Socialism. Beck had not scrupled to defend his subalterns, Scheringer and Ludin, when charged with the propagation of what was then (1931) the subversive doctrine of the Nazi Party. He had not been shaken in his belief that there was something good for Germany in all that Hitler promised until the first exhibition of bestial gangsterism in the blood-bath of June 30, 1934. Yet a few weeks later he had taken the Oath of Allegiance, albeit with grave and heart-searching reservations, to the man who had ordered this massacre. Not till the defiling hand of the Party was laid upon the sacrosanct privileges of the Army itself, four years later, was Beck roused to open opposition, but it must be stated that once he had been thus aroused he never looked back. From 1938 until now, on the sultry night of July 20, 1944, when he stood at the end of the road, he had fought and struggled to free Germany and the German Army from the fetters of National Socialism which he and many of his comrades had helped to rivet upon their wrists. He symbolized the best in German military resistance, the man who saw the error of his ways and did what he could, however futile and ineffective, to undo the harm which he had done. Beck was no bandwagon jumper, as had been Fromm and von Kluge and Rommel; he had watched inactive the Nazi circus go past him when all his world was following admiring in its train, but the years since 1938 had been a hell upon earth for him.

Something of all this must have been in his mind as he stood now, pistol in hand, confronting Fromm, with his fellow-conspirators, now his fellow-prisoners, about him. 'I recall the old days ...' he began, but Fromm interrupted him with crude brutality, increased by his own guilty anxiety for speed, and ordered him to get on with the business in hand. Beck gave him one contemptuous glance, and looked once in farewell to his friends. Then he put the pistol to his grey head and pulled the trigger.

His intention was better than his aim. The bullet grazed his temple, giving him a slight flesh wound, and buried itself in the ceiling. Beck staggered to a chair and collapsed into it, his head in his hands. 'You'd better give the old man a hand,' said Fromm callously, and left the room.

It was at this moment, shortly after ten o'clock, that Lieutenant Schlee and his detachment of the *Wachbataillon* made their unmolested entrance into the War Ministry. As they made their way up the deserted stairs and along the empty corridors, proceeding with caution lest a trap awaited them, they heard a single shot. There was silence for a moment, then a burst of voices and a general officer, whom Schlee recognized as Fromm, came into the passage. Schlee reported himself and placed his detachment under the General's orders. Fromm then went back to his office.

The scene was macabre. In a chair, supported by two officers, sat Beck, his face ashen and blood from his flesh wound running unchecked down his cheek. Half lying in another chair, attended by his brother and Werner von Haeften, was Claus von Stauffenberg, wounded from Bodo von der Heyde's bullet in the back. At the central table Olbricht and Hoepner were writing farewell letters to their families. Fromm looked evilly portentous: 'In the name of the Führer, a summary court-martial called by myself, has reached the following verdict: Colonel of the General Staff Mertz von Quirnheim, General Olbricht, the Colonel – I cannot bring myself to name him [von Stauffenberg] – and Lieutenant von Haeften are condemned to death.'

They were taken immediately to the courtyard below, von Haeften supporting the staggering von Stauffenberg. The headlights of the military trucks shone in their eyes, all but blinding them. The men of Schlee's detachment formed the firing-party. There was only one volley. Left alone with Beck and Hoepner, Fromm offered the latter a pistol but Hoepner was not prepared for this. He refused the way of suicide and allowed himself to be arrested. 'I am not a swine,' he said, 'that I should have to condemn myself.' It was a decision which he was doubtless later to repent.

'Now how about you?' Fromm asked Beck roughly, shaking him by the shoulder. Beck asked in a weak and weary voice for another pistol and it was given him. This time he was successful. The remaining prisoners who had been arrested in Olbricht's room, including Peter Yorck, Fritz von der Schulenburg, Eugen Gerstenmaier, Ulrich von Schwerin-Schwanenfeld, von Stauffenberg's brother Berthold and von Haeften's brother Bernd, were now herded down into the courtyard where, under Fromm's orders, a second firing-party had been ordered. But here Fate again intervened.

Before the second batch of executions could be carried out, thereby removing virtually the last traces of Fromm's complicity, there arrived at the Bendlerstrasse a group of Gestapo officials, escorting Kaltenbrunner and Skorzeny, with explicit orders that no further summary justice should take place. It was the first indication of the policy which Reichsführer-SS Heinrich Himmler, Minister of Interior and now Commander-in-Chief of the Home Army, was to pursue. The mere slaughter of the Führer's enemies was of no importance to him. They should die, certainly, but not before torture, indignity and interrogation had drained from them the last shred and scintilla of evidence which should lead to the arrest of others. Then, and only then, should the blessed release of death be granted them.

JOHN WHEELER-BENNETT

286.

In September 1944, an officer of the Coldstream Guards fighting in Belgium, where his battalion suffered heavy losses, learned that company commander Billy Hartington, heir to the Duke of Devonshire and husband of Kathleen 'Kick' Kennedy, JFK's sister, had been killed. His response captures the numbness and emotional confusion of many soldiers on many battlefields.

I said 'Oh.' There is nothing else one can say. I thought 'So in war the infantry just go on and on and in the end it's almost inevitable. If you are lucky you are hit; if you are unlucky you are killed. How few people know what it costs to live like that? Already there are other generations with lives to live. This must never happen to them. Oh, God, watch over us and keep us, and whatever is to be the end, it must not be in vain. Most people will forget because they have never known or understood, but if we who know the cost survive, what hope will there be when we forget? In my heart I knew I would forget as we all have – not the facts, but the misery and sacrifice which alone can give them meaning.

JOCELYN PEREIRA

287.

Any portrait of the condition of the soldier in war should include glimpses of its merciless squalor, as well as its peril. Gerald 'Val' Lamarque (c. 1920–78) served as a sergeant with the parachute Pathfinder Company at Arnhem in September 1944. He published an autobiographical novel of the experience under the pseudonym Zeno, while serving a post-war life sentence for murder – a crime passionelle. *Here, a section defending a Dutch house strives to tend a comrade badly wounded in the chest.*

'I want a shit.'

Liddon's voice was surprisingly strong, at the same time casual. He sounded the least concerned of the three of them, but he could not see his own back. 'You don't want a shit – not now, Del.' Mocock sounded like an elder brother reproving a younger one for having a call of nature at an inconvenient time. He sounded almost petulant. Bridgman said, 'Help me to get him up, and then get something for him to do it in.' They got Liddon into a squatting position, his hands resting against the wall, and while Bridgman undid his trousers, Mocock searched aimlessly about the kitchen. At last he came back towards them, a saucer held out in his hand.

'Jesus Christ! A saucer!'

Bridgman felt a wave of almost uncontrollable irritation sweep over him, and he had to close his eyes and breathe deeply, his lips clamped together against the explosive words building up in his throat. 'It doesn't matter. Hold him here. We can clean it up afterwards.' A draught broke through and made the candle gutter as the three

crouched together against the kitchen wall. Brogan and O'Neill arrived
as Liddon strained in their arms.

ZENO

288.

*In the last months of the war suicide became a pervasive theme in the
German officer corps, first among foes of the regime, later among
Nazis. Field-Marshal Erwin Rommel (1891–1944), hero of the
North African campaign, was not an active member of the anti-
Hitler conspiracy, but met his doom because the plotters regarded him
as a plausible figurehead to lead negotiations with the Western Allies.
Having been devotedly loyal to Hitler in his years of victory, Rommel
turned against the Führer only when he saw that Germany's defeat
was inevitable. His son, then fifteen, describes his last hours on 14
October 1944.*

When my parents arrived back at Herrlingen again after the long car
journey, they found a telephone message awaiting them to the effect
that two Generals were coming next day to talk to my father about his
'future employment'. My anti-aircraft battery, to which I had returned
several weeks before, had given me leave for the 14th October. I left the
gun position very early in the morning and arrived at Herrlingen at 7
a.m. My father was already at breakfast. A cup was quickly brought for
me and we breakfasted together, afterwards taking a stroll in the garden.

'At twelve o'clock today two Generals are coming to see me to
discuss my future employment,' my father started the conversation.
'So today will decide what is planned for me; whether a People's Court
or a new command in the East.'

'Would you accept such a command?' I asked.

He took me by the arm, and replied: 'My dear boy, our enemy in
the East is so terrible that every other consideration has to give way
before it. If he succeeds in overrunning Europe, even only temporarily,
it will be the end of everything which has made life appear worth
living. Of course I would go.'

Shortly before twelve o'clock, my father went to his room on the
first floor and changed from the brown civilian jacket which he usually

wore over riding-breeches, to his Africa tunic, which was his favourite uniform on account of its open collar.

At about twelve o'clock a dark-green car with a Berlin number stopped in front of our garden gate. The only men in the house apart from my father, were Captain Aldinger, a badly wounded war-veteran corporal and myself. Two generals – Burgdorf, a powerful florid man, and Maisel, small and slender – alighted from the car and entered the house. They were respectful and courteous and asked my father's permission to speak to him alone. Aldinger and I left the room. 'So they are not going to arrest him,' I thought with relief, as I went upstairs to find myself a book.

A few minutes later I heard my father come upstairs and go into my mother's room. Anxious to know what was afoot, I got up and followed him. He was standing in the middle of the room, his face pale. 'Come outside with me,' he said in a tight voice. We went into my room. 'I have just had to tell your mother,' he began slowly, 'that I shall be dead in a quarter of an hour.' He was calm as he continued: 'To die by the hand of one's own people is hard. But the house is surrounded and Hitler is charging me with high treason.

'"In view of my services in Africa",' he quoted sarcastically, 'I am to have the chance of dying by poison. The two generals have brought it with them. It's fatal in three seconds. If I accept, none of the usual steps will be taken against my family, that is against you. They will also leave my staff alone.' 'Do you believe it?' I interrupted. 'Yes,' he replied. 'I believe it. It is very much in their interest to see that the affair does not come out into the open. By the way, I have been charged to put you under a promise of the strictest silence. If a single word of this comes out, they will no longer feel themselves bound by the agreement.'

I tried again. 'Can't we defend ourselves …' He cut me off short. 'There's no point,' he said. 'It's better for one to die than for all of us to be killed in a shooting affray. Anyway, we've practically no ammunition.' We briefly took leave of each other. 'Call Aldinger, please,' he said. Aldinger had meanwhile been engaged in conversation by the General's escort to keep him away from my father. At my call, he came running upstairs. He, too, was struck cold when he heard what was happening. My father now spoke more quickly. He again said how useless it was to attempt to defend ourselves. 'It's all been prepared to the last detail. I'm to be given a state funeral. I have asked that it should take place in Ulm. In a quarter of an hour, you, Aldinger, will receive a telephone call from the Wagnerschule reserve hospital in Ulm to say that I've had a brain seizure on the way to a conference.' He looked at his watch. 'I must go, they've only given me ten minutes.' He quickly took leave of us again. Then we went downstairs together.

We helped my father into his leather coat. Suddenly he pulled out his wallet. 'There's still 150 marks in there,' he said. 'Shall I take the money with me?' 'That doesn't matter now, *Herr Feldmarschall*,' said Aldinger. My father put his wallet carefully back in his pocket. As he went into the hall, his little dachshund which he had been given as a puppy a few months before in France, jumped up at him with a whine of joy. 'Shut the dog in the study, Manfred,' he said, and waited in the hall with Aldinger while I removed the excited dog and pushed it through the study door. Then we walked out of the house together. The two generals were standing at the garden gate. We walked slowly down the path, the crunch of the gravel sounding unusually loud.

As we approached the generals they raised their right hands in salute. '*Herr Feldmarschall*,' Burgdorf said shortly and stood aside for my father to pass through the gate. A knot of villagers stood outside the drive. Maisel turned to me, and asked: 'What battery are you with?'

'36/7, Herr General,' I answered.

The car stood ready. The S.S. driver swung the door open and stood to attention. My father pushed his marshal's baton under his left arm and with his face calm, gave Aldinger and me his hand once more before getting in the car. The two generals climbed quickly into their seats and the doors were slammed. My father did not turn again as the car drove quickly off up the hill and disappeared round a bend in the road. When it had gone Aldinger and I turned and walked silently back into the house. 'I'd better go up and see your mother,' Aldinger said. I went upstairs again to await the promised telephone call. An agonizing depression excluded all thought.

I lit a cigarette and tried to read again, but the words no longer made sense. Twenty minutes later the telephone rang. Aldinger lifted the receiver and my father's death was duly reported. That evening we drove into Ulm to the hospital where he lay. The doctors who received us were obviously ill at ease, no doubt suspecting the true cause of my father's death. One of them opened the door of a small room. My father lay on a camp-bed in his brown Africa uniform, a look of contempt on his face.

It was not then entirely clear what had happened to him after he left us. Later we learned that the car had halted a few hundred yards up the hill from our house in an open space at the edge of the wood. Gestapo men, who had appeared in force from Berlin that morning, were watching the area with instructions to shoot my father down and storm the house if he offered resistance. Maisel and the driver got out of the car, leaving my father and Burgdorf inside. When the driver was permitted to return ten minutes or so later, he saw my father sunk forward with his cap off and the marshal's baton fallen from his hand. Then they drove off at top speed to Ulm, where the body was unloaded at the hospital; afterwards General Burgdorf drove on to Ulm Wehrmacht Headquarters where he first telephoned to Hitler to report my father's death and then on to the family of one of his escort officers to compose the menu for that night's dinner.

Perhaps the most despicable parts of the whole story were the expressions of sympathy we received from members of the German Government, men who could not fail to have known the true cause of my father's death and in some cases had no doubt themselves contributed to it, both by word and deed. I quote a few examples:

In the Field 16 October 1944
Accept my sincerest sympathy for the heavy loss you have suffered with the death of your husband. The name of Field-Marshal Rommel will be for ever linked with the heroic battles in North Africa.
    Adolph Hitler

Fuehrer's Headquarters 26 October 1944

The fact that your husband, Field-Marshal Rommel, has died a hero's death as the result of his wounds, after we had all hoped that he would remain to the German people, has deeply touched me. I send you, my dear Frau Rommel, the heartfelt sympathy of myself and the German Luftwaffe.

In silent compassion, Yours,

GOERING, Reichsmarschall des Grossdeutschen Reiches

MANFRED ROMMEL

289.

A British tank officer leaguers overnight in Germany in the closing weeks of the war.

Joe, Wally and I backed hard into the trees and heaved a sigh of relief. A tot or two of the blessed rum and so to bed. The following morning, as Briggsy reversed the tank back into business, there rose from literally under the left-hand track, with hands held well above his head, as dishevelled, grimy and altogether miserable a figure as anyone could imagine. His grey German uniform was scarcely recognizable under its coating of mud and oil. As we stared in amazement at this apparition, he grimaced and pointed to a narrow slit trench in which he had evidently been trapped under the tank track all night. There was something about that gesture which rang the very faintest of bells. I signalled to him to climb on to the turret. Sitting on top of the tank, we stared at each other in total disbelief.

In those long-ago sunlit days of the 1930s, when God was in his heaven and all was well with the world, my father had decreed that my brother and I should have a German tutor. His name had been Willie Schiller. Now the same Willie Schiller was facing me. There was nothing much either of us could do about it. He may have said '*Gott in Himmel*,' but I can't really remember. We just had a rum or two and smoked a cigarette. After the war I was telling my mother about this extraordinary affair. 'Nonsense,' she said firmly. 'It could not have been Willie. You must have been drunk.'

'I was not drunk,' I responded indignantly. 'Why do you say it could not have been Willie?'

'Because,' she said firmly, 'Willie was always so perfectly turned out.'

DOUGLAS SUTHERLAND

290.

Captain Ian Liddell (1922–45), a Coldstream Guards company commander, came through his military service unscathed until the closing month of the struggle, when many people thought it all over bar the victory celebration. The pathos of his experience epitomizes the injustice of war.

A gentle, humorous, quixotic young man with a history of schoolboy japes and ambitions to become a vet, Liddell loved animals. He served through 1941 with a company guarding the Royal Family, and was thrilled when George VI gave him a Labrador puppy. One day, he telephoned his mother to ask her to send him some cardboard Easter eggs. He explained that he was performing in a pantomime that the Coldstreamers were staging for their royal charges: 'I am playing the part of a chicken, and want to lay eggs in the laps of the King and Queen and little princesses!' The show is said to have won rave reviews from its audience.

Liddell saw no action until the invasion of Normandy. In January 1945, during a brief leave in London, he married a WAAF officer. Less than three months later, on 3 April, advancing through Germany, his battalion learned of an undestroyed bridge across the Ems river near Lingen, which his company was ordered to capture. Liddell had two hours in which to make his preparations: thus, what followed represented no momentary impulse, but instead a considered act. From a cottage overlooking the river, he and other officers studied the deserted crossing through their glasses. The eastern bank was defended by strongly-manned German fortifications. The western approach was blocked by a ten-foot-high barricade. They could see demolition charges, including 500lb aircraft bombs, with wires running to the enemy command post.

Liddell decided that the only chance of success was that he should himself disable the charges, before his men attacked. His lone dash in the face of the enemy lasted less than five minutes, but demanded a lifetime's courage, especially at a moment of the war when many allied soldiers were doing their utmost to escape becoming the last to die. In a grey drizzle, amid a storm of fire from both sides, he clambered alone over the barricade, and ran towards the far bank. Hundreds of watching British and Germans alike assumed that the bridge would erupt into rubble. Miraculously, however, Liddell was able to sever cables linked to two sets of charges with wire-cutters that he had borrowed from one of his platoon commanders, who begged him to take special care of them, because his father had carried them in World War I. Liddell then spotted more explosives under the bridge. He had just disabled these, too, when a German shot at him from close range. Having dropped his sten gun when climbing the barricade, the Englishman could respond only by hurling the heavy cutters in the man's face, causing him to slip and fall. Liddell then ran back to the west bank and waved forward his company, headed by a Sherman tank, which crashed the barricade before the infantry crossed and over-ran the German positions. Sixty enemy were captured in addition to those killed, for the loss of one Coldstreamer. Men marvelled, not only that day but for the rest of their lives, at Liddell's success and survival. One fellow-officer described Liddell as a simple man – meaning not stupid, but straightforward: 'he did what he did, because he thought he should'. Yet on 27 April near Rotenburg, Liddell faced the appointment in Samarra that had seemed his assured fate at the Ems bridge. A chance German bullet passed through a fellow-officer's head, then struck and killed Liddell also. He died without knowing that he had been awarded the VC for his action three weeks earlier, which his widow later received from the King.

MAX HASTINGS

291.

A British officer drives through Germany in the first days after its surrender.

I passed a farm wagon headed for the village. I glanced casually at the two men sitting up behind the horse. Both wore typical farmer head-gear and sacks were thrown over their shoulders protecting them from a light drizzle. We were just past them when something made me slam on the brakes and back up. I was right, the man who was not driving was wearing field boots. I slipped out from behind the wheel, pulled my revolver from its holster and told the corporal to cover me with his Tommy gun.

I gestured to the men to put their hands over their heads and told them in fumbling German to produce their papers.

'I speak English,' said the one with the field boots, 'this man has papers – I have none.'

'Who are you?' I asked.

He told me his name and rank – 'General.'

'We are not armed,' he added, as I hesitated.

Sandhurst did it – I saluted, then motioned to them to lower their hands.

'Where are you coming from, sir?'

He looked down at me. I had never seen such utter weariness, such blank despair, on a human face before. He passed a hand over the stubble of his chin.

'Berlin,' he said quietly.

'Where are you going, sir?'

He looked ahead down the road towards the village and closed his eyes.

'Home,' he said almost to himself, 'it's not far now ... only ... one more kilometre.'

I didn't say anything. He opened his eyes again and we stared at each other. We were quite still for a long time. Then I said, 'Go ahead, sir,' and added ridiculously ... 'please cover up your bloody boots.' Almost as though in pain, he closed his eyes and raised his head, then with sobbing intake of breath, covered his face with both hands and they drove on.

292.

*Scottish journalist George Macdonald Fraser (1925–2008) became
famous as the literary reviver of the fictional villain Flashman –
from Thomas Hughes' 1857 novel* Tom Brown's Schooldays *– and
the author of much memorable, lightly fictionalised history. Fraser
also wrote one of the finest private soldiers' memoirs of World War
II, an account of his own experiences in the Border Regiment with
Slim's 'Forgotten Army'. In old age he became a contemptuous critic
of the body armour and heavy equipment with which modern
soldiers encumber themselves, and sent me a photo of himself
'wearing the kit we wore to fight Johnny Jap': shorts, boots, rifle and
slouch hat. 'His' campaign in Burma continued for more than three
months after Hitler perished.*

We knew only too well that we were a distant sideshow, that our war
was small in the public mind beside the great events of France and
Germany. Oh, God, I'll never forget the morning in May 1945 when
we were sent out to lay ambushes, which entailed first an attack on a
village believed to be Jap-held. We were lined up for a company
advance, and were waiting in the sunlight, dumping our small packs
and fixing bayonets, and Hutton and Long John were reminding us
quietly to see that our magazines were charged and that everyone was
right and ready, and Nixon was no doubt observing that we'd all get
killed, and someone, I know, was muttering the old nonsense 'Sister
Anna will carry the banner, Sister Kate will carry the plate, Sister Maria
right marker, Salvation Army, by the left – charge!' when a solitary
Spitfire came roaring out of nowhere and Victory-rolled above us. We
didn't get it; on the rare occasions when we had air support the Victory
roll came after the fight, not before.

While we were wondering, an officer – he must have been a new
arrival, and a right clown – ran out in front of the company and
shouted, with enthusiasm: 'Men! The war in Europe is over!' There was
a long silence, while we digested this, and looked through the heat
haze to the village where Jap might be waiting, and I'm not sure that
the officer wasn't waving his hat and shouting hip, hooray. The silence

continued, and then someone laughed, and it ran down the extended line in a great torrent of mirth, punctuated by cries of 'Git the boogers oot 'ere!' and 'Ev ye told Tojo, like?' and 'Hey, son, is it awreet if we a' gan yam?' 'Gan yam' is Carlisle dialect for 'Go home'.

GEORGE MACDONALD FRASER

293.

Among the most shameful aspects of the British conduct of the war was the racial discrimination that was sustained in regard to hundreds of thousands of African colonial and Indian troops. They fought as mercenaries, but died in the Allied cause for a fraction of white troops' pay. The wife of a sepoy who had been absent for four years, serving with the Rajputana Rifles in the Middle East and Italy, caused her village letter-writer to pen this moving missive to his colonel, pleading her own destitution.

Sir, I most respectfully and humbly beg to state that my husband, NO.13312 Naik Boota Khan, has been serving the Crown on Overseas Service. Since his departure from India, no maintenance or family allotment is being paid or remitted to me. I am living at my parents' house who owing to indigent circumstances can hardly provide to maintain me. Owing to this trouble my father has run into debt as all the necessaries of life are too dear that can better be imagined than described. I venture to submit this my humble petition in the earnest hope that some measures may kindly be adopted to order my said husband to send me money from the date i.e. four years I am living at my parents' house. In the absence of any monetary help I will be sunk into a deplorable posture. [There is no record that her plea secured any better outcome than that of Thomas Hostelle (see no. 36) five hundred years earlier.]

NAZIR BEGUM

294.

Return of a prisoner of war – René Cutforth (1909–84), captured in the Western Desert in 1941, who later became a famous BBC foreign correspondent.

In the early summer of 1945 I climbed into a Dakota on an airfield in East Germany with all my worldly possessions – a pistol and ten rounds, and a few cigarettes. There were a dozen other men on the plane, all very dirty and most of them asleep. Nobody talked much, and at three in the afternoon we came down on a runway in a field full of buttercups in Oxfordshire. I had been away four years. It was England all right: there was that special kind of early summer light calming down the romantic distances and making the hedgerows into dark blue mists. Sweet reason was the prevailing atmosphere on the airstrip. Soothing girls gave us tea, rock cakes and cigarettes and asked soothing, ridiculous questions. They took down our names and units and addresses, doctors plied their stethoscopes, beat on our knees with little rubber hammers, and took blood samples. Quartermasters of a new kind to me, young and deploying a winsome charm, issued new underclothes and shoes. Orderlies showed us into bedrooms with beautiful white sheets and bathrooms attached. The first bath for three years was such a revelation that I made it last nearly two hours. Then I hacked off my beard and shaved.

They're softening you up, I said to myself from time to time. It's a trap. My neighbour in the next room, a bald major, put his head round the door with another version of events. 'I can only suppose,' he said, 'that the man who runs this place wants to get into Parliament.'

'I'm not stopping here,' I told him, 'no matter what they say or do. If I have to bust out I'm taking off tonight.'

'Oh, I don't know,' the major murmured, 'might give the bed a trial.' At this innocent speech such a fury rose in me that I had to move away at a run in case I fell upon him tooth and claw. So I went into the assembly hall where there was to be an announcement.

'Won't waste your time with speeches,' the commandant was saying, 'The sergeants sitting at tables down both sides of this room are experts in cutting red tape. They're here to see you get your advances on pay, temporary identity card, ration books, coupons, and all the bumf you need nowadays in the quickest possible time. Only one snag. Nobody leaves here until he is past the psychiatrist. The examination will take some time. There are 25 of you and three psychiatrists, so some of you will be staying the night. Any volunteers?' To my amazement and contempt he got 12.

The psychiatrist was Viennese and cat-like. 'I am not at all sure that you're fit to take your place in the civilian world. I ask you to stay. Will you stay?'

'No,' I said. 'I'll break out.'

'That would be very foolish, but also of some inconvenience to me,' the psych said. 'Have you a permanent address, a telephone number, somebody to look after you during your first three or four weeks?'

'My wife has taken a flat in South London,' I said, 'and she is expecting me.'

Quite suddenly I had to put on an absolutely blank face over one of those appalling rages which had been invading me ever since I got free from the prisoner-of-war camp. Deceit was the only thing. Bluff your way out. 'Oh, I think I'll be all right,' I said easily.

'Very well, you get three months' leave,' and he filled in a card. I hadn't taken him in, though: I saw the card later: 'Manic depressive type. Educated. High IQ. Possibly disturbed. Marked aggression.' Then two red asterisks.

'Have a drink,' said this abominable man, and I drank a large whisky very slowly to deceive him.

Twelve of us left for London by the 9.30 train, and if I was madder than most of them there wasn't much to choose between us. By midnight seven wives had a drama on their hands once more for better or for worse, and for the first time in history their dramas were being produced by the War Office. When I first heard this I was so furious I thought I was going to have a stroke. 'Do you mean to say,' I yelled, 'that the bloody army gave you lessons in how to be married to me?'

'Well, they told us what we might expect.'

'To hell with that,' I roared, and went off to the pub and stayed away for three days. When I got back I said: 'I took off because I will not have my life interfered with.' 'Oh yes,' my wife said, 'they told us about that, too.'

RENÉ CUTFORTH

295.

*Col. Richard Meinzerhagen (1878–1967), a retired officer and
passionate Zionist sympathizer, found himself aboard a British
troopship which docked briefly at Haifa in the midst of Israel's war of
independence. The author was already seventy. As he was also a
considerable fantasist, this account should be read with a
pinch of scepticism.*

23. IV. 1948.

We arrived at Haifa, on my way back from Arabia, at dawn and tied
up. There is a company of Coldstream Guards on board; some of the
officers knew Dan [the author's son] and we spoke about him; that
pleased me beyond measure; this company disembarked here to the
tune of a full-scale battle along the sea-front between the Arabs and
the Jews. By 4 p.m. the Arabs were on the run and the Jews in control
of the port except for the small depot still occupied by the British
Army. The Coldstreams walked right into the fight, their job being to
protect Government stores. Both rifle fire and the ping of bullets
continued throughout our stay in Haifa with an occasional mortar
shell or bomb. I realized that the Jews were going to be strained to the
utmost with seven Arab states invading them from all sides and with
an unsympathetic Great Britain.

I was quite determined to help if even in a small way. A private
soldier in the Coldstream detachment knew Dan and he was sick
and unable to land with the detachment. I went to his cabin and
asked him if he would lend me his complete equipment, rifle,
uniform and ammunition to which he agreed with a typical 'Take
the whole bloody lot, as far as I'm concerned.' So I took the 'whole
bloody lot' and marched ashore behind the Coldstream detachment,
unnoticed by the officers and falling-in in the rear. On leaving the
breakwater I broke away and walked to where I could hear firing. I
soon found the front line – about twenty Haganah well entrenched
in sand. So I ran up, scraped a hole and lay low for a bit. I had my
fieldglasses and sighted my rifle on an empty barrel, some 200 yards
away.

Some Arabs, about 600 yards away, had fired at me as I ran up to take position beside the Haganah, so I crawled forward by degrees taking every advantage of folds in the ground and eventually got a fine little scrape in the sand no more than 250 yards from some Arab snipers. Three Haganah did likewise and we four opened fire. I saw an Israeli looking at me and he said something which I did not hear; he then shouted to his friends and yelled at me in Hebrew. I paid no attention. My first shot was a bullseye and there was a cheer from the Haganah as the man rolled over and writhed about on the sand. I crept forward to a better position and found I could take the Arabs in the flank; which I proceeded to do.

Two more paid the penalty. The Haganah kept up with me and between us we got five more and one got up and ran towards us with his arms above his head so we captured him. There were now some twenty Arabs firing at us, at ranges between 300 and 600 yards; so we crept on and bagged four more: I had a narrow squeak when a bullet hit my little parapet and spluttered me with sand. We were doing well when one of the Haganah shouted at me, but not understanding I paid no attention. He then got up and ran back. About an hour later, after we had disposed of all the Arabs in front of us, a Coldstream officer came along, asked who the hell I was and on recognizing me, ordered me back to the ship, which I had to obey; but it mattered little as by then I had fired all my 200 rounds. So off I went, rather sheepishly, and without further incident got aboard, returned my uniform to my friend, cleaned his rifle and then ordered a large bottle of champagne which we shared in his cabin. I had had a glorious day. May Israel flourish!

RICHARD MEINZERHAGEN

296.

Douglas MacArthur (1880–1964) was relieved of his UN command in Korea by President Truman on 11 April 1951, after publicly and unilaterally threatening an extension of the war to mainland China.

Almost at the very moment yesterday that the news of General MacArthur's relief was coming over the radio at the divisional command post on the western front where I have been spending a few days, a terrific wind blew across the camp site, levelling a couple of tents. A few minutes later, a hailstorm lashed the countryside. A few hours after that, there was a driving snowstorm. Since the weather had been fairly springlike for the previous couple of weeks, the odd climatic goings on prompted one soldier to exclaim, 'Gee, do you suppose he really is God, after all?'

E. J. KAHN

297.

An officer of the Glosters recalls a moment during the battalion's doomed stand on 23 April 1951, during the Chinese offensive across the Imjin river in Korea. The battle entered the legend of the British Army for its indisputable heroism. Like most such sacrificial actions, however, it would have better served the Allied cause, as well as the participants, had it never taken place. The Glosters should instead have been withdrawn.

The dawn breaks. A pale, April sun is rising in the sky. Take any group of trenches here upon these two main hill positions looking north across the river. See, here, the weapon pits in which the defenders stand: unshaven, wind-burned faces streaked with black powder, filthy with sweat and dust from their exertions, look towards their enemy with eyes red from fatigue and sleeplessness; grim faces, yet not too grim that they refuse to smile when someone cracks a joke about the sunrise. Here, round the weapons smeared with burnt cordite, lie the few pathetic remnants of the wounded, since removed: cap comforters; a boot; some cigarettes half-soaked with blood; a photograph of two small girls; two keys; a broken pencil stub.

The men lounge quietly in their positions waiting for the brief respite to end. 'They're coming back, Ted.' A shot is fired, a scattered burst follows it. The sergeant calls an order to the mortar group. Already they can hear the shouting and see, here and there, the figures moving out from behind cover as their machine-guns pour fire from the newly occupied Castle Site. Bullets fly back and forth; overhead, almost lazily, grenades are being exchanged on either side; man meets man; hand meets hand. This tiny corner of the battle that is raging along the whole front, blazes up and up into extreme heat, reaches a climax and dies away to nothingness – another little lull, another breathing space.

[Lt. Philip Curtis] is called to the telephone at this moment; Pat [Angier]'s voice sounds in his ear. 'Phil, at the present rate of casualties we can't hold on unless we get the Castle Site back. Their machine-guns up there completely dominate your platoon and most of Terry's. We shall never stop their advance until we hold that ground again.' Phil looks over the edge of the trench at the Castle Site, two hundred yards away, as Pat continues talking, giving him the instructions for the counter-attack. They talk for a minute or so; there is not much more to be said when an instruction is given to assault with a handful of tired men across open ground. Everyone knows it is vital: everyone knows it is appallingly dangerous.

Phil gathers his tiny assault party together. They rise from the ground and move forward. Already two men are hit and Papworth, the medical corporal, is attending to them. They are through the wire safely – safely! – when the machine-gun in the bunker begins to fire. Phil is badly wounded: he drops to the ground. They drag him back through the wire somehow and seek what little cover there is as it creeps across their front. The machine-gun stops, content now it has driven them back; waiting for a better target when they move into the open again. 'It's all right, sir,' says someone to Phil. 'The medical corporal's been sent for. He'll be here any minute.' Phil raises himself from the ground, rests on a friendly shoulder, then climbs by a great effort on to one knee. 'We must take the Castle Site,' he says; and gets up to take it.

The others beg him to wait until his wounds are tended. One man places a hand on his side. 'Just wait until Papworth has seen you, sir –' But Phil has gone: gone to the wire, gone through the wire, gone towards the bunker. The others come out behind him, their eyes all on him. And suddenly it seems as if, for a few breathless moments, the whole of the remainder of that field of battle is still and silent, watching amazed, the lone figure that runs so painfully forward to the bunker holding the approach to the Castle Site: one tiny figure, throwing grenades, firing a pistol, set to take Castle Hill. Perhaps he will make it – in spite of his wounds, in spite of the odds – perhaps this act of supreme gallantry may, by its sheer audacity, succeed. But the machine-gun in the bunker fires directly into him: he staggers, falls, is dead instantly; the grenade he threw a second before his death explodes after it in the mouth of the bunker. The machine-gun does not fire on three of Phil's platoon who run forward to pick him up; it does not fire again through the battle: it is destroyed; the muzzle blown away, the crew dead. [*Lt. Curtis was awarded a posthumous VC for his action.*]

ANTHONY FARRAR HOCKLEY

298.

Simon Raven (1927–2001) was a gifted novelist and screenwriter who made a second career as rake, drunk, cad, bounder and sybarite. He was one of five soldiers featured in this book who attended my own school, Charterhouse – the others were Baden-Powell, Robert Graves, Orde Wingate and Billy Moss, a motley array of warriors. When I once invited Raven to lunch and asked whether he would prefer grouse or lobster, he promptly responded, 'Both.' Though enthusiastically gay, in 1951 he married a pregnant girlfriend, then abandoned her and their child to serve in Germany with the British Army.

As a married officer with a child Simon received an allowance which he was supposed to hand over to his spouse. This he had not so far done, his negligence resulting in a legendary exchange of telegrams. In response to Susan's WIFE AND BABY STARVING SEND MONEY SOONEST, Simon wired back SORRY NO MONEY SUGGEST EAT BABY. Susan was evidently not amused because she took her complaint to the War Office.

[*Having escaped from that scandal, and served with social if not battle-field distinction in Kenya, back at the regimental depot in 1957 Captain Raven incurred racing debts, and drew unmet cheques on the officers' mess funds, which terminated his army career:*] 'The Commandant reminded me that by incurring debts which I couldn't pay I was guilty of conduct unbecoming an officer and a gentleman. Once an official complaint was made I'd be court-martialled and in all probability cashiered, which would be very disagreeable for the regiment, for him and for me – in that order. He then remarked that in his youth they'd had an adage, "If you can't pay the bill, look for the fire escape". "You will send in your papers today, and I will undertake to get your resignation through. You'll still owe the bookies, but that will be no concern of ours. Meanwhile leave here and *lose* yourself until your resignation is gazetted. If you can't be found, you can't be court-martialled."'

299.

Michael Carver (1915–2001) was among the most formidable figures of the twentieth-century British Army. He was famously outspoken, as well as ruthless, naming in his memoirs each of the officers whom he had sacked for incompetence or loss of nerve in World War II, much to their distress. In 1984 he gave priceless support to me, a young author under attack for suggesting that the wartime German Army was better than its British and American counterparts, by publicly endorsing this opinion. After Carver served under Montgomery at NATO in 1953, the latter composed a confidential report that was as merciless as its subject, though some would say that for 'Monty' to pen such a critique represented a pot addressing a kettle.

This officer is outstanding for his rank. In the late war I gave him command of an armoured brigade when he was under 30. I must take the blame if it gave him a swollen head, which it did. He is very ambitious, but before he can rise to the heights to which his gifts entitle him, he has much to learn. He is widely read and has developed his military knowledge. But he must also develop human qualities, which at present he lacks. It is important for him to understand that the higher the mountain the more fiercely blow the storms about it; as he gains in rank he will need the support and sympathy of others. He will not get that support unless he can learn to be less critical in his general outlook. If he can take advantage of this hint, he should rise to the highest ranks. If he cannot, but merely regards me as one of the many 'bloody fools' who stand in the way of his quick promotion to high rank, he will crash badly. I have given Colonel Carver a copy of these remarks. I want him to accept the hint! [Carver never really did so, but he became a field-marshal and chief of defence staff anyway.]

VISCOUNT MONTGOMERY

300.

The French GCMA – Groupement de Commandos Mixtes Aeroportes – *who fought behind Vietminh lines in Indochina during the early 1950s evolved from World War II maquis experience. Each unit was composed of native tribesmen led by French officers and NCOs, and supplied by air. One of the many tragic stories of that war derived from the fate of surviving GCMA groups which could not be evacuated after the French withdrawal. Their fate was here described by one of the conflict's most vivid if sometimes romance-prone chroniclers, the French-Austrian war correspondent and later academic Bernard Fall (1926–67), writing in his 1961 book* Street Without Joy.

The cease-fire of July 1954 brought an end to G.C.M.A. operations. Frantic efforts were made by the French to broadcast messages to all the groups operating behind Communist lines to fall back to Laos, the 17th parallel, or to the shrinking Haiphong perimeter, before the Bamboo Curtain rang down on them for good. But for many, the broadcasts came too late, or the T'ai or Meo could not reconcile themselves to leave their families exposed to the Communist reprisals which were sure to come. And the Frenchmen with them, who could not possibly make their way back across hundreds of miles of enemy territory, stayed with the tribesmen to the end.

This was a fight to the finish, and no quarter was given on either side. One by one, as the last commandos ran out of ammunition, as the last radio sets fell silent, the remnants of the G.C.M.A. died in the hills of North Viet-Nam. There was no fuss: France did not claim the men, and the Communists were content to settle the matter by themselves. French officers recalled with a shudder the last radio message picked up from somewhere in North Viet-Nam nearly two years after the fighting had officially stopped. The voice was French, the message was addressed to the French. It said: 'You sons-of-bitches, help us! Help us! Parachute us at least some ammunition, so that we can die fighting instead of being slaughtered like animals!'

But the cease-fire was in effect and the last French troops left Indochina in April 1956, in compliance with the demands of the Vietnamese nationalists. The few remaining GCMAs kept on fighting. The Communists' own weekly *Quan-Doi Nhan-Dan* ('People's Army') of September 3, 1957 reported that from July 1954 to April 1956 their forces in the mountain areas east of the Red River had, 'in spite of great difficulties and hardships', killed 183 and captured 300 'enemy soldiers', while inducing the surrender of 4,336 tribesmen and capturing 3,796 weapons. Some of the luckier tribesmen, such as the Muong and Nung who were closer to the French lines, made their way to South Viet-Nam and are now resettled in the southern hills near Dalat, in a setting and climate very close to that of their beloved T'ai country. Others continue to trickle into neighboring Laos, whose own mountain tribes are their close relatives.

By 1959, the struggle was over. Whatever Frenchmen there had been left among them were now dead or captured. Only one, Captain C—, who was thoroughly familiar with several mountain dialects, is known to have made his way out of the Communist-occupied zone after a harrowing 500-mile trek through the mountains.

BERNARD FALL

301.

The 1956 Anglo-French Suez invasion was marked by military confusion as great as the political misjudgement. In the weeks following mobilization, the British expeditionary force assembled with embarrassing sluggishness.

When the units started to collect their vehicles and stores from depots which were spread around the country and manned mostly by civilians who took the weekend off, it was found that mobilization scales were out-of-date, that numerous items were out-of-stock and that many of the vehicles were, to say the least, decrepit. To move the tanks from Tidworth to Portland and Southampton, where they were to be loaded into Landing Ships, help had to be sought from Pickfords [removals company] to supplement the few tank-transporters still left to the Army. Pickford's men were subject to trade-union rules and civilian regulations, and their massive transporters took a week for a journey which an army convoy could complete in three days; behind each bunch of their vehicles trailed a number of empty spares, as was required by the Regulations of British Road Services. It took four weeks to move and load ninety-three tanks.

When ships eventually offloaded at Port Said, there were surprises. At the fishing harbour a senior staff officer noticed a 3-ton lorry, so overladen that its rear springs were all but concave and stuck fast on the ramp. 'Who the bloody hell are you,' he enquired kindly, 'and what are you doing?' 'I, sir,' responded a voice of much dignity, 'am the mess-sergeant of Her Majesty's Life Guards, and I have with me the officers' mess silver and champagne.'

ROY FULLICK AND GEOFFREY POWELL

302.

A young tank officer recalls the experience of spearheading the assault on Port Suez.

As we trundled through, suddenly the driver tugged my trousers and said 'Sir, do you know if they drive on the right hand or the left in this country?' We hadn't gone much further when he gave another tug. I said, 'What on earth's the matter now?' and he said, 'Sir, there's a traffic light at red. Do we stop?'

DEREK OAKLEY

303.

In June 1958 Evelyn Waugh's son Auberon (1939–2001) was serving in Cyprus as a troop commander with the Royal Horse Guards, part of the UN peacekeeping force, when a mishap befell him.

On patrol, we always travelled with a belt in the machine-gun, but without a bullet in the breech. I had noticed an impediment in the elevation of the gun on my armoured car, and used the opportunity of our taking up positions to dismount, seize the barrel from in front and give it a good wiggle. A split second later I realized that it had started firing. No sooner had I noticed this, than I observed with dismay that it was firing into my chest. Moving aside pretty sharpish, I walked to the back of the armoured car and lay down, but not before I had received six bullets – four through the chest and shoulder, one through the arm, one through the left hand. My troop corporal of horse swore horribly at my driver, whom he imagined to be responsible. In fact nobody had been in the armoured car, as I explained from my prone position.

I was rather worried and thought I was probably going to die, as every time I moved the blood pouring out of holes in my back, where the bullets had exited, made a horrible gurgling noise. To those who suffer from anxieties about being shot I can give the reassuring news that it is almost completely painless. Although the bullets caused considerable devastation on the way out, the only sensation at the time was of a mild tapping on the front of the chest. I also felt suddenly winded as they went through a lung. But there was virtually no pain for about three-quarters of an hour, and then only a dull ache before the morphine began to take effect.

The machine-gun had shot nearly the whole belt – about 250 rounds – into the Kyrenia road, digging an enormous hole in the process, before being stopped. In the silence which followed, Corporal of Horse Chudleigh came back to me, saluted in a rather melodramatic way as I lay on the ground and said words to the effect that this was a sorry turn of events. He looked so solemn that I could not resist the temptation of saying: 'Kiss me, Chudleigh.' Chudleigh did not spot the historical reference, and treated me with some caution thereafter. At least I think I said 'Kiss me, Chudleigh.' I have told the story so often now that I honestly can't remember whether it started life as a lie.

The Blues medical officer accompanied me in the ambulance to hospital and read the De Profundis to me on the way. This struck me as rather gloomy, although there was nothing else to do, but not nearly as gloomy as the surgeon, Colonel John Watts, who prepared to operate immediately. Feeling quite happy as the morphine took effect and fortified, as they say, by the last rites of the Church in the shape of an

Irish priest to whom I took an instant dislike, I said to Colonel Watts in as nonchalant a tone as I could muster: 'Tell me, Colonel, what chance do you actually think I have of pulling through?' He fixed me with his cool blue eyes and said: 'I think you've got a very good chance.' I felt as if an icy hand had been placed over my heart, but more even than terror I felt fury. 'What do you fucking mean I've got a good chance? You're supposed to say I will be out of bed the day after tomorrow.'

I later learned that none of them thought I would survive, but Colonel Watts performed brilliantly, taking out the lung, spleen and two ribs in a hospital whose general standard of equipment, as I later realized, was somewhere around that of a cottage hospital in the Soviet Union. I woke up after the operation in stifling heat, delighting Colonel Watts, when he told me that he had removed my spleen, by asking whether that would improve my temper.

AUBERON WAUGH

304.

The French army's operations against Algerian insurgents in the early 1960s were conducted with notable ruthlessness, such as this one in the Némentchas mountains, described by a Foreign Legionnaire.

For three days, two companies had been blockading a cliff. They were unable to dispose of a very strong rebel band, solidly entrenched in some caves. They tried everything, even flame-throwers. Nothing doing. The caves were impregnable. Every legionnaire who tried to force his way in was machine-gunned at point-blank range, after a dozen steps in the dark. They decided to reduce the besieged men by smoking them out. But there was no wood. They would have needed hundreds of smoke grenades, because the caves were so vast. Then, tired of waiting – the rebels seemed to have the caves filled with provisions – they had a typical legionnaire idea: they sent for quicklime and gravel by helicopter: and they calmly sealed all the caves in the cliff, one by one, hermetically, under covering fire of a battery of machine-guns! After which they went away. A few months later, they went back for a look. They carefully dismantled their untouched walls. A horrifying stench of death emerged. The stiffened corpses of rebels massed behind them fell into their arms.

PIERRE LEULLIETTE

305.

A British member of the Legion described his experience in Algeria on 23 April 1962, when his battalion's commander committed the unit to an attempted generals' coup against President Charles de Gaulle.

In the middle of the night we were aroused and ordered to collect our equipment and board the trucks. In a long column the regiment drove due west towards Algiers, three hundred miles away. Dawn came slowly through, cold and slate grey – and then suddenly it was day and sun. Thousands and thousands of Europeans lined the roads madly cheering us through. Car horns hooted and hooted – three short blasts and two long (*Al-gé-rie fran-çaise*) – again and again. Teeming crowds of wild faces laughing and yelling, pushing and shoving in a great heaving morass of bodies. Our trucks were barges floating in a sea of faces and waving arms. They loved us, they raved for us. They have never shown this emotion before and I suspect its depth is not substantial, but we must seem to them to be their saviours.

In all of this there is one amazing and incomprehensible fact. One question has not yet been answered at all clearly and that is, what side are we really on? We crawl through the streets of Algiers in the late evening and occupied barracks of French regulars who appear to have temporarily disappeared – still unbelievably no attempt at explanation from an officer. This credits us with nil intelligence. They assume we will just obey orders as we have always done; so we remain at the mercy of rumour based on snatches of information from transistor radios.

[Early next morning] we drove to the airport and our role became immediately clear. We were to occupy the airport, which was held by French marines, and they weren't having any of it. So we were given wooden batons, heavy, with sharp points, and in one long line we slowly eased into the marines, pushing them forward like bolshie rams. They frequently turned and attacked with aggression. This was met in many cases with savage beatings and it became a sad and shoddy business. Marine officers were pushed around by our officers – there were scenes of officers yelling at one another with questions of loyalty and accusations of traitor and so on. L'Hospitalier bust his baton on the head of one of the marines. Gradually they were herded out of the airport premises and we were in control of the base from which we will apparently make the [planned mass parachute] drop on Paris.

The evening has come. A kind of stalemate appears to have been reached. De Gaulle has brought up tanks in France and threatened to shoot parachutists out of the air if a drop is made. This dampened some of yesterday's thoughts of dancing in Paris and put prospects of dinner at Maxim's tomorrow night slightly further away. [The mutiny fizzled out shortly afterwards.]

SIMON MURRAY

306.

Commanders often give uplifting addresses to their troops before setting forth for war. This one was delivered to the men of US 2/4th Marines, 'The Magnificent Bastards', on 23 April 1965, when the battalion was about to deploy to Vietnam. Their colonel was a much-decorated veteran of Korea, and especially of the December 1950 Chosin reservoir campaign.

Men, this stuff flying around about Americans not wanting to fight is a lot of crap. Americans love a challenge, when you Marines here were kids you all admired the champion marble players, the smartest kid in the class, the fastest runner, the handiest kid with his fists, the big league home run king, the All-American Football player, AMERICANS LOVE A WINNER!! They cannot tolerate a loser, Americans play to win every time and all the time. I wouldn't give a hoot in hell for any Marine who lost and laughed. That's why Marines have never lost a battle. For the very thought of losing is despicable to any Marine.

If we do go to fight in South-East Asia, you are not all going to die, only about two percent of us would die in a major battle. Death must never be feared. Death in time comes to all. Yes, every man is scared in his first battle; if he says he isn't he is a damn liar. Some men are cowards, yes, but they fight just the same or they get the hell slammed out of them watching men fight who are just as scared as they are. The real hero is the Marine who fights even though he is scared. Some get over their fright in a few minutes under fire, for others it takes days. But the real Marine never lets fear of death overpower his sense of honor, his sense of duty to his Country, Family and his Manhood.

All through your careers, you men have bitched about what you call 'CHICKEN SHIT DRILL!!' that, like everything else in the Marine Corps, is for a definite purpose, obedience to orders and to create a sense of alertness! This must be bred into every Marine. I don't give a good damn about any Marine who is not always on his toes. The Battalion is ready for whatever may be in store for us. A Marine to continue breathing has got to be alert at all times if not, some day in combat some Commie Son of a Bitch will sneak up behind him and beat him to death with a sock full of you know what! Somewhere's near Hagaru in North Korea there are over 750 neatly marked graves, 'All because one man went to sleep on the job', but they are Chinese Communist graves for we caught the Bastards asleep before his own Officers did.

This Marine Battalion is a team; we live, sleep, eat and fight as a team. This individual heroic stuff is a lot of junk. The Bilious Bastards who write that kind of stuff for 'TRUE' magazine didn't know any more about real fighting than they do about screwing. We have the best food, the best equipment, the best spirit and Marines in this battalion that there are in the whole world. Why by GOD!!! I actually pity any Viet Cong sons of bitches this battalion ever goes up against!!! By GOD I do! My Marines don't surrender – I don't ever want to hear of any under my command being captured – even if you get hit you can still fight. That's just not junk either. The kind of Marines I want in this Battalion is like 'MAYOR' of Birmingham Ala. – who even though hit in the chest at the [Chosin] reservoir crawled out, took a tommy-gun and ammo off a dead chink and shot the hell out of 20 more Chinese before he was evacuated – or my M.G. PLAT CMDR Lt John Miller, who even though he had been shot in the arm said 'Skipper I'm not leaving because at least with my one good hand and arm I can load magazines for your BAR and you can continue to blast the bastards to the sea' – and 'Vinegar John' was right beside me all the way down!! All the real heroes are not storybook combat fighters, every single Marine in this Battalion plays a vital role. DON'T EVER LET DOWN!!, thinking your role is unimportant.

Every Marine in this Battalion is a link in a chain. What if every mule or truck-driver decided he didn't like the whine of shells overhead, turned yellow and jumped in a ditch – this bird would say to himself, hell they'll never miss me just one guy in a thousand – what would our country, our loved ones, our homes, even the world be? No, thank GOD Marines don't think like that, every fire team, every Marine does his job. Every section, every squad is highly important to winning any campaign. Our supply section is vital to provide the chow, bullets and machinery of war to keep us rolling. Our motor transport to bring up the food, bullets, for where we may be called upon to fight there isn't a hell of a lot to steal. Every Marine in our mess section even the one who heats the water for mess gear so we don't get the G.I.'s has a job to do. Yes, even our Chaplain – is important, for if we got killed and he wasn't there to bury us we'd all go to HELL!!

LT.-COL. JOSEPH 'BULL' FISHER USMC

307.

Phil Caputo (b. 1941), author of the celebrated 1977 Vietnam memoir A Rumor of War, *describes his introduction to war, as a young officer posted to Da Nang in March 1965.*

It was a peculiar period in Vietnam, with something of the romantic flavour of Kipling's colonial wars. It was not so splendid for the Vietnamese, of course, and in early April we got a hint of the nature of the contest that was being waged in the bush. Two Australian commandos, advisers to an ARVN Ranger Group, walked into Charley Company's area. They were tough-looking characters, with hatchet-hard faces, and were accompanied by an even tougher-looking Ranger, whose eyes had the burned-out look of a man no longer troubled by the things he had seen and done. The Aussies looked up Sergeant Loker, Tester's platoon sergeant, who had once served as adviser with them. There was a noisy reunion.

A few of us, curious about these strangers, gathered nearby to listen. The Australians were describing a fire-fight they had been in that morning. The details of this clash have vanished from my memory, but I recall the shorter of the two saying that their patrol had taken a 'souvenir' off the body of a dead VC, and pulled something from his pocket and, grinning, held it up in the way a fisherman posing for a photograph holds up a prize trout. Nothing could have been better calculated to give an idea of the kind of war Vietnam was and the kind of things men are capable of in war if they stay in it long enough. I will not disguise my emotions. I was shocked by what I saw, partly because I had not expected to see such a thing and partly because the man holding it was a mirror image of myself – a member of the English-speaking world. Actually, I should refer to 'it' in the plural, because there were two of them, strung on a wire; two brown and bloodstained human ears.

PHILIP CAPUTO

308.

Nicholas Tomalin (1931–73) made several trips to Vietnam for the London Sunday Times. *This was his most famous dispatch, filed in June 1966.*

After a light lunch last Wednesday, General James Hollingsworth, of Big Red One, took off in his personal helicopter and killed more Vietnamese than all the troops he commanded. The story of the General's feat begins in the divisional office, at Ki-Na, twenty miles north of Saigon, where a Medical Corps colonel is telling me that when they collect enemy casualties they find themselves with more than four injured civilians for every wounded Viet Cong. The General strides in, pins two medals for outstanding gallantry to the chest of one of the colonel's combat doctors. Then he strides off again to his helicopter, and spreads out a polythene-covered map to explain our afternoon's trip. The General has a big, real American face, reminiscent of every movie general you have seen. He comes from Texas, and is forty-eight.

His present rank is Brigadier General, Assistant Division Commander, 1st Infantry Division, which is what the big red figure one on his shoulder flash means.

'Our mission today', says the General, 'is to push those goddam VCs right off Routes 13 and 16. Now you see Routes 13 and 16 running north from Saigon toward the town of Phuoc Vinh, where we keep our artillery. When we got here first we prettied up those roads, and cleared Charlie Cong right out so we could run supplies up. I guess we've been hither and thither with all our operations since, an' the ol' VC he's reckoned he could creep back. He's been puttin' out propaganda he's goin' to interdict our right of passage along those routes. So this day we aim to zapp him, and zapp him, and zapp him again till we've zapped him right back where he came from. Yes, sir. Let's go.'

The General's UH18 helicopter carries two pilots, two 50-calibre machine-gunners, and his aide, Dennis Gillman, an apple-cheeked subaltern from California. It also carries the General's own M16 carbine (hanging on a strut), two dozen smoke bombs, and a couple of CS anti-personnel gas-bombs, each as big as a small dustbin. Just beside the General is a radio console where he can tune in on orders issued by battalion commanders flying helicopters just beneath him, and company commanders in helicopters just below them. Under this interlacing of helicopters lies the apparently peaceful landscape beside Routes 13 and 16, filled with farmhouses and peasants hoeing rice and paddy fields.

So far today, things haven't gone too well. Companies Alpha, Bravo and Charlie have assaulted a suspected Viet Cong HQ, found a few tunnels but no enemy. The General sits at the helicopter's open door, knees apart, his shiny black toecaps jutting out into space, rolls a filter-tip cigarette to-and-fro in his teeth, and thinks.

'Put me down at Battalion HQ,' he calls to the pilot.

'There's sniper fire reported on choppers in that area, General.'

'Goddam the snipers, just put me down.'

Battalion HQ at the moment is a defoliated area of four acres packed with tents, personnel carriers, helicopters and milling GIs. We settle into the smell of crushed grass. The General leaps out and strides through his troops.

'Why General, excuse us, we didn't expect you here,' says a sweating major.

'You killed any 'Cong yet?'

'Well no General, I guess he's just too scared of us today. Down the road a piece we've hit trouble, a bulldozer's fallen through a bridge, and trucks coming through a village knocked the canopy off a Buddhist pagoda. Saigon radioed us to repair that temple before proceeding – in the way of civic action, General. That put us back an hour …'

'Yeah. Well Major, you spread out your perimeter here a bit, then get to killin' VCs will you?'

Back through the crushed grass to the helicopter.

'I don't know how you think about war. The way I see it, I'm just like any other company boss, gingering up the boys all that time, except I don't make money. I just kill people, and save lives.'

In the air the General chews two more filtertips and looks increasingly forlorn. No action on Route 16, and another Big Red One general has got his helicopter in to inspect the collapsed bridge before ours.

'Swing us back along again,' says the General.

'Reports of fire on choppers ahead, sir. Smoke flare near spot. Strike coming in.'

'Go find that smoke.'

A plume of white rises; in the midst of dense tropical forest, with a Bird Dog spotter plane in attendance. Route 16 is to the right; beyond it a large settlement of red-tiled houses. Two F-105 jets appear over the horizon in formation, split, then one passes over the smoke, dropping the trail of silver, fish-shaped canisters. After four seconds' silence, light orange fire explodes in patches along an area fifty yards wide by three-quarters of a mile long. Napalm. The trees and bushes burn, pouring dark oily smoke into the sky. The second plane dives and fire covers the entire strip of dense forest.

'Aaaaah,' cries the General. 'Nice. Nice. Very neat. Come in low, let's see who's left down there.'

'How do you know for sure the Viet Cong snipers were in that strip you burned?'

'We don't. The smoke position was a guess. That's why we zapp the whole forest.'

'But what if there was someone, a civilian, walking through there?'

'Aw come son, you think there's folks just sniffing flowers in tropical vegetation like that? With a big operation on hereabouts? Anyone left down there, he's Charlie Cong all right.'

I point at a paddy field full of peasants less than half a mile away. 'That's different son. We know they're genuine.'

The pilot shouts: 'General, half-right, two running for that bush.'

'I see them. Down, down, goddam you.'

In one movement he yanks his M16 off the hanger, slams in a clip of cartridges and leans right out of the door, hanging on his seatbelt to fire one long burst in the general direction of the bush.

'General, there's a hole, maybe a bunker, down there.'

'Smokebomb, circle, shift it.'

'But General, how do you know those aren't just frightened peasants?'

'Running? Like that? Don't give me a pain. The clips, the clips, where in hell are the cartridges in this ship?'

The aide drops a smoke canister, the General finds his ammunition and the starboard machine-gunner fires rapid bursts into the bush, his tracers bouncing up off the ground round it.

We turn clockwise in ever tighter, lower circles, everyone firing. A shower of spent cartridge cases leaps from the General's carbine to drop, lukewarm, on my arm.

'I … WANT … YOU … TO … SHOOT … RIGHT … UP … THE … ASS … OF … THAT … HOLE … GUNNER.'

Fourth time round the tracers flow right inside the tiny sand-bagged opening, tearing the bags, filling it with sand and smoke.

The General falls back off his seatbelt into his chair, suddenly relaxed, and lets out an oddly feminine, gentle laugh. 'That's it,' he says, and turns to me, squeezing his thumb and finger into the sign of a French chef's ecstasy.

We circle now above a single-storey building made of dried reeds. The first burst of fire tears the roof open, shatters one wall into fragments of scattered straw, and blasts the farmyard full of chickens into dismembered feathers.

'Zapp, zapp, zapp,' cries the General. He is now using semi-automatic fire, the carbine bucking in his hands.

Pow, pow, pow, sounds the gun. All the noises of this war have an unaccountably Texan ring.

'There's nothing alive in there,' says the General. 'Or they'd be skedaddling. Yes there is, by golly.'

For the first time I see the running figure, bobbing and sprinting across the farmyard towards a clump of trees dressed in black pyjamas. No hat. No shoes.

'Now hit the tree.'

We circle five times. Branches drop off the tree, leaves fly, its trunk is enveloped with dust and tracer flares. Gillman and the General are now firing carbines side by side in the doorway. Gillman offers me his gun: No thanks.

Then a man runs from the tree, in each hand a bright red flag which he waves desperately above his head.

'Stop, stop, he's quit,' shouts the General, knocking the machine-gun so tracers erupt into the sky.

'I'm going down to take him. Now watch it everyone, keep firing round-about, this may be an ambush.'

We sink swiftly into the field beside the tree, each gunner firing cautionary bursts into the bushes. The figure walks towards us.

'That's a Cong for sure,' cries the General in triumph and with one deft movement grabs the man's short black hair and yanks him off his feet, inboard. The prisoner falls across Lieutenant Gillman and into the seat beside me.

The red flags I spotted from the air are his hands, bathed solidly in blood. Further blood is pouring from under his shirt, over his trousers.

Now we are safely in the air again. Our captive cannot be more than sixteen years old, his head comes just about up to the white name patch – Hollingsworth – on the General's chest. He is dazed, in shock. His eyes calmly look first at the General, then at the Lieutenant, then at me. He resembles a tiny, fine-boned wild animal. I have to keep my hand firmly pressed against his shoulder to hold him upright. He is quivering. Sometimes his left foot, from some nervous impulse, bangs hard against the helicopter wall. The Lieutenant applies a tourniquet to his right arm. 'Radio base for an ambulance. Get the information officer with a camera. I want this Commie bastard alive till we get back … just stay with us till we talk to you, baby.'

The General pokes with his carbine first at the prisoner's cheek to keep his head upright, then at the base of his shirt. 'Look at that now,' he says, turning to me. 'You still thinking about innocent peasants? Look at the weaponry.' Around the prisoner's waist is a webbing belt, with four clips of ammunition, a water bottle (without stopper), a tiny roll of bandages, and a propaganda leaflet which later turns out to be a set of Viet Cong songs, with a twenty piastre note folded in it. Lieutenant Gillman looks concerned. 'It's OK, you're OK,' he mouths at the prisoner, who at that moment turns to me and with a surprisingly vigorous gesture waves his arm at my seat. He wants to lie down.

By the time I have fastened myself into yet another seat we are back at the landing pad. Ambulance orderlies come aboard, administer morphine, and rip open his shirt. Obviously a burst of fire has shattered his right arm up at the shoulder. The cut shirt now allows a large bulge of blue-red tissue to fall forward, its surface streaked with white nerve fibres and chips of bone (how did he ever manage to wave that arm in surrender?).

When the ambulance has driven off the General gets us all posed round the nose of the chopper for a group photograph like a gang of successful fishermen, then clambers up into the cabin again, at my request, for a picture to show just how he zapped those VCs. He is euphoric. 'Jeez I'm so glad you was along, that worked out just dandy. I've been written up time and time again back in the States for shootin' up VCs, but no one's been along with me like you before. I'll say perhaps your English generals wouldn't think my way is all that conventional, would they? Well, this is a new kind of war, flexible, quickmoving. Us generals must be on the spot to direct our troops. The helicopter adds a new dimension to battle. There's no better way to fight than goin' out to shoot VCs. An' there's nothing I love better than killin' Cong. No sir.'

NICHOLAS TOMALIN

309.

Karl Marlantes (b. 1944), a 1969–70 US Marine officer,
is the author of Matterhorn, *one of the great Vietnam novels,*
published in 2010. He wrote this in a later work, What It Is
Like to Go to War.

People lie. They lie in business, they lie in universities, they lie in marriages, and they lie in the military. Lying, however, is usually considered not normal, an exception. In Vietnam, lying became the norm, and I did my part. Kill ratios and body counts were the prime examples of how 'normal' became abnormal. Or was it vice versa? Take a typical squad-level firefight on a routine security patrol, Lance-Corporal Smithers in charge. All sorts of shit breaks loose. Teenagers get killed and maimed. The radio nets go crazy with artillery missions, sit-reps, mortar missions, anxious platoon commander, anxious company commander, anxious battalion commander, S-2, S-3 – all wanting to know what's happening. When the shooting stops, the pressure begins for the most important piece of information, the sole justification in the Vietnam war for all this sorrow: the body count. Smithers, what's the score?

The teenage adrenaline-drained patrol leader has to call in the score so analysts, newspaper reporters and politicians back in Washington have something to do. Smithers's best friend has just been killed. Two other friends are missing pieces of their bodies and going into shock. Smithers is tired and has a lot of other things on his mind. With scorekeepers often 25 kilometres away, no one is going to check on the score. In short, Smithers has a great incentive to lie. He also has a great need to lie. His best friend is dead. 'Why?', he asks himself. This is where the lying in Vietnam all began. It had to fill the long silence following Smithers's anguished 'Why?'

Imagine the scene in ancient Greece if the Greeks had the same attitude. Word has just reached Greek headquarters that Leonidas and the three hundred Spartans have all died defending the pass at Thermopylae. 'Good Zeus! They're all dead? What was the body count? What do you mean you didn't think it was important? If you think I'm going to take this news to Themistocles with nothing for show for it but a bunch of dead Spartans, you'd better think again.'

KARL MARLANTES

310.

Field-Marshal Lord Guthrie (b. 1938) confesses to an attack of cowardice in 1966, while leading an SAS patrol in the remotenesses of then-British Aden, modern Yemen.

As dusk fell one evening we were holed up in a wadi making the best of our evening meal when four local boys appeared, no more than 15 or 16 years of age. One of them said in English 'We look for English doctor.' I raised my hand, to which the boy, reverting to Arabic, asked if he could be circumcised. This prompted a lot of hilarity among my troopers, who said I should 'give it a go'. One said 'Don't worry Boss, I'll hold it for you, you just cut it away.' Rather to my shame, I flunked it. The poor boys broke down in tears, knowing that their tribal elder would perform the operation with a bit of flint.

CHARLES GUTHRIE

311.

On 19 July 1972, a nine-man SAS training team, commanded by Captain Mike Kealy (1945–79), was deployed in a mud-walled house at Mirbat in Oman, amid an insurgency in which British troops were covertly supporting the Sultan.

A few hundred yards from the house stood a fort, beside it a pit in which stood a 25-pounder manned by one of the regiment's best-beloved characters, a huge Fijian named Sergeant Talaiasi Labalaba, aged thirty, together with an Omani soldier, Walid Khafan. At 0530, just as dawn was breaking, incoming shelling and machine-gun fire signalled the start of an attack by several hundred insurgents, which raged through the next four hours, as wave after wave pressed forward through the encircling wire, to within thirty yards of the defenders. When the gunpit temporarily fell silent Mike Kealy, a small, fair-haired man of twenty-three, quiet and self-effacing, raced across seven hundred yards of open ground with Trooper Tobin, to go to the aid of the crew.

They found Labalaba with his jaw blown off, still trying to fire the 25-pounder at point-blank range over open sights. Hardly had they reached the pit when he was shot dead. Tobin was mortally wounded, but continued to fight off the enemy, as did Kealy. Their superhuman courage – and that of all the defenders – enabled the defenders to hold out, but only just. I should have liked to recommend Kealy for the Victoria Cross, but this was politically impossible, for a VC would have attracted far too much attention and publicised our presence in Oman. Instead he received a DSO, while Corporal Bob Bradshaw was awarded a Military Medal, Trooper Tobin a posthumous Military Medal, and Sergeant Labalaba a posthumous Mention in Despatches – although for security reasons, no announcement of any of the medals was made until four years after the event.

PETER DE LA BILLIÈRE

312.

A platoon commander of the Parachute Regiment describes a July 1973 night in Belfast, at the height of Northern Ireland's Troubles. His account is taken from a book which prompted shock on first publication in 1983, because of its frankness about the uglier aspects of 'peacekeeping'. Here, the army's attentions were directed against Protestant paramilitaries.

Standing around in the Ops Room when a call comes through on the radio. A panic voice, indistinct amid the crackle of static. Immediately the whole atmosphere changes. The stand-by section commander, half-asleep in his chair, suddenly wide awake and half-way out of the door to get the section ready and Saracen [armoured vehicle] fired up. Within two minutes I've joined them with the information that a patrol has been ambushed. 'Straight down to the Shankill, turn left and up to the Agnes Street junction. Go like hell.' I'm shouting into the driver's ear above the roar of the engine. 'When we get there, I'm dropping three of you off short of the turn and want you down in fire positions covering the rear. Jimmy, take the other three round the immediate area. There's a back-up on its way as soon as they get a Pig [armoured vehicle] out to pick them up.' Shouted orders. Thumping heart. Eyes wide in expectation. Cocked weapons. Holy fuck, this is it!

There's a grey Morris 1800 in the middle of the street. Confusion. A Military Policeman kneeling beside the front wing. Another appearing from a doorway.

'What the fuck are you doing here?'

'We were just doing a routine car patrol, checking on stolen vehicles.'

'You were what?' I'm incredulous. I don't believe my ears. Two M.P.s casually driving down the Shankill in the middle of the night in a civilian car in full uniform.

'Oh, forget it. What happened?'

'As we were driving along, we were fired at from down there.' Pointing away to the south of the road. 'We stopped and returned the fire.'

'Did you hit anything?'

'No.'

'How many rounds?'

'They fired about four or five, I think. My mate loosed off his magazine.'

'What, the whole lot?'

'Yes.'

Thirty-odd rounds sprayed into the darkness. What would they have done if the gunmen had stayed around for a second go? Crap-hats.

Jimmy returned from his snoop around and found nothing. Not even a sign of life in the vicinity. Anti-climax. Another fucking anti-climax. The tension screw winds a little tighter. The lads are giving the monkeys hell. The back-up patrol has arrived along with the O.C., the C.O., the Battalion Ops officer and God knows what else. If only they would all stay out of the way and let us get on with it.

I manage to get my guys back into the Saracen and we are meandering slowly back to Leopold Street. The high-pitched whine in low gear is shattering, and guaranteed to wake even the heaviest of sleepers.

Wake up you bastards! If we don't sleep you don't. A couple of guys walking down the street with a girl. Great! Pile out, rifles levelled.

'Assume the position. Hands on the wall, fingers spread, now get those legs apart. Afraid you'll drop something?'

The girl is standing still, not saying a word. All our questions meeting no response. They're more afraid of what the 'boys' will do to them than what we have to offer. Soon change that, sunbeam. The dull thud of an idly swung baton up between the legs. Gasp of pain.

'Who told you to move, cunt.' Baton swings again, cracking hard onto a kneecap. Shrug from the tom in my direction. 'Looked as if he was going to hit you, boss.' A body search with hands up hard between the legs. Squeezing testicles. Tears in the eyes. Fear in the eyes. Hopelessness in the face, turning slowly to anger and hard resolve. That's good, mate, get angry, try something. Better still, go tell your mates.

'What are you doing with these two specimens, love? Why not come back with us? Bet they don't know what it's for. Nice girl like you.' Hungry-eyed toms, with open crudeness, visually undressing the girl.

'Is she a good fuck, mate? Do a good blow job, does she?' Chuckles and laughter.

'OK lads, you've had your fun, back in the vehicle. Let's go.'

Back to Leopold Street. Back to the cocoon. Back to the freaky non-talk of people grouped together for too long, to a private world behind a blanket stretched over the opening of a bunk, to the constant banter and false bravado. The 'street' is the reality and the unreality. The centre of the universe, the beginning and the end of time. The whole spectrum of all human existence in full living colour written on the walls, scored with the thump of explosive, photographed in the mind of a diseased body. Replayed every day, relived with boredom. Selling our morality to ourselves over and over again, with the help of war books, films, TV and sleep. Don't think of the rights and wrongs, just let the beast rise and enjoy the primeval passion.

An Army sniper who had just shot and killed a terrorist on the street was asked by a woman reporter what it was like to shoot someone. 'You just squeeze this little thing here,' he said. She went away with the

wrong impression, but no doubt just what she wanted to hear. The
soldier sadly shaking his head. We are here to create the news for a
hundred poised pens and ready cameras. To provide a nation with its
quota of violence, to give people the chance to shake their heads,
others to organise marches, pressure groups and all the other paraphernalia of a well-organised growing industry. Northern Ireland is an
industry, providing reporters with the opportunities to further their
already stagnant careers, for social workers to martyr themselves on the
unsympathetic conscience of an unimaginative nation. An entertainment without interlude. To hell with the lot of you.

A. F. N. CLARKE

313.

*On 6 October 1973, the Syrian and Egyptian attack on the Israeli
army achieved devastating surprise, and in its first hours came close to
achieving a decisive breakthrough. But the Israeli genius for battlefield
improvisation proved sufficient – just sufficient – to stem the assault.
The defence of the Golan Heights against overwhelming odds stands as
an epic of military endeavour, fragments of which I myself witnessed
from a relatively safe distance. Chaim Herzog, writing soon after the
Yom Kippur war, identified its heroes by pseudonyms.*

By the afternoon of 9 October, the position of the Israeli 7th Brigade
had become critical. Lt.-Col. Yossi had become world-famous in June
1967, when his photograph appeared on the cover of *Life* magazine
swimming in the Suez Canal after the triumphal climax of the Sinai
campaign. Now, he found himself in battle once again, under very
different circumstances. They had been fighting for four days and three
nights, without a moment's rest or respite, under constant fire. On
average each tank was left with three or four shells. At the height of
battle Avigdor turned and spoke to his operations officer. The officer
began to reply but suddenly in the middle of his sentence slid to the
floor of the armoured carrier, fast asleep. Avigdor spoke to Raful and
told him that he did not know if he could hold on. Already in a daze,
he described the condition of his brigade. Raful, as ever quiet, calm
and encouraging, pleaded with him, 'For God's sake, Avigdor, hold on!

Give me another half an hour. You will soon be receiving reinforcements. Try, please, hold on!'

At this critical moment, Lt.-Col. Yossi, leading remnants of the Barak Brigade with a force of eleven tanks, entered the divisional area and was directed by Raful to Avigdor. Yossi had handed over command of his battalion in the Barak Brigade on 4 September and decided that his honeymoon would be a non-conventional one. So with his newlywed wife, Naty, he flew to the Himalayas. On Yom Kippur eve they rode by motorbike to the Chinese frontier. Back in Katmandu for Yom Kippur, the receptionist in the hotel said to him, 'You're from Israel, aren't you? Something is happening in your area. You ought to listen to the news.' Racing against time Yossi and Naty, using every form of subterfuge, managed to fly back to Israel via Teheran and Athens. From Athens Yossi phoned his family to bring his uniform and equipment to the airport.

As he rushed northwards, little did he realize that he would receive command of the remnants of his former brigade. He hurried to Hofi's advanced headquarters and heard what had happened to the Barak Brigade. It was Tuesday morning. When Dov had reached the Barak Brigade centre, remnants began to arrive in dribs and drabs. Oded had in the meantime evacuated from the area of Tel Faris, taking with him some 140 infantry men who arrived on foot down the Gamla Rise. Dov and the other officers organized technical teams and began to recover abandoned tanks in the field, while ordnance units began to repair them. At noon on Tuesday a psychiatrist arrived from the medical centre of Tel Hashomer to take care of the soldiers of the Barak Brigade. He stood and looked at the dishevelled, unshaven, gaunt-eyed soldiers, some of them burnt and most of them blackened by the smoke and flames, working silently on the damaged tanks and putting them in shape. It was a moving and sobering sight. He asked them what they were doing and they explained that they were preparing the tanks to take them back into battle again. 'If they are going into battle again, I had better forget everything I ever learnt,' he remarked.

Dov notified command headquarters that he already had thirteen tanks ready for battle. He organized crews, brought in ammunition, begged some mortars and then he heard from command headquarters that Yossi was arriving to take command. The news of Yossi's arrival spread and Shmulick, who had been Yossi's second-in-command and who had been wounded in the first day of battle, escaped from the hospital in Safed and came to rejoin him and go back into battle. Conscious that they were to avenge the comrades of their brigade, Dov led Yossi's force to the front in a jeep. As they approached and received orders to join the 7th Brigade, Yossi heard on the radio that Tiger on the southern sector of the brigade front was out of ammunition and unable to hold out on the slopes of the 'Booster' against the Syrian advance.

Tiger's force was by now left with two shells per tank. 'Sir,' he radioed in a tone of desperation to the brigade commander, 'I can't hold on.'

'For heaven's sake hold on for only ten minutes,' implored Avigdor. 'Help is on the way.' When Tiger ran out of shells completely, he began to fill his pockets with hand grenades and withdraw. At this moment Yossi moved up to the 'Booster', opened fire and in the initial clash destroyed some thirty Syrian tanks. He had arrived just as the 7th Brigade, left with 7 running tanks out of an original total of approximately 100, was on the verge of collapse. Both sides had fought to a standstill. Avigdor had told Raful that he could not hold the Syrian attack, but suddenly a report came in from the A3 fortification (surrounded by Syrians and well behind the Syrian advance forces), that the Syrian supply trains were turning round and withdrawing. The Syrian attack had been broken; their forces broke and began to withdraw in panic.

The remnants of the 7th Brigade, including Yossi's reinforcements, totalled some twenty tanks. Exhausted, depleted to a minimum, many wounded, with their tanks bearing the scars of war, they now began to pursue the Syrians, knocking out tanks and armoured personnel carriers as they fled. On the edge of the anti-tank ditch, they stopped: the brigade had reached the limits of human exhaustion.

Avigdor stood in a daze looking down on the Valley of the Tears. Some 260 Syrian tanks and hundreds of armoured personnel carriers and vehicles lay scattered and abandoned across this narrow battlefield between the Hermonit and the 'Booster'. In the distance he could see the Syrians withdrawing in a haze of smoke and dust, the Israeli artillery following them. Raful's quiet voice came through on the earphones as he addressed them on the network of the 7th Brigade. 'You have saved the people of Israel.'

CHAIM HERZOG

314.

Lt. Hiroo Onoda (1922–2014) of the Imperial Japanese Army provided one of the strangest postscripts to World War II. Along with many other Japanese, when his position in the Philippines was overrun by the Americans in 1945, he retreated into the wilderness. In the years that followed, first the Americans, then the Philippines army and police, and finally Japanese well-wishers attempted to hunt down the fugitives. Almost all were killed or induced to surrender. But when a young Japanese traveller, Norio Suzuki, finally made contact with Onoda in his fastness early in 1974, the soldier declined to surrender, or even to concede the reality of Japan's defeat, until he had received orders to this effect from his former commanding officer, Major Yoshimi Taniguchi. At last, the profoundly wary Onoda approached an agreed rendezvous.

I hid in the bushes, waiting for the time to pass. It was a little before noon on March 9, 1974, and I was on a slope about two hours away from Wakayama Point. My plan was to wait until the time of the evening when it is still just possible to tell one face from another and then approach Wakayama Point rapidly, in a single manoeuvre. Too much light would mean danger, but if it were too dark, I would not be able to make sure that the person I was meeting was really Major Taniguchi. Also, late twilight would be a good time for making a getaway, if I should have to. Just after two in the afternoon, I crept cautiously out of my hiding place and crossed the river.

Making my way through a grove of palms, I soon came to an area where the islanders cut trees for building. At the edge of a clearing, I stopped and looked the place over. I could see nobody around, but to be on the safe side, I camouflaged myself with sticks and dried leaves before dashing across the shelterless area. I reached a position about three hundred yards from the appointed spot. This was the place where I had met and talked with Norio Suzuki, just two days earlier. A message from Suzuki asking me to meet him here again had been left in an agreed message box. I was still afraid it might be a trap. The enemy might be waiting for me on the hill.

I proceeded with the utmost caution but saw no signs of life.

At the top of the hill, I saw a yellow tent. I could make out a Japanese flag waving above it, but I could not see anybody. Were they resting in the tent? Or were they hiding somewhere else waiting for me to show up? After thirty tense minutes, I decided they must be in the tent and settled down to wait for sunset. I inspected my rifle and retied my boots. I was confident. I felt strong because I had rested while keeping watch. I jumped over a barbed-wire fence and made for the shade of a nearby bosa tree, where I paused, took a deep breath, and looked at the tent again. The time came. I gripped my rifle, thrust out my chest, and walked forward into the open.

Suzuki was standing with his back to me, between the tent and a fireplace they had rigged up by the riverbank. Slowly he turned around, and when he saw me, he came toward me with arms outstretched. 'It's Onoda!' he shouted. 'Major Taniguchi, it's Onoda!' In the tent, a shadow moved, but I went forward anyway. Suzuki, eyes bursting with excitement, ran up to me and with both hands clasped my left hand. I stopped about ten yards from the tent, from which there came a voice.

'Is it really you, Onoda? I'll be with you in a minute.'

I could tell from the voice that it was Major Taniguchi. Motionless, I waited for him to appear. Suzuki stuck his head in the tent and brought out a camera. From inside, the major, who was shirtless, looked out and said, 'I'm changing my clothes. Wait just a minute.' The head disappeared, but in a few moments Major Taniguchi emerged from the tent fully clothed and with an army cap on his head. I barked out, 'Lieutenant Onoda, Sir, reporting for orders.'

'Good for you!' he said, walking up to me and patting me lightly on the left shoulder. 'I've brought you these from the Ministry of Health and Welfare.' He handed me a pack of cigarettes with the chrysanthemum crest of the emperor on them. I accepted it and, holding it up before me in proper respect for the emperor, fell back two or three paces. Major Taniguchi said, 'I shall read your orders.' I held my breath as he began to read from a document that he held up formally, 'Command from Headquarters, Fourteenth Area Army ... Orders from the Special Squadron, Chief of Staff's Headquarters, Bekebak, September 19, 1900 hours.

'1. In accordance with the Imperial Command, the Fourteenth Area Army has ceased all combat activity.

'2. In accordance with Military Headquarters Command No. A-2003, the Special Squadron in the Chief of Staff's Headquarters is relieved of all military duties.

'3. Units and individuals under the command of the Special Squadron are to cease military activities and operations immediately and place themselves under the command of the nearest superior officer. When no officer can be found, they are to communicate with the American or Philippine forces and follow their directives.

'Special Squadron, Chief of Staff's Headquarters, Fourteenth Area Army, Major Yoshimi Taniguchi.'

After reading this, Major Taniguchi paused slightly, then added, 'That is all.' I stood quite still, waiting for what was to follow. I felt sure Major Taniguchi would come up to me and whisper, 'That was so much talk. I will tell you your real orders later.' After all, Suzuki was present, and the major could not talk to me confidentially in front of him. I watched the major closely. He merely looked back rather stiffly. Seconds passed, but still he said no more. The pack on my back suddenly seemed very heavy.

Major Taniguchi slowly folded up the order, and for the first time I realized that no subterfuge was involved. This was no trick – everything I had heard was real. There was no secret message. The pack became still heavier. We really lost the war! How could they have been so sloppy? Suddenly everything went black. A storm raged inside me. I felt like a fool for having been so tense and cautious on the way here. Worse than that, what had I been doing for all these years? I pulled back the bolt on my rifle and unloaded the bullets.

HIROO ONODA

315.

*Launched from bases 2,000 miles away, the rescue of almost all the
102 hostages held by terrorist hijackers of an Air France Airbus at
Entebbe airport, Uganda, on the night of 3 July 1976, was probably
the most daring special forces operation of all time. The Sayeret
Matkal – 'General Staff Reconnaissance unit' – the Israeli equivalent
of Britain's SAS – which spearheaded the assault was commanded by
Lt.-Col. Yoni Netanyahu (1946–76), elder brother of Binyamin, the
later prime minister. With his deputy, Major Muki Betser, he and his
men landed in darkness aboard a Hercules transport, then set off in a
Mercedes followed by two Land Rovers, to drive the mile to the Old
Terminal where the hostages were held, hoping to pass as a Ugandan
senior officer and his escort. The plan almost came disastrously
unstuck.*

Just as they turned onto the approach road, the occupants of the
Mercedes could see in its headlights two Ugandan sentries. The one
on the left quickly vanished into the darkness, but his comrade aimed
his rifle at the Mercedes and called out 'Advance!' Betser assumed
from his time in Uganda that it was a routine challenge and that the
soldier was unlikely to fire. They could drive by him. 'Eighty, seventy,
sixty,' he said under his breath, counting down to the moment when
they would leave the car. 'Amutzur,' said Yoni, 'cut to the right and
we'll finish him off.' Betser was horrified. 'Leave him, Yoni,' he said
quietly but emphatically. 'It's just his drill.' After the briefest pause
Netanyahu repeated his order, convinced that the sentry was suspicious. They had to take him out. Netanyahu and Zussman drew and
cocked their silenced Berettas. Betser felt they were making a fatal
mistake. 'Yoni, no!' he urged. 'Don't shoot!' But his warning was
ignored.

At a range of ten yards they opened fire from the moving car, the
silencers turning the crack of their shots into a 'bare whisper'. The
sentry tottered and fell. Suddenly the silence was shattered by a long
burst of automatic fire. Betser jerked his head round to see the sentry,
who must have got back on his feet, crumple in a hail of Kalashnikov

bullets from one of the Israeli Land Rovers. The sound of the gunshots had put the whole operation in jeopardy. The element of surprise had been well and truly lost. Betser, for one, expected the terminal building to vanish at any moment 'in a fireball of explosions as the terrorists followed through with their threats to blow up the hostages'. 'Drive!' shouted Netanyahu at Kafri, who had braked. 'Fast!' Kafri stamped on the accelerator but had only covered a short distance when bullets spat out of the darkness. Yoni and Betser both yelled 'Stop!' The car screeched to a halt, as did the Land Rovers behind.

They had stopped at least fifty yards from the Old Terminal, rather than the five they had planned for. Flinging his door open and shouting at the others to follow, Betser began running. He could hear the thumping of boots behind him and knew that his break-in team were close behind. From the darkness came a burst of fire. Behind him the rest of the assault team were bunched up. It was a 'complete contradiction of the battle plan'. Netanyahu shouted at Betser to move forward. [The colonel] was almost opposite the first entrance, and not far from the point where he planned to set up his command post, when he stopped. By pausing in open ground, he was vulnerable to Ugandan snipers. Shots rang out from the control tower, and Netanyahu was struck in the chest and lower right arm. Sighing, he collapsed to the ground. [The Israeli CO was mortally wounded, but Betser and the rest of the assault team stormed the Old Terminal, killing the seven terrorists who, by supreme good fortune, sought to shoot it out with the attackers, rather than exploit the warning given by the exchanges of fire with the Ugandans to massacre the hostages.

Almost nine hours later, the pilot of one of the Hercules carrying the French, American and Israeli passengers rescued at a cost of five dead, including Netanyahu, broadcast to the soldiers and hostages crammed in the hold.] 'Ladies and gentlemen,' he said in Hebrew, 'we are now flying above Sharm El Sheikh.' They had reached Israeli territory. The hostages knew that their long and traumatic ordeal was finally over. They let out a great cheer and embraced those near to them. Some were in tears. It was now that Eran Dolev, head of the surgical team, revealed his dry sense of humour. 'I have to tell you,' he said over the intercom, 'that last night Israel introduced VAT for the first time at eight per cent. If you don't like it, you're free to return to Entebbe.' No one laughed.

SAUL DAVID

316.

In December 1979, the Kremlin ordered the overthrow and killing of Afghanistan's president Hafizullah Amin. Soviet troops and KGB personnel were committed to night assaults on the presidential palace and other key installations. These attained the black comic quality that has characterized all foreign interventions in Afghanistan through two centuries.

The most important and difficult target was the General Staff building. Fourteen special forces, accompanied by Abdul Wakil, a future foreign minister, were assigned to deal with it. A deception plan was devised to ease the odds. That evening General Kostenko, the Soviet adviser to Colonel Yakub, the Chief of Staff, took a number of Soviet officers to pay a formal call, including General Ryabchenko, the commander of the newly arrived Guards Air Assault Division. They discussed questions of mutual interest with the unsuspecting Yakub, a powerful man who had trained in the Ryazan Airborne School and spoke good Russian. Ryabchenko had no difficulty in behaving naturally, since he knew nothing of what was about to take place. Meanwhile other Soviet special forces officers were spreading through the building, handing out cigarettes and chatting to the Afghan officers working there.

When the explosion at the communications centre took place, they burst into Yakub's office. Yakub fled to another room after a scuffle in which his assistant was killed, but then surrendered and was tied up and placed under guard. The fighting lasted an hour. As it died away, Abdul Wakil appeared in Yakub's office. He talked in Pushtu to the general for a long time, and then shot him. Twenty Afghans were killed. A hundred were taken prisoner, and as they so heavily outnumbered the attackers, they were herded into a large room and tied up with electric cable.

There was an unpleasant moment when a company of Soviet paratroopers, who had arrived forty minutes late, advanced on the General

Staff building in armoured personnel carriers and opened up a heavy fire, forcing [their compatriots] inside to take cover as tracer bullets flew across the room, glowing like red fireflies. Order was restored and the paratroopers helped to secure the building. The Russians needed the Radio and Television Centre to broadcast [the new puppet president] Karmal's appeal to the people. They reconnoitred it very carefully throughout 27 December, some of them posing as automation experts. In the assault seven Afghans were killed, twenty-nine wounded, and over a hundred taken prisoner. One Soviet soldier received a minor wound. There was no serious resistance at the Interior Ministry building, though one Russian soldier was wounded. The attackers had orders to arrest the Interior Minister, S. Payman, but he had fled in his underwear and sought refuge with his Soviet advisers.

By morning the firing had more or less died down. But not quite. As they drove into town in their Mercedes, the senior Soviet officers who had directed the attack on the palace were fired on by a nervous and trigger-happy young paratrooper. The bullets hit the car but not the occupants. A colonel jumped out and gave the soldier a sharp clip round the ear. General Drozdov asked the young lieutenant in charge, 'Was that your soldier? Thank you for not teaching him to shoot straight.'

Once the fighting in the Taj Bek Palace had stopped, Colonel Kolesnik set up his command post there. The victorious Soviet soldiers were dropping with fatigue. Since it was possible that Afghan troops in the area might try to retake the palace, they set up a perimeter defence, their nerves still at full stretch. When they heard rustling in the lift shaft, they assumed that Amin's people were launching a counter-attack, sprang to arms, fired their automatic weapons, hurled grenades.

It was the palace cat.

RODRIC BRAITHWAITE

317.

Medical support for the Soviet forces fighting in Afghanistan was as wickedly inadequate as everything else about their 1979–89 campaign, in which 57,000 men were wounded and 416,000 fell sick.

The medical services almost lost the battle not against war wounds, but against infectious disease. More than three-quarters of those Russians who served in Afghanistan spent time in hospital. Some 11 per cent were wounded or injured. The rest – 69 per cent of all those who served during the war – suffered from serious sickness: 28 per cent from infectious hepatitis, 7.5 per cent from typhoid fever, and the rest from infectious dysentery, malaria, and other diseases. Units were often far below strength because so many of the soldiers were sick. The main scourge was hepatitis. The joke among the soldiers was that soldiers got jaundice, officers got Botkin's disease, and generals were treated for hepatitis.

There were stories that some evaded duty by getting medical orderlies in the hospitals to give them urine from infected patients: if you drank it, you got the disease. By the end of 1981 every fourth soldier in the 5th Guards Motor-rifle Division had been laid low. The commander, his deputies, and all the regimental commanders were down at the same time. At any given moment up to a quarter, perhaps even a third, of the 40th Army might be incapacitated by disease. At the height of the epidemic, there was only one nurse for every three hundred patients.

Hepatitis was not the only problem. In the summer of 1985 a patrol from the Independent Motor-rifle Brigade in Jalalabad drank the water from a roadside spring as they were returning from patrol. A few days later three of them collapsed on parade with cholera. More than half the brigade fell sick. A rumour circulated that the water had been deliberately infected by 'two Europeans dressed in local clothes'. To prevent further infection, people said, the bodies were being cremated – something almost unheard of in what was still at heart an Orthodox country. The sick were isolated behind barbed wire, the doctors and nurses were isolated with them, and additional medical staff had to be flown in from Moscow.

The situation of the 40th Army thus resembled that of the British and French armies in the Crimean War, and for much the same reasons: dirty water, appalling sanitation, dirty cooks, dirty canteens, dirty clothes, poor diet. The Soviets prided themselves on the number of hospitals and orphanages they built in Afghanistan. But they filled more hospitals and orphanages than they constructed, and the bigger the hospital, the worse the sanitary conditions. Soviet medical services in the Second World War were much better.

RODRIC BRAITHWAITE

318.

An SAS trooper describes the storming of the Iranian Embassy in London's Princes Gate just before 8 p.m. on the night of 5 May 1980, six days after Arab terrorists occupied it, taking hostage its staff and visitors. The author, who served eighteen years in the regiment, also took part in the siege of Mirbat, described above.

We took up a position behind a low wall as the demolition callsign ran forward and placed the explosive charge on the Embassy french windows. It was then that we saw the abseiler swinging in the flames on the first floor. It was all noise, confusion, bursts of submachine-gun fire. I could hear women screaming. Christ! It's all going wrong, I thought. There's no way we can blow that charge without injuring the abseiler. Instant change of plans. The sledge-man ran forward and lifted the sledge-hammer. One blow, just above the lock, was sufficient to open the door. They say luck shines on the brave. We were certainly lucky. If that door had been bolted or barricaded, we would have had big problems.

'Go. Go. Go. Get in at the rear.' The voice was screaming in my ear. The eight callsigns rose to their feet as one and then we were sweeping in through the splintered door. All feelings of doubt and fear had now disappeared. I was blasted. The adrenalin was bursting through my bloodstream. I got a fearsome rush, the best one of my life. I had the heavy body armour on, with high-velocity plates front and back. During training it weighs a ton. Now it felt like a T-shirt. Search and destroy! We were in the library. There were thousands of books. As I adjusted my eyes to the half-light – made worse by the condensation on my respirator eyepieces – the thought occurred to me that if we had blown that explosive charge we might have set fire to the books. Then we would really have had big problems: the whole Embassy would have been ablaze in seconds.

The adrenalin was making me feel confident, elated. My mind was crystal-clear as we swept on through the library and headed for our first objective. I reached the head of the cellar stairs first, and was quickly joined by Sek and two of the callsigns. The entry to the stairs was blocked by two sets of step-ladders. I searched desperately for any signs of booby-traps. There wasn't time for a thorough check. We had to risk it. We braced ourselves and wrenched the ladders out of the way. Mercifully there was no explosion. The stairs were now cleared and we disappeared into the gloom of the basement.

I fished a stun-grenade out of my waistcoat and pulled the pin. Audio Armageddon, I thought as I tossed the grenade down into the darkness. We descended the stairs, squinting into the blinding flashes for any unexpected movement, any sign of the enemy, and then we were into the corridor at the bottom. We had no sledge, no Remington with us, so we had to drill the locks with 9-milly, booting the doors in, clearing the rooms methodically as we went along. Minutes turned into seconds; it was the fastest room clearance I'd ever done.

It was when I entered the last room that I saw the dark shape crouched in the corner. Christ! This is it, I thought. We've hit the jackpot. We've found a terrorist. I jabbed my MP5 into the fire position and let off a burst of twenty rounds. There was a clang as the crouched figure crumpled and rolled over. It was a dustbin! Nothing, not a thing. The cellars were clear. I was now conscious of the sweat. It was stinging my eyes, and the rubber on the inside of the respirator was slimy. My mouth was dry and I could feel the blood pulsing through my temples. And then we were off again, no time to stop now, up the cellar stairs and into the Embassy reception area. As we advanced across the hallway, there was smoke, confusion, a tremendous clamour of noise coming from above us. The rest of the lads, having stormed over the balcony at the front and blasted their way into the first floor of the building, were now systematically clearing the upper rooms, assisted by a winning combination of the stunning effect of the initial explosion, the choking fumes of CS gas, the chilling execution of well-practised manoeuvres and the sheer terror induced by their sinister, black-hooded appearance. We were intoxicated. Nothing could stop us now.

Through the gloom I could see the masked figures of the other team members forming into a line on the main staircase. My radio earpiece crackled into life. 'The hostages are coming. Feed them out through the back. I repeat, out through the back.' I joined a line with Sek. We were six or seven steps up from the hallway. There were more explosions. The hysterical voices of the women swept over us. Then the first hostages were passed down the line. I had my MP5 on a sling around my neck. My pistol was in its holster. My hands were free to help the hostages, to steady them, to reassure them, to point them in the right direction. They looked shocked and disorientated. Their eyes were streaming with CS gas. They stumbled down the stairs looking frightened and dishevelled. One woman had her blouse ripped and her breasts exposed. I lost count at fifteen and still they were coming, stumbling, confused, heading towards the library and freedom.

'This one's a terrorist!' The high-pitched yell cut through the atmosphere on the stairs like a screaming jet, adding to the confusion of the moment. A dark face ringed by an Afro-style haircut came into view; then the body, clothed in a green combat jacket, bent double, crouched in an unnatural pose, running the gauntlet of black-hooded figures. He was punched and kicked as he made his descent of the stairs. He was running afraid. He knew he was close to death. He drew level with me. Then I saw it – a Russian fragmentation grenade. I could see the detonator cap protruding from his hand. I moved my hands to the MP5 and slipped the safety-catch to 'automatic'.

Through the smoke and gloom I could see callsigns at the bottom of the stairs in the hallway. Shit! I can't fire. They are in my line of sight, the bullets will go straight through the terrorist and into my mates. I've got to immobilize the bastard. I've got to do something. Instinctively, I raised the MP5 above my head and in one swift, sharp movement brought the stock of the weapon down on the back of his neck. I hit him as hard as I could. His head snapped backwards and for one fleeting second I caught sight of his tortured, hate-filled face. He collapsed forward and rolled down the remaining few stairs, hitting the carpet in the hallway, a sagging, crumpled heap. The sound of two magazines being emptied into him was deafening. As he twitched and vomited his life away, his hand opened and the grenade rolled out. In that split second my mind was so crystal clear with adrenalin it zoomed straight in on the grenade pin and lever. I stared at the mechanism for what seemed like an eternity, and what I saw flooded the very core of me with relief and elation. The pin was still located in the lever. It was all over, everything was going to be okay.

I crossed the room to my holdall and as I began pulling off my assault equipment I could feel the tiredness spreading through my limbs. It wasn't just the energy expended on the assault, it was the accumulation of six days of tension and high drama, of snatched sleep in a noisy room, of anxiety and worry over the outcome. I looked to my left. The Toad had just returned. He looked tired, his face was flushed and he was out of breath. He looked at me and shook his head. 'I'm getting too old for this sort of thing.' 'So am I,' I replied.

PETE WINNER

319.

When, in April 1982, 3 Commando Brigade, Royal Marines received the order to embark for the Falkland Islands, it seemed wildly implausible that they would fight as they were to do six weeks later. But Nick Vaux (b. 1936), most impressive of the formation's commanding officers, one of whose battles I attended in June, had a nice sense of the melodramatic, as reported here by the Marines' local newspaper in Plymouth.

A bleak wind sliced across the parade ground at Bickleigh Barracks as 42 Commando paraded before their departure for the South Atlantic and a possible battle for the Falkland Islands. The Dewerstone glowered in the background under dark clouds as Maj.-Gen. Jeremy Moore, who commands Commando Forces, spoke to the hundreds of officers and men on parade in full fighting order. The parade moved off the square after Lt.-Col. Vaux had given the final order '42 Commando, to the South Atlantic – Quick March!'

WESTERN MORNING NEWS

320.

My own account, writing as a war correspondent, of the last day of the Falklands War, 14 June 1982.

I awoke from a chilly doze to find a thin crust of frozen snow covering my sleeping-bag and equipment in the dawn. Around me in the ruined sheep pen in which we lay, a cluster of white-tinted ponchos and rucksacks marked the limits of battalion headquarters. The inexhaustible voice of Major Chris Keeble, second-in-command of 2 Para, was holding forth into a radio handset as fluently as it had been two hours earlier, when I lost consciousness. All firing in front of us, where the battalion's rifle companies had stormed a succession of enemy positions in the darkness, was ended. Desultory Argentine shells were falling on untenanted ground some 600 yards to the right. We could hear heavy firing further south, where the Guards and Gurkhas were still fighting for their objectives.

Might I go forward, I asked? By all means, said the energetic major. He pointed across the hill to the new positions of the rifle companies, and detailed a soldier from the defence platoon to show me the way. We left the headquarters group huddled around their radios and bivouacs, and began to stride across the frozen tussock grass, my guide chattering about men he knew who had been hit during the night, and the amazing helicopter pilot John Greenhalgh who flew in his Gazelle without benefit of night-vision aids to bring up ammunition and recover the wounded.

We began to pass abandoned enemy positions, strewn with weapons and ammunition, clothing and food. 'Not short of much, was they?' said the Para laconically. 'So much for the navy's blockade.' We reached 'A' Company, a few hundred yards frontage of unshaven scarecrows surrounded by arms and equipment, their positions dotted with flickering flames from the little hexamine cookers on which they were brewing hot chocolate and porridge from their Arctic ration packs. Nearby stood the Scorpion and Scimitar light tanks which had supported the attack. Lt. Lord Robin Innes-Ker, who had seemed at times less keen than some of his more homicidal comrades about passing the summer soldiering in the South Atlantic, had at last entered into the spirit of the thing. 'Did you see us?' he enthused. 'It was tremendous. We fired a hundred and fifty rounds. Once we saw somebody light a cigarette and that was it ...'

Everyone agreed that the night had been a huge success, not least because those doing the talking had come through it alive. I sat down in an untenanted bivouac – a poncho flapping uneasily above a peat hag – to write a dispatch. A gunner forward observation officer came and sat beside me. He asked if I had the faintest idea what was going on. I said I only knew that I was due to join 3 Para late in the afternoon, in time to see them launch the next – and everybody hoped final – attack of the war that night. A cheeky eighteen-year-old private soldier put his head under the poncho and demanded a cigar, which I was not cheeky enough to refuse him. 'We did pretty good, didn't we? You get paid extra for doing this? Why does BBC always tell everybody where we're going to attack? How many more days do you reckon?'

Suddenly we heard men calling to each other through the snow shower outside: 'They're running away! It's on the radio! The Argies are running everywhere!' The company commander, Major Dare Farrar-Hockley, shouted to his platoon commanders to be ready to move in five minutes. I gathered my own equipment. 'Do you want a lift?' called Roger Field of the Blues and Royals. One of the most pleasant parts of this war was that, after so many weeks together, so many people knew each other. I scrambled clumsily up onto the hull of his Scimitar, and clung to a smoke projector as we bucketed across the hillside. We halted for a few moments by one of the enemy positions captured during the night. The men picked a path through plausible souvenirs. We speculated about the identity of a sad corpse covered with a poncho, its feet encased in British-issue rubber boots.

We clattered down the rock-clad ridge alongside the long files of 2 Para. As we approached the skyline, we saw soldiers lying, standing, crouching along it, fascinated by the vision below. They were gazing upon the wreckage of a cluster of large buildings three hundred yards beneath us – the former Royal Marine base at Moody Brook. Two or three miles down a concrete road eastward, white and innocent in the winter sunshine, stood the little houses and churches of Port Stanley.

Suddenly the focus of all our ambitions, which at dawn had been apparently as distant as the far side of the moon, lay open for the taking. The soldiers, three nights without sleep, began to chatter like schoolboys. The battalion commander, David Chaundler, was giving his orders: '… I'm not having anybody going down that road unless the high ground is covered, so I'm getting "B" Company up there. The Scimitars will stay here and provide a firebase.'

The first men were already threading their way through Moody Brook and up the opposite hillside. The Blues and Royals took up positions among the rocks from which they could cover the entire route to Port Stanley. 'A' Company was to march straight up the road. I trotted after Dare Farrar-Hockley. We crossed Moody Brook amidst orders shouted down the line to stay rigidly in file on the track because of the danger of mines. With deplorably selfish professional ambitions in my mind, I started to reflect aloud on the risk that 45 Commando might already be approaching Stanley from the other flank; that if negotiations started, our advance would be stopped in its tracks. The word was called forward to quicken the pace. Dare Farrar-Hockley, signaller and correspondent trailing in his wake, hastened past the files of the leading platoon to take position with the point section, each man praying silently for the radio to remain mute.

We passed a burning building opposite the seaplane jetty; abandoned vehicles; loose ammunition littering the road like sweet papers in Hyde Park. Then we were among the first demure little bungalows of the Stanley seafront. 'We've got to stop, sir, and wait for the CO!' shouted a signaller. There was a groan, then reluctant acquiescence. The men dropped into a crouch by the roadside, peering ahead towards the town centre. Then an NCO at the head of the company called: 'I think I can see a Panhard moving in front.' The soldiers hastily adopted tactical positions on either road verge, searching the distance for the threatened armoured car. A man with a rocket-launcher doubled forward, just in case. Nothing happened except that a trawler moved out across the harbour, showing a white flag. Through binoculars, I studied men standing watching our progress from the hillside perhaps a mile across the bay – SAS troopers, who had landed there during the night.

Then there was more excited chatter around the signallers: 'The Argies have surrendered! No one to fire except in self-defence.' Up the road from behind us strode a knot of officers led by the colonel. 'Get in behind the Colonel's party, "A" Company,' ordered Dare Farrar-Hockley urgently. 'Nobody but "A" is to get in behind the Colonel.' Every man who had not lost his red beret was wearing it now, passionately conscious that a unique opportunity for regimental glory was within their grasp. The battalion's officers had advanced perhaps two hundred yards beyond our initial halting place when a new signal was brought to the Colonel's attention: no British soldier was to advance beyond the racecourse, pending negotiations.

There was a bitter mutter of disappointment. Where was the race-course? Beside us now. There was a chorus of urgings about Nelsonian blind eyes, rapidly stifled. The Colonel ordered 'A' Company to turn aside. Suddenly, the tiredness of the men seeped through. They clattered on to the little wooden grandstand and sat down, still draped in weapons and machine-gun belts, to cheer one of their number as he clambered out on to its roof and, after some technical difficulties, tore down the Argentine flag on the flagpole, raised that of 2 Para. At their urging, I photographed this memorable gathering of desperadoes on the stepped benches. Then men began to brew up and to distribute a few cases of Argentine cigarettes they found in the starter's hut, first booty of the battle.

I wandered down to the road. It stretched empty, the cathedral clearly visible perhaps half a mile ahead. It was too good a chance to miss. Pulling off my web equipment and camouflaged jacket, I handed them up to Roger Field in his Scimitar, now parked in the middle of the road and adorned with a large Union flag. Then, wearing a civilian anorak and clutching a walking-stick that I had carried since we landed at San Carlos Bay, I set off towards the town, looking as harmless as I could contrive. 'And where do you think you are going?' demanded a para NCO in the traditional voice of NCOs confronted with prospective criminals. 'I am a civilian,' I said firmly, and walked on unhindered.

Just round the bend in the road stood a large building fronted with a conservatory that I suddenly realized was Government House. Its approaches were studded with bunkers, whether occupied or otherwise I could not tell. Feeling foolish, I stopped, grinned towards them, raised my hands in the air, and waited to see what happened. Nothing moved. Still grinning and nodding at any possible spectres within, I turned back on to the road and strode towards the cathedral, hands in the air. A group of Argentine soldiers appeared by the roadside. I walked past them with what I hoped was a careless 'Good morning'. They stared curiously, but did nothing.

Then, ahead of me, I saw a group of obviously civilian figures emerging from a large, official-looking building. I shouted to them: 'Are you British?' and they shouted back 'Yes.' Fear ebbed away, and I walked to meet them. After a few moments conversation, they pointed me towards the Argentine major on the steps of the administration block. I introduced myself to him as the correspondent of *The Times* newspaper, on the basis that it was the only British organ of which he might have heard. We talked civilly enough for a few minutes. He kept saying that most of my questions could be answered only after four o'clock, when the British general was due to meet General Menendez. Could I meanwhile go and talk to the British civilians, I asked? Of course, he said.

Years later, I met the major again and asked why he had not ordered his men to shoot as I approached their lines. He responded rather wittily: 'You did not look very military, and as far as I knew there was no lunatic asylum on the island. So I assumed that you must be a journalist.' I walked towards that well-known Stanley hostelry the Upland Goose, down a road filled with file upon file of Argentine soldiers, assembling ready to surrender. They looked cowed, drained of hostility. Yet I did not dare to photograph their wounded, straggling between comrades, because in defeat men are bitter and dangerous.

It was only when I saw officers peering curiously at me from their vehicles that I realized that my efforts to look civilian were compromised by my face, still blackened with camouflage cream – the previous evening, we had all 'disguised fair nature with hard-favoured rage'. Walking into the hotel was the fulfilment of a dream, a fantasy that had filled our thoughts for almost three months. 'We never doubted for a moment that the British would come,' said the proprietor,

Desmond King. 'We have just been waiting for the moment.' It was like liberating an English suburban golf club.

MAX HASTINGS

321.

Russian soldiers serving in 1990s Chechnya endured many privations, even before the enemy entered the story. So did their pets.

He came to us when we had only two days of food supplies left. A handsome, smart face, fluffy coat and a tail that curled in a circle. Amazing eyes, one orange, the other green. He was well fed, but not as much as the other Grozny dogs that went out of their minds as they gnawed on corpses in the ruins. This one was good-natured. We warned him. We talked to him like a person and he understood everything. Here, at war, everyone and everything seems to be at one with their surroundings, be it a person, a dog, a tree, a stone, a river. It seems everything has a spirit. When you dig a foxhole in the stony clay with a sapper's shovel, you talk to it as if it were a loved one: 'Come on my dear, just one more shovel, just a tad more,' and it yields to your entreaties, gives another chunk, hiding your body deeper. Everyone and everything understands and knows what their fate will be. And they are entitled to make their own decisions – where to grow, where to flow, where to die. We didn't need to reason with him, just one word and it was all clear. We warned him and off he went. But later he came back anyway, because he wanted to be with us. It was his choice, no-one forced him.

Then our food started to run out. We stretched it another day, thanks to some beef given to us by the 15th regiment. Then it was all gone. 'I'll skin and gut him if someone else kills him,' said Andy, our cook, stroking Sharik behind the ear. 'I won't kill him. I love dogs, and all animals really.' No-one wanted to do it. We agonized for another half day while Sharik sat at our feet and listened to us discuss who would kill him. Finally Andy took it upon himself. He led Sharik off to the river and put a bullet behind his ear. It killed him outright, not even a whimper, and his skinned body was soon strung from a tree branch. Sharik had plenty of meat on him and the fat on his side glistened. 'That has to be cut off, dog fat tastes bitter,' Andy told us. I did as he said and then cut up the warm flesh. We boiled it for two hours in a pot and then stewed it in some ketchup left from the dry rations. It tasted pretty good.

Next morning they brought us supplies of oats.

ARKADY BABCHENKO

322.

The kit that a modern American warrior carries to war, as rehearsed by an officer packing to deploy in Afghanistan in 2002. Nathaniel Fick (b. 1977) from Baltimore, Maryland, joined the Marine Corps after attending Dartmouth College.

Rucksack, flak jacket, ceramic plates to stop AK-47 fire, helmet, M-16 rifle, twelve magazines of 5.56 mm ammunition, M9 pistol with five magazines of 9 mm ammo, ten quarts of water, sleeping bag and Gore-Tex liner, fleece jacket, wool hat and gloves, face paint, first-aid kit, maps, blood chit, grease pencils, compass, GPS (Global Positioning System) receiver, toilet paper, eight-inch dive knife, two pairs of underwear, five pairs of socks, three T-shirts, one rain jacket, Pashto and Dari translation guides, disposable cameras, calculator, plastic transparencies, case of PowerBars, iodine tablets, earplugs, entrenching tool, picture of my family taken the Christmas before, camp stove, signal mirror, Angle of Repose, atropine injector, sunglasses, headscarf, toothbrush, electric razor, American flag, and a thousand dollars in twenties, just in case.

NATHANIEL FICK

323.

In March 2003, 5ft 4in Jane Blair (b. 1973), daughter of a New York lawyer, was a lieutenant serving at the Kuwait base of a US UAV unit, from which drones overflew Iraq. Its personnel saw themselves as serving on the battlefield, even as they viewed it on a monitor far out of harm's way. This is a quintessential glimpse of future war, from which the human component will be increasingly distanced.

'This is your first night flight, isn't it, ma'am?' he said. 'It is, Staff-Sergeant.' 'Well, you better stand back, the bird makes a lot of noise.' As he said it, 'Freddy' took off in an exhilarating motion, cutting the air like an arrow heading for its mark. Its loud motor roared as it sped off into the distance, and as the sound grew fainter, the UAV was lost to our eyes in the night sky. 'That was a textbook takeoff!' Staff-Sergeant Moore observed. 'Damn, that looked near perfect with my untrained eyes,' I replied. 'Hopefully this'll set the tone for the entire op.' In the ground control station the internal and external pilot flew the bird, monitoring its altitude and path along the checkpoints. The external pilot executed the initial takeoff and the loitering around the camp until the UAV acquired a signal and was passed to the control of the internal pilot. We flew up to eight thousand feet and soon would cross the Iraqi border. At that altitude, it would be hard for the enemy to hear the bird passing overhead. Once we lowered to six thousand, the bird could be heard by the average person.

Surreal lunar images were displayed on the monitor inside the combat operations center. Most of the Marines in the squadron who weren't working piled into the COC to watch this first flight. We were all stoked. Here we were, looking at imagery we had seen in pictures. Now, we were seeing it firsthand. For the moment, we were the tip of the spear. In this tent, we shared the knowledge that no one knew what was going on in Iraq but us. Sergeant Leppan, looking very intense, zoomed in on some black dots on the screen. The night vision equipment in the bird could see heat-emitting images as well as terrain features of the cooling landscape below. Because it was night, this is the camera we would use. During the day, we could alternate between our infrared camera and our regular one.

'What's that?' I asked.

'Well, ma'am, it looks like a flock of sheep.' He was a little miffed at my ignorance, but he smiled.

'Oh, so what's that?' I asked, indicating a fast-moving dot encircling the sheep. I thought for sure I had spotted something noteworthy. 'That would be the crazed sheepdog who's barking at the UAV!' We laughed. I couldn't figure out how my imagery analysts could distinguish one thing from another at the height we were flying. But they did. The UAV crept slowly north into the lines that marked the border between the two countries. The CO came on the ICS again.

'Marines ... well, this is the moment. We're in Iraq.' We all cheered, but inside we were nervous. A strange disquieting feeling came over me. We had just passed into a world we had only read about – the mythic world of Iraq. We were making history. The terrain from six thousand feet didn't look very different from Kuwait. We were seeing very little save the numerous revetments, or tank ditches. We finally came upon the gas oil separation plants. Down below, we saw men hurriedly throwing boxes into the backs of parked pickup trucks when they heard the bird.

Motorcycles, possibly the command vehicles, raced around, passing word. Within forty-five minutes, they were gone. No more than three hundred meters away was a flaming oil trench, the kind that was used during the Gulf War and the Iran–Iraq War to prevent aircraft from reaching a strategic target. They dug large trenches around the area and filled them with crude oil. They would then ignite them, forming both an igneous pit and an obscurant. All we could see was a line of burning oil and the heat waves produced by the infrared imagery. Then, suddenly, we saw static on the screen. Apparently the signal from the bird to the radio link was occasionally lost. However, the bird had a pre-programmed route and would follow the route back if a signal was not re-established.

'Goddam it!' Where the hell is the bird?' The mission commander, Captain Hamill cursed, yelling in the headset. 'She'll be comin' round the mountain, sir, just watch,' Sergeant Young, another imagery analyst, said. He started humming 'She'll Be Coming 'Round the Mountain.' He was always smiling and singing. I couldn't understand how he was always so cheerful. He was from Indiana but sounded like he was from Georgia. Sergeant Young was quite different from the average Marine. Back in garrison, his singing was welcome; out here, Sergeant Leppan was getting annoyed. We had lost our first aircraft. She wasn't 'coming around the mountain,' I told Sergeant Young. Two hours had passed, and there was no response. If it didn't acquire a signal, it would eventually crash from lack of fuel. Several of the squadron members left. I walked over to the GCS.

'I'm guessing we should write this one off,' I said to Captain Hamill, who was now reading a technical manual, trying to figure out a solution. 'This is not good. It should have returned an hour ago.' Sergeant Young chimed in, 'Looks like we're going to have to go into Iraq with a couple vehicles and a .50 cal and pick up her remains!'

'Yeah, great freakin' plan! We'll just tell the Iraqi forces that we need to pick up our aircraft.'

'It's the gremlin watchdogs,' said Sergeant Leppan.

'The what?' I asked.

'Ma'am, they're the little hand-sized demons that jump on top of the UAV and rewire it for a little while so they can take it wherever they want. Then they return it to us after they're finished playing around.'

'I see … interesting theory, Sergeant Leppan.' Our watchdog squadron patch looked like a gremlin demon with wings. At the moment, Sergeant Leppan looked crazed. By now it was around midnight and the mission commander, Captain Hamill, was writing this off as a combat loss. As we started packing things away, we thought we heard our sister squadron's bird returning home.

'Is that their UAV?'

'They're not flying right no – my God …'

'It's Freddy!' The external pilot was jumping up and down, waving his cover, or what civilians call a hat. Squadron members near the flight line started cheering. 'She's come home!' 'How crazy is that, she's back!' one Marine yelled. 'Miracles never cease to happen,' Sergeant Young added emphatically. The external pilot regained his composure and guided the UAV back in safely. 'See ma'am, she came around the mountain.' Sergeant Young smiled.

JANE BLAIR

324.

A US unit cautiously enters Saddam Hussein's capital. The author, born in Hong Kong, attended St Paul's School and University College London, before enlisting as a private in the US Marine Corps — his mother was American. During the ensuing thirty-one years he rose to colonel, serving in war zones across the world.

Reports on the radio made things sound worse than they were, as reports often do, and it was by no means apparent that the occupation of Baghdad was proving to be a cake walk. As we approached the Ministry of Information, a series of shots raised spurts of dust in the road to our front, sending the crowd scrambling for cover and eliciting an ear-splitting clatter of return fire from every gun in the patrol. That was about it as far as resistance went. Aside from that, our biggest challenge was dealing with the increasing confidence of the crowds who surged forward to clasp our hands. 'Thank you!' they shouted, 'Welcome to Baghdad!' and, 'Bush, Bush, Buuuuuush!'

Jeff and I linked back up with RCT-7. It was a scene of organized chaos as Marines swarmed everywhere, parking vehicles, putting up tents, starting generators and carrying equipment in the now familiar drill of setting up camp. As Jeff maneuvered our Humvee towards a gap in the hastily laid concertina wire circling the perimeter, a couple of bearded individuals in filthy camouflage utilities flagged us down. I climbed out and approached them cautiously – noting that they appeared to be Westerners. They carried weapons and one was wearing flip-flops. 'Hello mate, welcome to Baghdad.' I shook their outstretched hands. 'I'm Ivan – this is Tim, we're with a UK SOF det.' – [the SAS]. 'How long have you been here?' asked Jeff incredulously.

'A couple of days – you caught us up earlier than we expected.'

'It's lucky that we got to you before the Iraqis.'

'We were actually most worried about getting drilled by you guys.'

Tim chimed in. 'That's no joke – we pulled this side of the river because the army is firing up everything in their path.' True by all accounts. Later that day Jeff and I assisted a distraught Spanish camera crew who had lost one of their team when a US tank fired on the Palestine hotel – despite warnings about the presence there of Western media. I led Ivan and Tim into the newly erected operations tent and introduced them to Nick who, wincing at the beards, flip-flops, and first name introductions, was uncharacteristically sceptical of their story. Ivan persisted politely, and when it became apparent that he and Tim had a better picture of Iraqi positions in the city than any of us, Nick's usual hospitable self returned and he invited them to bring the rest of their small detachment, six soldiers in all, into the 7th Marines' fold.

Later that afternoon [9 April], Jeff and I were in Firdos Square where we had linked up with the headquarters section of 3/4. Hundreds of Iraqis packed the square, swarming around the giant statue of Saddam in scenes of riotous celebration. They engulfed the few dozen Marines present with displays of gratitude, bouncing up and down with joy. The crowd made a path for Marines to back a tank retriever up to the base of the statue, and helped them wrap a cable around Saddam's concrete torso. As the statue toppled, there were cries of warning, and the crowd parted just in time to allow the dictator's hapless effigy to crash to earth – bouncing upwards in fragments. An American flag appeared for a moment on Saddam's face, and then disappeared as someone remembered the prohibition on such displays of triumphalism and whipped it away. The crowd danced and bayed their delight. It was a moment of pure euphoria, the likes of which I had never seen before nor since.

ANDREW MILBURN

325.

After a roaming, fatherless childhood, Kayla Williams (b. 1976) abandoned the punk scene to join the US Army in 2000 and train as an Arabic linguist. Her advancing intelligence unit was bemused by their convoy's reception.

Locals come out to look at us. Waving white flags (as in 'Don't shoot me' or 'Go America,' we can't be sure). Giving us the thumbs-up. Always the thumbs-up. The people we see seem so happy. Waving and smiling and pumping their fists in the air. But at the same time I recognize pro-Saddam graffiti on some of the buildings, though no one comments on the disconnect. Mainly we are relieved that the people appear to welcome us. Not hate us. We go on. Suddenly they're waving dead white chickens. None of us has ever encountered anyone waving a dead white chicken. None of us knows what it means. Are they offering us something to eat? Are they making an obscure reference to something about Americans, the United States, our invasion?

Is it a gesture of resistance or of mockery? What the hell is going on?
I'm not talking one dead white chicken. Or ten. Or fifty. As we
continue to drive, we are seeing locals waving dead white chickens
everywhere. And now that you mention it, we are seeing dead white
chickens in the road and on the roadsides by the hundreds. We drive
and drive, and there are thousands of dead white chickens.

We have been trained how to recognize a chemical attack. And one
of the first signs of a chemical attack is dead animals littering the roads.
And so we consider the scenario and we ask ourselves: Should we suit
up? Don our chemical protection gear? Are we under chemical attack?
But the locals are absolutely unfazed. They are smiling and laughing as
they wave the dead chickens. That's the first problem. The second
problem is the perfectly healthy-looking black and brown chickens
scampering in our path. If this is a chemical attack, it has killed only
the white chickens. The non-white chickens are all fine. Has the evil
dictator of Iraq devised a racist chemical weapon that only kills white
chickens? Soon someone on our team starts to laugh. And then we're
all laughing. Laughing hysterically. There's something so blatantly
bizarre about this situation. So manifestly hallucinatory. We don't feel
threatened. We know we are not the target of a chemical attack. Just a
cosmic joke. And given the scheme of things – our fatigue and frustration at having gotten so utterly lost – it's a huge relief.

A couple of weeks after the dead white chickens incident, the *News-
week* reporter traveling with our unit does some independent research.
He has an answer, and it's nothing like what we expect. Turns out that
one of Saddam's sons, Uday or Qusay – no one can say for sure – had
a chicken factory where he liked to breed white chickens. With the
invasion, the factory had been abandoned and the locals had overrun
it. They'd decided to take out their anger at Saddam by releasing the
chickens into the wild where they had no survival skills. And promptly
died by the thousands. The locals, wishing to demonstrate their enthusiasm for the US troops, waved the chickens to demonstrate their
hostility to the toppled dictatorship.

KAYLA WILLIAMS

326.

A correspondent embedded with Maj.-Gen. David Petraeus's (b. 1952) 101st Airborne Division, describes the similar astonishment of the general and his officers, on the same day, at finding themselves welcomed by local people.

'You come in here looking for a fight and you find guys hugging and kissing you,' Linington said. 'You hit him hard, don't take any shit from him, and he just melts away.'

'What a difference a day makes,' Petraeus said. He cautioned against gloating or trophy-taking. 'We want to be a benevolent and humble presence.' Dwyer surveyed the festive crowd gathering in the plant driveway. 'We're being pelted with love,' he said. A distinguished elderly man in a white robe emerged from the throng and identified himself as Abdul-Razzaq Kasbi, a teacher at the Saddam Secondary School. He spoke English with slow formality, enunciating each syllable. Schools in Hilla had closed on March 17, when war became inevitable, Kasbi said, and many had been converted into armouries. The water plant had shut down earlier in the week for lack of fuel. 'Mr General,' he said to Petraeus, 'we are afraid you will control us, as he has done.' There was no need to specify who 'he' was.

'No,' Petraeus said, 'we won't.'

Freakley leaned over and told Petraeus in a stage whisper, 'He's a show-me guy.'

'Any honest Iraqi person,' Linnington said, 'wants to see a good man control Iraq, not one bad man replaced with another.'

'We are still afraid of that,' Kasbi said. He kneaded his worry-beads.

'We will show you by our actions,' Petraeus said. He offered the teacher a brass division coin to seal the bargain, but Kasbi politely refused the token.

'I can't have anything of you unless I am sure you have come for the sake of our people. We want to live in peace. We don't want to substitute one bad person for another bad person.'

Petraeus took the rejection gracefully. Another man, sporting a trim goatee, tapped the elbow of a staff officer and offered to help. 'Okay,' the officer replied, 'tell us where all the fuckers are.' I edged over to Freakley, who was entangled in a political debate with a French radio reporter. 'Would you rather be known as the person who stood for the moral right,' Freakley asked, 'or would you rather be known as a person who went with popular opinion?'

'I understand the intellectual aversion to nation-building,' Petraeus mused. 'On the other hand, I don't see how you avoid it.'

RICK ATKINSON

327.

A woman member of a US intelligence unit experiences technical difficulties in Iraqi mountain country.

Humvees are remarkable machines. They work almost everywhere and can do almost anything an off-road vehicle needs to do. But this mountain was rocky and it was steep. Very steep. We were going slow, maybe five to ten kilometres an hour, and I was creeping us up, holding tight to the wheel.

Then we slipped a little sideways. 'Hey,' Quinn said, popping the passenger door. 'Let me ground-guide you.' A sensible call. The goal here was to help avoid the larger rocks and steer us past them. But the wheels began to slip some more no matter where I turned, and there was a feeling in the front like the wheels were lifting slightly when I gunned the engine. 'Hey,' Reid said, popping open the door in the back. 'I'm getting out of this thing.' So now I was alone in the Humvee. This was unbelievable. The guys in my team out there walking up the mountain. Me in the Humvee feeling pretty confident the truck was about to flip over. 'You guys are fucking pussies!' I yelled. No one contradicted me. No one volunteered to get back in the Humvee, either. Whatever happened to moral support?

My legs started to tremble, and I clutched the steering wheel. Sweaty palms made a firm grip theoretical. Quinn was in front of the truck, waving right, waving left, doing something useful. I couldn't see Reid. Maybe when the truck flipped, it would flip right onto his lousy ass. Now, that would be poetic. Things kept on in this vein for longer than

I care to imagine – or remember. Up and up the damned mountain, my two team members safely out of harm's way. I'd downshifted into low-lock, and we made it up eventually. Quinn yanked big rocks from in front of the wheels sometimes. I swear the front tires lost touch with the ground once or twice when I pressed hard on the pedal.

Honestly, I thought this was the end of me. Finally, though, when we arrived at the site, the FISTers [Fire Support Teamers] were grinning. Said they'd been watching us through binos the whole way up. Said they were betting for sure we'd flip it. Surprised to find a girl behind the wheel. 'You going to help set up?' Sergeant Quinn asked. 'Are you fucking kidding?' I said. I was shaking so bad I could barely stand. The other team cracked up laughing – but I could tell right away that they were laughing with me, not at me. I had won their respect by driving while the guys walked. 'Can I have a cigarette?' I asked them. I had been trying to quit smoking for a few weeks, but this drive tore my resolve. 'Boobs,' a FISTer said, like it was some genuine insight. 'Look, this one's got boobs.'

KAYLA WILLIAMS

328.

A US officer fighting the Iraq insurgency tries to make the best of its intractable moral as well as tactical challenges.

I would find myself soundlessly repeating the same quote from *Hamlet*: 'Nothing is good or bad but thinking makes it so,' which, in retrospect, is not true. Standing in the doorway of a house occupied by medieval fanatics waiting to kill you is bad, whether you think about it or not. A more apt quote from the same play would have been: 'For conscience doth make cowards of us all'. It certainly does. If the door was locked, you'd pound at the handle with a rifle butt – all the while waiting for the wood to splinter in a storm of outgoing fire. As soon as it swung open, you'd be moving with it, knowing that your life depended on getting your silhouette clear of the fatal funnel. Not running, because then you couldn't shoot, but moving as quickly as you could at a half crouch, peering into the gloom over the sights of your rifle, finger tense against the trigger. And when shooting started it was absolute pandemonium. Usually it began with the ear-splitting stutter of an automatic weapon, flashes and smoke and dust shaken out of the ruptured walls turning the gloom into darkness. And you were firing – without conscious decision or thought – mind subject to impulse. Sometimes you could make out targets – crouched, furtive figures behind the muzzle flashes – but usually you were just shooting blindly. There were shouts, the harsh guttural sounds of Arabic – your men or the enemy it was never clear – and once a scream that froze the blood. And then it would be over. The air thick with cordite smoke, the sound of coughing, groans, urgent demands from those behind, and a loud hammering that you realized was your heart. And a back slap from the Marine behind you: 'You OK?'

Going upstairs was even worse because it meant feeling your way up the stairwell, upper body contorted so that you could cover the next floor with your rifle, at a clear disadvantage to anyone lurking on the landing with grenades in hand. On occasion, Elesky or I would glance back to find that no one was following us. And perhaps this was just as well. A day into the battle, one of our number was killed while clearing a stairwell. When the other Americans on his team recovered his body they discovered that he had been shot in the back of the head. From that moment on, Elesky and I took it in turns to sleep.

Everyone has his own priority of fear, and what scares one man witless may not worry another unduly, and vice versa. My own special antipathy was entering a darkened house knowing that there was a good chance that there were men inside waiting to kill you … Our target that night looked like any other house in Mosul, a small, squat building set in a courtyard the size of a suburban driveway surrounded by a 6ft wall. The squad assigned to the cordon peeled off from the column as we approached, and with a few whispered instructions, Ali directed two soldiers to climb the outer wall, which they did, carefully using their rifle butts to remove shards of broken glass embedded in the concrete at the top of the wall, and then helping each other shimmy over. One of them gave the signal that all was clear, and the rest of the squad followed, boosting each other up onto the wall and disappearing in quick succession into the darkness beyond. On previous raids, we'd have had at least one advisor enter the building with the first jundi, but now the Iraqis seemed to be running the show, another good sign – or so I thought.

Rocco and I followed, though somewhat less gracefully, dropping heavily into the courtyard and pausing to get our bearings. As we checked our weapons and started to move towards the house there was a burst of firing: a brief stutter then the sound of several weapons on full automatic, amplified to an ear-splitting din in the close confines of the courtyard. We sprinted across the compound and into a cluster of soldiers on the front steps of the house. The firing had stopped. Rocco and I hurled ourselves against the door which swung open, spilling us into a darkened hallway. I stepped quickly to one side, pulled out my flashlight, and swung the beam along the floor, revealing a dark puddle, and a smeared trail that led back into the darkness. Someone flicked on a light, and in the sudden harsh glare I saw that there was a body lying in the hallway face down.

A jundi advanced towards it, rolling it onto its back and kicking across the floor an AK-47 that had been lying under it. It was a middle-aged man, gray-haired and unshaven wearing a pajama top and baggy pants. His eyes were closed and his chest, a mass of blood and pulp, was still. Rocco knelt over him to check the pulse on his neck, Gene and I pushed past them towards a light at the end of the hallway, our weapons at the ready. We stepped from the hallway into the kitchen and a sight that I will never shake loose. A middle-aged woman was sheltering under the table holding two children – a young girl and boy, probably around 3 and 5 years old. A third child, an older girl in her early teens, was standing behind the table staring at us with a mix of fear and hatred.

'It's OK,' I said. 'We're not going to hurt you,' suddenly realizing the awful import of what had just happened. Before Gene or I could react, the girl darted past us into the hallway, and let out a piercing scream. Rocco returned her to the room, holding her shoulders with both hands and pushing her gently in front of him. He was saying 'It's alright, it's alright.' But it wasn't alright – she had just seen her father's corpse and was shrieking uncontrollably. The mother started to sob, her children staring at her open-mouthed, uncomprehendingly. We had an interpreter but there was little for us to say beyond the fact that we were very sorry and that we would send a vehicle the next morning to take them to the morgue where they could make funeral arrangements.

We still weren't sure what had happened, but I was beginning to fear that it had been a horrible mistake. Many families in Mosul kept a weapon for self-protection. The father must have heard the jundi banging on the door; maybe he saw the weapons, and assumed that they were Muj, or maybe he himself was really an insurgent. We had no way of knowing. In any case, he had opened fire, and they had shot him through the door. I walked out the front door, two jundi followed me dragging the corpse, its head thudding down the steps.

'Pick him up for fuck's sake,' I yelled. And then, as they put him down, I saw his chest rise. 'Rocco – he's alive!' We bent over him and Rocco reached for his pulse again but there was none. 'He's been shot in the heart, there's no way that we can bring him back. Those must have been some kind of post-mortem contractions,' Rocco said, rinsing his hands with a bottle of water. This faint hope now dashed, Gene, Rocco and I trailed the pall-bearers back to the convoy. On the drive back to the hospital no one spoke. As soon as we came to a halt back in the compound, Gene dropped down from the turret, a stricken look on his face. 'We killed an innocent man.'

'Let's not jump to conclusions,' Rocco said. 'We need to talk about this.' The three of us gathered in the room that Rocco and I shared, sitting on camp beds facing each other by the dim light of a flashlight suspended by cord from the ceiling. 'We will never know if he was really an insurgent or not,' Rocco began, 'you guys know how this works. We encourage them to collect intelligence and most of the time it's decent but sometimes it's not. It's all single-source, and sometimes they make mistakes or someone with a grudge passes them false info. We do what we can to check that they're taking the right steps but that's not always possible, and in the end, we want it to be their war – right?'

We said nothing. I couldn't shake the images of the family from my mind. Rocco went on: 'If it's anyone's fault it's mine. But you know even if one of us had been up there with Ali it wouldn't have made a difference. Once that guy fired through the door, he was a dead man.'

'John?' Rocco looked at Gene.

'Yes sir. I feel better,' Gene said in a monotone.

'Get to bed, gents.'

After they had left, Rocco rubbed his hands over his face.

'I do believe that we are a force for good here, Andy,' he said – as much to himself as to me. He climbed onto his rack and stared at the ceiling. 'But some of the shit that happens here is going to stay with me for a long time,' he said, and closed his eyes.

ANDREW MILBURN

329.

Some of the travails of a woman warrior in Iraq had nothing to do with the enemy, as Kayla Williams describes, after confronting her sergeant, Quinn, before he rotated out of the country.

'Why the hell won't the guys in our platoon ever talk to me anymore?' I ask. 'What the fuck's going on? Why do they act like they all hate me?' And Quinn tells me because he knows he won't have to deal with the consequences. 'They think you're a big whore,' Quinn says, looking away. 'They think you're a slut. And they don't want to have anything to do with you. Because they think you're a slut.'

'I don't get it.'

'Listen,' Quinn says. 'I think I know what this is all about. Remember? Right before we deployed? You were having sex with Connelly, right? The guys think that makes you a big slut.' This makes zero sense to me. And I am getting very angry about it. 'Connelly was the only guy I ever fucked at Fort Campbell before we deployed,' I say to Quinn. 'From when I got there in July until we left, he was the only one. So what the hell ...'

Then I put it together. The only thing I can think is that I tried briefly dating this guy while I had been at Campbell. But it hadn't worked out. I'd told the guy I was messed up in the head because I'd just gotten divorced. 'I'm not in a place where I can give you an emotional connection,' I told this guy. 'At all.' And the guy said: 'That's fine.' So we were just dating. We never even had sex. Then at a party at Sergeant Biddle's house, I got really drunk and fucked Connelly. So this other guy I'm dating, his feelings got hurt, and he got upset about

it. For three days. And then he got over it. And we remained friends.
We are still friends today.

But the guys in my platoon acted all pissed off. They decided I was
a slut or a whore, that I cheated on the guy I was dating. They probably did not even know I never slept with this other guy. They just
assumed that if a man and a woman are together, they must be having
sex. So they decided to make this big bullshit judgment on me. And it
was really painful.

KAYLA WILLIAMS

330.

*Rory Stewart (b. 1973) is a slightly built, puckish Scottish Old
Etonian who has been variously a royal tutor, soldier, spook, roaming
adventurer, Tory MP and government minister. He and Aubrey
Herbert, cited in extract no. 211, would have found much to say to
each other. Implausibly, in 2004–5 he served as a deputy province
governor in turbulent Iraq. In* Occupational Hazards, *an exuberant
account of that experience, he described difficulties in persuading local
leaders in Amara to abandon a hostile demonstration. He learned a
lesson in crowd control after a delegation asserted its demands.*

'Can I answer these one by one?' I asked. 'Not yet,' said the small man
who had so far acted as the conciliator. He said that he would never
cooperate with the Coalition or take money from them; they were an
evil force. He was an educated man; he had graduated from secondary
school; and now look at what he was wearing. He gestured to his track
pants, which had 'Reeboock' printed on the side. He said that without
immediate full employment the entire province would go up in flames.
'Can I answer your complaints now?' 'No. We have heard enough
from you,' said the unshaven man. 'In which case,' I said, sounding
angrier than I felt, 'perhaps you should leave.' And I stood up. And so
did they. 'We will gather ten thousand in the streets,' the unshaven
man shouted. The others took him by the arm. 'Thank you,' said the
short conciliator. 'You are Hitler,' added the unshaven man.

The short man asked me if we could have a private word. I took him
aside into a room. He sat down opposite me, fixed me with his dark

eyes and began to mutter something quickly in Arabic. As he did so, he stroked the knuckles of his right hand, down his cheek and twitched his hand in front of his crotch, loudly sucking in his breath and groaning. I said I didn't understand. Again he stroked his cheek, gestured at his genitals and made what I took to be a sound of pain, this time miming taking off his 'Reeboock' pants.

I told him to wait and called in an interpreter. 'He says, sir, that when he was shaving his pubic hair this morning … na'am … na'am,' the interpreter said, nodding and waiting for the next phrase, 'with the razor blade – he cut his penis … the tip of his penis. Off. And he wonders what he should do.' 'Has he tried the hospital?' 'He is going to take down his trousers. He wants to show you his penis now.' 'Thank him for the offer. Tell him it would be better to go immediately to the hospital.' 'He says he has no money. Can you give him money for the doctor?' 'Tell him the CPA is not authorized to provide money to individuals. Our funds are for communities.' I took a twenty-dollar bill from my wallet. 'Tell him it is a present. And tell him in return to make sure everyone goes home.' The short man took the money with a smile and walked outside. Within ten minutes the crowd had gone.

RORY STEWART

331.

While Stewart strove to direct a restoration of local governance, violence against British troops increased daily, but intelligence proved imperfect, as is often the case during insurgencies.

We learned that there was activity in a ruined sugar-cane factory. Men gathered there late every night, arriving in conditions of great secrecy and leaving singly. It was on the edge of the neighbouring city of Majar, which, surrounded by shantytowns of displaced Marsh Arabs, had been a hotbed of resistance against Saddam and was now one of the most extreme anti-Coalition communities in the south. Six British military policemen had been caught and murdered in the centre of Majar in the summer, the largest number of British troops killed in a single enemy attack since the Falklands War. There were reports of an al-Qaeda cell. Some intelligence analysts suggested that the factory had been adapted for the manufacture of bombs; others thought that since the narcotics trade was flourishing it had something to do with the manufacture of heroin.

The general was consulted. The covert-observation team of the battle group – which had been trained to watch IRA safe-houses, hiding for days in hedgerows, sleeping, eating, and performing all other bodily functions where they lay, with the help of plastic bags – was deployed with expensive night-vision sets. Just after dusk the team observed Iraqi vehicles arriving with their headlights off. Some men entered the factory. The soldiers advanced with their image-intensifying equipment. And then suddenly the radio net was filled with rough voices saying, 'Extract, extract.' For a moment it must have seemed to the operations room as if they had stumbled on a Martian landing. It was only the next morning that word began to spread that what they had seen, highlighted in the green glow of their night-image goggles, were men gathering for gay sex.

RORY STEWART

332.

Bullying is historically endemic throughout the Russian Army, manifested in a code of hierarchy and behaviour closely resembling that of English public schools in my boyhood. It attained a nadir during the Russians' brutal and unpopular 1999–2009 campaign in Chechnya.

Strictly speaking, there is no *dedovshchina* bullying in our regiment. *Dedovshchina* is a set of unofficial rules, a kind of a code of laws which, if violated, incur corporal punishment. For example, your walk. Your walk is determined by the amount of time you have served. The 'spirits' – those who have just been called up – are not supposed to walk at all, they are supposed to 'flit' or 'rustle'. Those in their second six months – the 'skulls' or 'bishops' – are entitled to a more relaxed mode of walking but their gait is nonetheless supposed to reflect humility. Only the 'lords', who are about to be demobilized, can walk with a special swagger that is allowed to the older recruits alone; a leisurely pace, their heels scraping the floor. If I had even thought about walking like that in training I'd immediately have been showered with punches. 'Up for demob now, are you?' they'd have asked, and then they'd have given me hell. If I stuck my hands in my pockets I'd also get a thump on the head: that is the privilege of the older soldiers. A 'spirit' should forget about his pockets entirely. Otherwise they fill them with sand and sew them up: the sand chafes the groin and two days later you have weeping sores.

If a 'spirit' doesn't show respect in his conversation with an older soldier, a 'granddad', he'll get beaten up. If he talks too loudly or goes about the barracks clattering his heels, he'll get beaten up. If he lies on his bed in the day, he'll get beaten up. If the people back home send him good rubber slippers and he decides to wear them to the shower, he'll get beaten up and lose his slippers. And if a 'spirit' even thinks of turning down the tops of his boots or walking around with his top button undone, or if his cap is tipped back on his head or to one side, or he doesn't do his belt up tightly enough, they'll thrash him so hard he'll forget his name. He is a 'spirit', the lowest dregs, and it's his job to slave until the older soldiers have been discharged.

But at the same time the older soldiers jealously guard their rights over their 'spirits'. Every self-respecting granddad has his own, a personal slave, and only he is allowed to beat and punish him. If someone else starts to harass this 'spirit' then he'll go straight to the 'granddad' and then there are conflicts: 'You're bugging him so you're bugging me …' It's also good for a 'spirit' to have his personal 'granddad'. First of all, only one person beats you. Then you can always complain to him if someone else makes claims on you, and he'll go and sort it out. If a 'skull', a soldier in his second six months, thumps a 'spirit's head or takes money from him, then he'll get a good sound beating – only 'granddads' are allowed to rob the young ones. A 'spirit' is only obliged to rustle up money, cigarettes and food for his own 'granddad', and he can ignore anyone else's demands. The only exception is a 'granddad' who's stronger than yours.

But there is none of this in our regiment. All of that stuff – the unbuttoned tunics, the belt and the walk – is all just child's play. It's the big league here. I can walk how I like and wear what I like and it doesn't bother anyone. They beat us for completely different reasons.

Our older conscripts have already killed people and buried their comrades and they don't believe they'll survive this war themselves. So beatings here are just the norm. Everyone is going to die anyway, both those doing the beating and their victims. So what's the big deal? There's the runway, two steps away from here, and they keep bringing back bodies by the dozen. We'll all die there.

Everybody beats everyone. The 'dembels', with three months service to go, the officers, the warrant officers. They get stinking drunk and then hammer the ones below them. Even the colonels beat the majors, the majors beat the lieutenants, and they all beat the privates; and 'granddads' beat new recruits. No-one talks to each other like human beings, they just smack each other in the mouth. Because it's easier that way, quicker and simpler to understand. Because 'you're all going to snuff it anyway, you bitches'. Because there are unfed children back home, because the officer corps is addled with impoverishment and hopelessness, because a 'dembel' has three months left, because every second man is shell-shocked. Because our Motherland makes us kill people, our own people, who speak Russian, and we have to shoot them in the head and send their brains flying up the walls, crush them with tanks and tear them to pieces.

Because these people want to kill you, because your soldiers arrived yesterday straight from training and today they are already lying on the airstrip as lumps of charred flesh, and flies lay eggs in their open eyes, and because in a day the company is reduced to less than a third, and God willing, you'll stay among that third. Because the one thing that everyone knows is how to get drunk and kill, kill and kill some more. Because a soldier is a stinking wretch, and a 'spirit' doesn't have any right to live at all, and to beat him is to actually do him a favour. 'I'll teach you what war is about, you pricks! You can all have a smack in the mouth so you don't think life is too rosy, and thank your mother that she didn't have you six months earlier or you'd all be dead now!'

Everyone hates everyone else in this regiment – the hatred and madness hang over the square like a foul black cloud, and this cloud saturates the young boys with fear, just like pieces of barbecue meat being marinated in lemon juice, only they get stewed in fear and hatred before they get sent off to the meat-grinder. It will be easier for us to die there. I stand on sentry duty as Timokha walks past. He swings round and kicks me in the chest so hard I fly off my feet and hit the wall, knocking down the wooden sign with the timetable of the company's activities: 'The company is engaged in fulfilling a government assignment.' They call this war a government assignment. In death notices sent to families they could just as well write: 'Decapitated while on government assignment.' The sign falls and its corner hits me hard on the back. I crumple up with pain. Timokha keeps walking.

ARKADY BABCHENKO

333.

Some of the methods used in Iraq by the Americans when interrogating insurgent suspects shocked some of those who carried them out, including Kayla Williams.

One day, a HUMINT interrogator approaches me and asks whether I might be willing to assist in interrogations as a female Arabic linguist. I assume he asks because he wishes to interrogate a female prisoner. Or because he simply needs my skills as a linguist. But these assumptions turn out to be wrong. I'm familiar with the cages. I know we are playing loud rock music day and night to irritate the prisoners. Anything to keep them awake. I know we make prisoners participate in chants of 'I love Bush' or 'I love America.' Anything to piss them off. When the interrogator and one or two other HUMINT guys coach me on my role for these interrogations, it is not what I expect. Once we get down to the cage area and I cover my nametapes and rank with tape (this is standard practice to prevent retribution), I am told what they will want me to do. 'We are going to bring these guys in. One at a time. Remove their clothes. Strip them naked. Then we will remove the guy's blindfold. And then we want you to say things to humiliate them. Whatever you want. Things to embarrass them. Whatever you can say to humiliate them.' I am surprised by this, but I don't turn around, either. I don't get myself out of there. I want to help so I do as I am told.

So I enter the interrogation room. Some HUMINT guys are there along with some other MI guys as guards. There is a civilian interpreter present as well. The prisoner enters the room with a blindfold on and his hands tied behind his back. Things happen like they said they would. They remove his clothes. They position him so he is facing me. When they remove the blindfold, I am the first person he sees. The civilian interpreter and the interrogator (who also speaks Arabic) mock the prisoner. Mock his manhood. Mock his sexual prowess. Ridicule the size of his genitals. Point to me. Remind him that he is being humiliated in the presence of this blond American female. Anything. Anything that comes to mind.

Degrade the prisoner. Try to break him down. Try to break his spirit. Occasionally they also ask questions on topics that might have some possible Intel value. I am watching this, and it is my perception that the Intel value of this prisoner is very limited. However, I am not in a position to judge. I haven't read the files. I am prompted to participate. To mock this naked and crying man. What do I say? What can I say? 'Do you think you can please a woman with that thing?' I ask, gesturing. I have no aptitude for this work. I prove almost immediately that I am no good at this. I tell him he had better tell us what we want to know, or we won't stop. But I am almost feeling pity. What do you say to someone to make them feel like shit? It's just not something I've ever really practiced in my personal life. It is not anything I've ever studied how to do.

I'm sure there must be plenty of women who would probably know exactly what to say, but I discover that I am actually not one of them. It is all odd and uncomfortable. But I don't know enough about what it is the HUMINT people do to know whether what I am seeing is what is supposed to happen. Soldiers flick lit cigarette butts at the prisoner. It's one thing to make fun of someone and attempt to humiliate him. With words. That's one thing. But flicking lit cigarettes at somebody – like burning him – that's illegal. It's a violation of the Geneva Conventions. They smack the prisoner across the face. These actions definitely cross a line.

While I am watching them do these things to this prisoner, I think a lot about Rick [her Palestinian ex-husband]. I imagine what it would be like for him in a situation like this. Especially with a woman there to watch. How much it would distress him. The face is not the same, but the prisoner's eyes look a lot like Rick's. The same shape of eyes, the same eye color. The same lashes. What would it be like for Rick if he ever went home to Palestine and got picked up by the Israelis and treated this way? I imagine Rick in this room. That becomes about the only thing I think about while everything else is going on. I'm no longer trying to contribute. I'm not insulting the prisoner or trying to mock him. I've fallen silent. But no one notices, because I am still a useful prop. No one seems to mind that I have nothing to say. When it's over after one more prisoner and a couple of hours, I tell the interrogator that I do not want to do this again.

For months afterward, I think about this episode … I wonder if my own creepy sense of pleasure at my power over this man had anything to do with being a woman in this situation – the rarity of that enormous power over the fate of another human being. But maybe it has nothing to do with being female. I've talked to a number of people – both guys and girls – who did these sorts of interrogations of detainees for months. People who relished their sense of power. They loved doing this job, though I've come to wonder what it must be like (maybe especially for a guy) to come back from doing this work for six months and live again with a wife and a small child. What kind of readjustment is that? What kind of psychological damage does this kind of work do? How long does it take to recover from a situation where he gets used to being suspicious of everyone and where he uses threats and intimidation to get what he wants? And then comes home to a wife and a three-year-old?

All of us, guys and girls, were in a situation in Iraq where we were powerless much of the time. Powerless to change what we did. Powerless to go home. Powerless to make any real decisions about how we were living our lives while deployed. And then we found ourselves in this situation where we had all this power over another person. And suddenly we could do whatever the fuck we wanted to them.

KAYLA WILLIAMS

334.

Heather Paxton served in Iraq in 2003 as a US Army Civil Affairs specialist, working closely with local people.

Hussein stood by himself that morning, lurking in the corner of the guard shack. Before he said hello, he landed me a box wrapped in a cheap blue plastic bag. I stared at the bag, not quite sure what to do with it. I shielded my eyes from the never-ending sun in the clear Iraqi sky. 'What's this?'

'A present. Perfume. Women should smell like women, not men.' On his face was a mischievous grin.

'You need to think of me as a soldier, not a woman,' I said. This wasn't the first time he had given me a gift, and I was torn between feeling flattered and horrified. His crush on me only seemed to get worse as time went by. His two wives didn't approve, and nor did my commander. Hussein was the local Sheikh's first-born son, and a critical asset in catching insurgents and gunrunners in Diyala province.

My job required that I transport him every day from the front gate to the operations centre. This made my attempts to ignore him difficult. 'I can't accept this, and you know it,' I thrust the bag back into his hands. The smirk on his face vanished, and he stared at me with his dark eyes. 'Why? You not accept my gift because you a soldier, not a woman? Take it. You a woman too. You make me happy if take gift.' I snatched the bag from his outstretched hands. 'Get in the vehicle,' I barked. 'We're running late.' After I dropped him off with my superiors, I stole away for a moment to my room. I untied the knot in the plastic bag and took out the box containing the perfume. Inside was a beautiful oblong glass bottle, a mixture of clear and smooth, milky and tough, like fine sandpaper. A gold pendant hung from the neck of the bottle, Parfum D'Or.

The only scents I'd smelled for the past four months were sand, sweat, gunpowder and the overpowering cologne that our Iraqi interpreters poured on every day. I pressed the pump and a spray of perfume shot out, saturating the air around me. I savoured its spicy bouquet. My heart ached for the world I left behind. I was tired of the stench of fear that clung to every pore of my body. I dreamed, just for a moment, that the fragrance of the perfume could bring me back home where I was safe. But no amount of perfume could cover my fear. So I put the bottle into my trunk, washed my face and went back to work.

Two months later Hussein was dead. Shot in the chest five times while driving home from work. The day I learned of his death, I took the perfume bottle out of my trunk, I pictured his mangled body on the side of the highway. I pulled the cap off and inhaled, trying to recapture the joy his present gave me, but it only deepened my grief. I was sorry I'd never thanked him. He gave me a sense of home, where I felt safe and where I was loved.

HEATHER PAXTON

335.

Johnson Beharry was a 24-year-old Grenadian, one of eight children. Nicknamed Paki by his army comrades, he was awarded the VC for extreme courage under fire in Iraq in 2004. In his autobiography, he vividly describes the struggles of many soldiers to make good after unhappy upbringings. To his surprise, he found his platoon sergeant describing a personal history no less troubled than his own.

'I'm from Dover,' he tells me. 'The bit of Dover that makes the docks look like a holiday camp. My dad was in the army, but he wasn't ever around much, and him and my mum split when I was a kid. I was running around with a crowd of losers, always in and out of trouble. Smashing up a bus shelter was the absolute highlight of my day.' 'But why the army?' 'I needed a job, mate. Easiest thing to do if you live in Dover is work on the ferries. I had a shitload of interviews, but got turned down for every job I applied for. Being colour-blind meant I couldn't be a deckhand, for some bloody reason, and they weren't exactly queuin' around the block to employ me at the captain's table.' He takes a sip of coffee. 'So I walked into a recruitin' centre and signed up.

'I was seventeen. First thing we did was get sent to Northern Ireland. To me, then, the army was a job; a way to put a bit of money in my pocket and keep me off the streets. But try tellin' that to the twelve-year-old kids chuckin' fridges at us from a block of flats on some God-forsaken estate in West Belfast.

'The first time I'm back home on leave, I go and see my mates, who are either down the pub, on the dole or kiddin' themselves that they're Arthur bloody Daley – plastic Londoners, the lot of them. When I start tellin' them about my day in South Armagh I might as well be talkin' Polish. That's when I decided, no more pissin' around. Make the time that I'm here the real deal.' When I hear him speak this way I realise that Broomstick's story is not a million miles from my own. We're all running from something. The army has given us a place to hide.

JOHNSON BEHARRY

336.

Twenty-two-year-old British gunner Lee Thornton was mortally wounded on patrol in Iraq on 5 September 2006, three days after his closest comrade was killed by a bomb. Like many warriors in many wars, he had written a 'last letter', to be delivered 'in case of my death'. Thornton's was addressed to his 21-year-old fiancée, Helen O'Pray. An extract deserves a place here not because it is exceptional, but because it is typical.

Hi babe,

I don't know why I am writing this because I really hope that this letter never gets to you, because if it does that means I am dead. It also means I never had time to show you just how much I really did love you. You have shown me what love is and what it feels like to be loved. Every time you kissed me and our lips touched so softly I could feel it. I got the same magical feeling as our first kiss. I could feel it when our hearts get so close they are beating as one. You are the beat of my heart, the soul in my body; you are me because without you I am nothing. I love you Helen. You are my girlfriend, my fiancée and my best friend. You are the person I know I could turn to when I needed help, you are the person I looked at when I needed to smile and you are the person I went to when I needed a hug.

When I am away it is like I have left my soul by your side. You have shown me so much while you have been in my life that if I lost you I could not live. You have shown me how to live and you have shown me how to be truly happy. I want you to know that every time I smile you have put it there. You make me smile when others can't, you make me feel warm when I am cold. You have shown me so much love and so much more. You are my whole world and I love you with all my heart. You are my happiness. There is no sea or ocean that could stop my love for you. It is the biggest thing I have ever had.

You make time to talk to me and listen to what I have to say. I know God put me and you on this earth to find each other, fall in love and show the rest of the world what true love really is. I know this is going to sound sad but every night I spent away I had a photo of you on my headboard. Each night I would go to bed, kiss my fingers, then touch your face. Well, now it is my turn to look over you as you sleep and keep you safe in your dreams. I will always be looking over you to make sure you're safe. Helen, I want to say something and I mean this more than I ever did before. You were the love of my life, the girl of my dreams. Just because I have passed away does not mean I am not with you. I'll always be there looking over you keeping you safe. So whenever you feel lonely just close your eyes and I'll be there right by your side. I really did love you with all I had. You were everything to me. Never forget that, and never forget I will always be looking over you.

Love always and forever, Lee.

LEE THORNTON

337.

Some American men serving in Iraq found it hard to treat women with the respect due to them as comrades, or to accept their authority as officers, as Lt. Jane Blair describes.

We had finally gotten showers, or at least the Wing finally told us we could use theirs. Since it was now almost a month since my last, I was already past the point of caring. I had gotten used to my smell. I took the group of female Marines to the shower area. As we pulled up in our high-back Humvee, a group of infantrymen began gawking at us. I gave them the evil eye, but some of the other females were not giving them the same look. They were flirting back. For some of them, this was the most attention they got from males in a long time. It wasn't just that the balance between males and females was so uneven, but also that most places Marines were stationed didn't have that many females, either. Female Marines usually have the pick of the litter, which often led to severe misjudgements and divorce or single parent-hood. You get a nineteen-year-old away from home for the first time in his life, and his hormones are going to override his rational sensibil-ities. That's why, despite the attention females constantly get, the excessive interest shouldn't fool them. You put these same guys around college co-eds, and they're going to turn their backs on some of the female Marines they had been paying attention to.

Being a prior enlisted and having my own share of failed dating experiences, I knew this was true. But at the Basic School things had been different. Thousands of prescreened, physically fit, college-educated males were placed next to about one hundred females for training. But the stakes were against us there; we were discouraged from dating other lieutenants, and if we did, we had to be extremely discreet about it. I think that's how I came to realize that my fellow squadron member, now my husband, was the right one for me. We fought against our friendship developing into anything more. Despite the fact that he had seen me at my worst — wearing mud-drenched cammies with a camouflaged face, hair matted with dirt, and sometimes fighting a serious lack of sleep — he still fell in love with me. Indeed, the Marine Corps had tested the strength of our relationship.

For more than half the time we had been together, we had been separated. You have to have a strong sense of resolve to be able to hold out against those odds, but more important, you really have to be in love. The Marine Corps has a disastrous failure rate for marriages, but Peter and I were trying to beat the odds. They say absence makes the heart grow fonder, but I could shoot the idiot who told that lie. Sepa-ration just makes you realize how little the rest of the bullshit in the world matters. I had been heartbroken these last few months. I felt I could stay in Iraq forever, just as long as he was here.

My mind wandered back to the present, to the hormonally charged infantrymen who were ogling my girls. One, an average-looking Joe

Corporal, had the balls to approach them when I turned away. 'Hey, how's it goin' ladies?' he said. 'You've gotta excuse me for staring, ladies, but it's been almost three months since we've seen chicks, and I just wanted to talk to say hi and to welcome you all to Iraq. We've been out there for at least a month without anything. By the way, you don't have a clean pair of socks I can have, do you?' We had been out here just as long as they had, under the same conditions. True, we weren't infantrymen, but we had done everything and gone everywhere they had. Apparently this guy thought we were fresh off the boat. I walked around the vehicle just in time to intercept the sock transaction and said, 'What are you doing, Corporal?'

'I'm trying to get a pair of socks,' he responded with a complete lack of military courtesy. 'We haven't been able to go to the PX in over a month.'

'Really, where do you think we're coming from?'

'Uh, I'm not really sure. Kuwait?'

'We've been out there just as long as you have, no PX, no showers. So you're done. Move. Now.'

He looked at me strangely and turned around.

'Oh, and by the way, Corporal …'

'Yeah?'

'I'm an officer, so you better go to your officer and get some training on the proper customs and courtesies for addressing officers.'

'Yes.'

'Unless you want me to do it for you?'

'No, ma'am.'

He walked back toward the vehicles, saying quite audibly to his fellow squad members, 'The Lieutenant said no and to fuck off!' I heard them all laughing. One of the females, Lance-Corporal Gonzalez, was shocked, holding her only pair of clean socks in her hands. She had been the one flirting with them. 'So can I still give him the socks, ma'am?' she asked me. 'No, Lance-Corporal,' I said. 'They don't really want clean socks, but I want them to respect you.'

JANE BLAIR

338.

In 2007–10 British diplomat Emma Sky (b. 1968) was seconded as political adviser to Gen. Ray Odierno, senior US Army officer in Iraq, with whom she experienced a feisty relationship.

The military could be so frustrating to work with. They had categorized Iraqis, on three-by-five-inch cards, as good guys (pro-America, Kurds) and bad guys (anti-America, Arabs). Good guys got full support: money and love. Bad guys got wrath: 'kill or capture'. [During a helicopter ride] I mentioned to General O the graffiti I had seen on the wall of a building: The hero, the martyr Saddam Hussein. He commented that the dead dictator was a mass murderer. I said, 'We still don't know who killed more Iraqis: you or Saddam, sir.' There was total silence on the helicopter. Everyone froze. I thought this time I really had gone too far. Then General O shouted, 'Open the doors, pilots. Throw her out!'

[After the two met again in the US, however], we sat for a while in silence. 'You made me a better general,' he said. It was the greatest compliment he had ever paid me. 'You changed my life, sir. I can never thank you enough for the opportunities you gave me.' 'What are you going to do when it is all over? I've never asked you this but are you going to get married? Have kids?' 'I think I've left it a bit late, sir,' I replied, somewhat uncomfortable to be having this conversation with him. 'I don't think that will happen for me.' I went on, 'When guys come back from war, they can sit in bars and impress girls with their stories. It's a bit different for women. No guy is going to want to hear my war stories.'

EMMA SKY

339.

Australian former soldier David Kilcullen (b. 1967) has become an expert on counter-insurgency, following experience in East Timor, Iraq and Afghanistan. He coined a notable phrase – 'the accidental guerrilla' – to explain the fashion in which bored young men across

*the world can take up arms not from ideological commitment, but in
a mere search for excitement and purpose. Here, he describes a firefight
in Afghanistan on 19 May 2006 between Taliban and US special
forces, which cost the Americans one dead and seven wounded, and in
which local people behaved no differently from their nineteenth-
century forebears, armed with jezails.*

The most intriguing thing about this battle was not the Taliban, it was
the behaviour of the local people. One reason the patrol was so heavily
pinned down was that its retreat was cut off by a group of farmers
who had been working in the fields and, seeing the ambush began,
rushed home to fetch their weapons and join in. Three nearby villages
participated, with people coming from as far as 5 kilometres away,
spontaneously marching to the sound of the guns. There is no
evidence that the locals cooperated directly with the Taliban; indeed,
it seems they had no directly political reason to get involved in the
fight. But, they said, when the battle was right there in front of them,
how could they not join in? Did we understand how boring it was to
be a teenager in central Afghanistan? This was the most exciting thing
that had happened in their valley in years. It would have shamed
them to stand by and wait it out, they said.

DAVID KILCULLEN

340.

*An incident in Afghanistan, described by a British officer who argues,
convincingly and importantly, that popular perception is as relevant
as supposed military reality in deciding who is winning or losing
modern battles.*

In July 2010 Major Shaun Chandler was commanding A Company 1
Royal Gurkha Rifles in central Helmand. He knew the insurgents were
telling villagers that we were cowards because we did not fight man to
man, but used aircraft and artillery. Major Chandler shaped percep-
tions by very rarely using helicopter support, never using aircraft, and
never using artillery. He repeatedly outmanoeuvred the insurgents
using basic infantry tactics and snipers. At one point he invited the

insurgents, through the villagers, to a 'fair' fight in a field at a given time in which he would bring only twelve men, and not use air support or artillery. He showed up on time, the villagers showed up to watch, but the insurgents refused to fight. After thirty minutes Major Chandler walked into the middle of the field by himself and looked around. The events were perceived by all parties – us, the villagers and the insurgents – as a defeat for the insurgents' narrative.

EMILE SIMPSON

341.

A Welsh Guards officer's attitude to mortal peril differs somewhat from that of an Afghan comrade.

[Rob] Gallimore was discovering differences between the British and Afghan armies. On one operation Gallimore, wearing body armour and helmet, squeezed himself into the back of a Ford Ranger beside Lt. Col. Abdul Hai, the Afghan army commander he was mentoring. Hai, a graduate of the American staff college at Fort Leavenworth, Kansas and fluent in English, looked at him and sighed. 'Rob, Rob, Rob, you are always worrying about these IEDs going off,' he said. 'With your armour plating, ballistic matting and body armour you worry: 'Will I lose a leg? Will I lose an arm? Will I die? Will I still look pretty for the girls?' He smiled. 'In a Ford Ranger, there is no doubt whatsoever that we are all dead, and if that is God's will then so be it. So take off your silly armour and helmets and let us advance towards the Taliban, air conditioning on and Afghan music blaring.' A little further down the road, Ali, Hai's loyal bodyguard and driver, spotted and defused an IED ahead of them. Hai turned to Gallimore and said: 'There, Rob. We need no protection. I have Ali.' The Afghan army method of searching for IEDs was dubbed by Gallimore's men '*Barma inshallah*' – *barma* being army slang for mine clearance and *inshallah* being the Arabic word for 'God willing'.

TOBY HARNDEN

342.

General Sir Rupert Smith (b.1943) commanded the UK armoured division in the 1991 first Gulf war, then in 1995 the UN forces in Bosnia. Ten years later, he published the most significant book written by a British soldier of his generation, The Utility of Force. *In it, he sought to define a new paradigm of conflict underlying 'wars amongst the people', a phrase that he coined.*

War amongst the people is characterised by six major trends:

1. *The ends for which we fight are changing* from the hard absolute objectives of interstate industrial war to more malleable objectives to do with the individual and societies that are not states.
2. *We fight among the people,* a fact amplified literally and figuratively by the central role of the media: we fight in every living-room in the world as well as on the streets and fields of a conflict zone.
3. *Our conflicts tend to be timeless,* since we are seeking a condition, which then must be maintained until an agreement on a definitive outcome, which may take years or decades.
4. *We fight so as not to lose the force,* rather than fighting by using the force at any cost to achieve the aim.
5. *On each occasion new uses are found for old weapons:* those constructed specifically for use in a battlefield against soldiers and heavy armaments, now being adapted for our current conflicts since the tools of industrial war are often irrelevant to wars among the people.
6. *The sides are mostly non-state* since we tend to conduct our conflicts and confrontations in some sort of multination grouping, whether it is an alliance or a coalition, and against some party or parties that are not states.

Our new form of war is no longer a single massive event of military decisions that delivers a conclusive political result.

RUPERT SMITH

343.

The distinguished Canadian historian Margaret MacMillan caused controversy among those who know nothing of soldiers, when she asserted in her 2018 Reith Lectures and subsequent book the harsh truth that more than a few, through the ages, have relished their work, including the process of killing.

This enjoyment of war is something that makes civilians in peaceful societies uncomfortable. A Canadian general I was once interviewing spoke about the excitement of war, but only after I had turned off my tape recorder. It was, he said, like riding a very fast motorcycle. The knowledge that you could crash and die at any moment added to the thrill. Even more disturbing is the realisation that some combatants actually enjoy having the power of life and death over another and take pleasure in killing and destruction. The Australian Llewellyn Idriess, who was a sniper in World War I, said that he felt 'only pride' that in fair warfare I had taken the life of a strong man' … Julian Grenfell's letters to his mother are equally open and are full of references to how lucky he is to be in the war and to what fun he is having: 'I adore war. It is like a big picnic without the objectlessness of a picnic. I have never been so well or happy. Nobody grumbles at one for being dirty.' The excitement of fighting 'vitalises everything, every sight and word and action'. He was shot in the head and died in May 1915 … As Alexievich says, 'War remains, as it has always been, one of the great human mysteries.'

MARGARET MACMILLAN

344.

A personal last word.

In 1980 I interviewed a wartime SAS veteran named Corporal Sam Smith, who vividly described experiences with the regiment in the Mediterranean and France. He told many stories that illustrated the toughness of his fellow-soldiers. One of Smith's anecdotes especially lingers in my mind. In southern Italy in the autumn of 1943, for some arid hours his troop was pinned down on a hillside by German fire. A brick-hard former boxer lying beside him noticed that fifty yards away, a cluster of Italian women was sheltering under a bridge. The man suddenly leapt to his feet, scorning shouted protests and a hail of enemy fire, 'belted for the bridge, had one of the women and was back inside three minutes'. Smith, then twenty-one, described the deed to me with admiration almost forty years later: 'I was in love with the SAS,' said the old soldier. 'It was my life. Our greatest fear was that we might be sent back to our units.'

Yet the story he told about the ex-boxer in Italy was that of a rape. Such things have been done by every army, in every conflict in history. But they should be seen for what they are, just as several recent SAS excesses in Afghanistan should properly have been investigated and not excused, even as the regiment's gallant deeds were applauded. 'Private armies' must never be permitted private ethics. Though General Monck in 1645 asserted the honourable nature of the soldier's profession, he also acknowledged that some men are attracted by a lust 'to do evil'. I have always honoured soldiers, but am also sensitive to the wickedness of which some are capable.

Societies rightly respect their warriors, but must beware of fostering a cult in which virility, courage, toughness, ruthlessness and barbarity become confused. The units of Hitler's Waffen-SS were among the most effective fighting organisations the world has ever seen. But their killings of civilians and prisoners annul any claims they might otherwise possess upon the regard of posterity. Their perverted ethic celebrated might as the sole legitimate determinant of human affairs; they despised and trampled upon weakness and indeed right.

All armies periodically commit atrocities – some have been described in these pages. The Allies in 1944 Normandy sometimes killed captured German tank crewmen, because they wore black overalls that caused them to be confused with the SS. Likewise US generals in North-West Europe licensed their men to shoot captured snipers, an illegal practice that merely indulged soldiers' dislike of selective enemy marksmen. But the Waffen-SS, and in some measure the entire Wehrmacht and wartime Japanese army, institutionalised atrocities, while in 1945 Germany Stalin's Red Army institutionalised rape. In 1984, when I wrote a book about the Normandy campaign, it was among the first works to confront the harsh truth that, man for man, the German army was a more effective fighting force than its British and American counterparts. But I have since come to understand, and to believe immensely important, that if the Western Allied armies had matched the spirit of the Waffen-SS, this would have annihilated the very purpose, the very values, for which the war was being fought.

The challenge, for all armies in all ages, is to cherish warrior virtues, while rejecting warrior excesses. This is never easy, especially in the twenty-first century's 'wars among the people'. But those of us who love and respect soldiers – men, and now also women, many of whom possess virtues that the rest of us, civilians, lack – recognize the importance of upholding the rule of law even at war, and of sustaining the values of a civilized society even as warriors perform that most uncivilized of all human actions, the killing of one another. If we ever allow a spirit to prevail either on the battlefield or, worse still, back at home or in Parliament, that shrugs over a dead civilian or prisoner, 'What's one less towelhead or Irish terrorist?', then we would lose the most important war of all.

MAX HASTINGS

345.

Many men and women go to wars in expectation of adventures. The fortunate come home alive and uninjured, having learnt something about conflict, which a World War II Norwegian Resistance hero characterized as 'wholly evil, and not redeemed by glory'. This is not a declaration of pacifism; instead it is mere recognition of a fundamental truth. In a 2015 autobiographical novel, a former British cavalry officer described a parade before the funeral of a subaltern killed by an IED in Afghanistan, which reflects the special grief surrounding the death of a very young man – as soldiers almost always are.

They lined up in three ranks behind the church, immaculate in their service dress. Their drill boots were polished harder than they had ever been. The whole squadron was there; they had come down by bus from the barracks that morning. They had had their medals parade the day before, and after today would be going on leave. As the rest of the congregation went in the front in dribs and drabs the younger soldiers thrust out their chests to show off their first medals.

Regimental Sergeant-Major Brennan went down the ranks. His long rack of medals gleamed in the sunshine, and when he spoke to one of the boys it was not with parade ground harshness, but quietly and kindly. He came to Dusty and Dav. Davenport's arm was still in a sling, but he had struggled into his uniform for the service. Brennan looked down at them, both even shorter than he was. Dusty's eyes blazed defiance, as if willing tears to dare to come. Brennan smiled: 'You all right, boys? Stay strong for me in there. Stay strong for the boss.'

'Yeah. We're OK, sir.' Davenport spoke for both of them. Dusty found that no words would come out of his mouth. 'We'll be fine. Just, just I never been at a funeral before.'

Brennan nodded. 'I understand. They're shit, fellas. Not going to lie. People take them the wrong way. People stand up and try to speak as though it's a celebration of a life. Which is bollocks. The whole thing's a fucking tragedy.'

BARNEY CAMPBELL

Sources and Acknowledgements

The editor and publisher gratefully acknowledge permission to reproduce copyright material in this book.

Samuel Ancell: from his 'Circumstantial Journal of the Long and Tedious Blockade and Siege of Gibraltar', published in 1782.

Lord Anglesey: from *One Leg: The Life and Letters of Henry William Paget, 1st Marquess of Anglesey, K.G.* (1961).

Ralph Arnold: from *A Very Quiet War* (Hart-Davis, 1962).

Rick Atkinson: from *The British Are Coming: The War for America 1775–1777*, William Collins, 2019, copyright © Rick Atkinson, 2019. Reproduced by permission of HarperCollins Publishers Ltd and Macmillan Publishers International Limited on behalf of the author.

John Aubrey: from *Brief Lives* (Boydell Press, 1982).

Arkady Babchenko: from *One Soldier's War in Chechnya*, translated by Nick Allen, Grove Press, copyright © Arkady Babchenko, 2006, translation copyright © Nick Allen, 2007. Reproduced by kind permission of the translator.

Michael Barber: from *The Captain* (Duckworth, 1996).

Sing Basnet: quoted in *History of the 9th Gurkha Rifles* by Lt. Col. G. R. Stevens, Vol. 3 1921–48 (Aldershot, 1952).

Charles Bean: from *Australian Official History*, Vol.1.

Antony Beevor: from *The Battle for Spain*, Orion, 2007, copyright © Ocito Ltd, 2006. Reproduced by permission of The Orion Publishing Group, London.

Nazir Begum: from French Papers held at the National Army Museum. Reproduced with permission.

Johnson Beharry: from *Barefoot Soldier*, Sphere, 2006, copyright © Johnson Beharry, 2006. Reproduced by permission of Little, Brown Book Group, London.

Henry Belcher: from *The First American Civil War* (Macmillan, 1911).

Mark Bence-Jones: from *The Cavaliers* (1976).

The Bible: Authorised Version. All rights in respect of the Authorised King James Version of the Holy Bible are vested in the Crown in the United Kingdom and controlled by Royal Letters Patent.

Grace Bignold: quoted in Lyn Macdonald, *The Roses of No Man's Land* (Michael Joseph, 1980).

Peter de la Billière: from *Looking for Trouble*, HarperCollins, 1994, Copyright © Sir Peter de la Billière, 1995. Reproduced by permission of HarperCollins Publishers Ltd, and Curtis Brown Group Ltd, London on behalf of Sir Peter de la Billière.

Matthew Bishop: from *Life and Adventures* (London, 1744).

Jane Blair: from *Hesitation Kills: A Female Marine Officer*, Rowman & Littlefield Publishers, 2011, copyright © 2011 by Jane Blair. Reproduced by permission of Rowman & Littlefield Publishers and the author Jane Blair-Stokes.

John Blakiston: from *Twelve Years' Military Adventures in Three Quarters of the Globe*, vol. I (Henry Colburn, 1829).

Lesley Blanch: from *The Game of Hearts: Harriette Wilson and her Memoirs* (Gryphon, 1957).

Russell Braddon: from *The Naked Island*, Michael Joseph, 1952, copyright © Russell Braddon, 1952. Reproduced by permission of Curtis Brown Ltd, London, on behalf of The Estate of R.R. Braddon.

Rodric Braithwaite: from *Afgantsy: Thessians in Afghanistan, 1979–89*, Profile, 2011, copyright © Rodric Braithwaite, 2011, 2012. Reproduced by permission of the publisher.

Gordon Brook-Shepherd: from *November 1918*, Collins Publishers, 1981. Reproduced by kind permission of the Estate of the author.

Richard Brooke: from *Visits to Fields of Battle in England of the Fifteenth Century* (London, 1857).

Dee Brown: from *Bury My Heart at Wounded Knee: An Indian History of the American West*, Holt, Rinehart & Winston, 1971. Reproduced by permission of SLL/Sterling Lord Literistic, Inc.

W. S. Brownlie: from *The Proud Trooper* (Collins Publishers, 1964).

The Diaries of Private Horace Bruckshaw, edited by Martin Middlebrook (1979).

Sir Robert Bruce Lockhart: from *The Marines Were There* (Putnam, 1950).

Dr John Butter: from *Autobiography* (London, 1847).

Julius Caesar: from *Gallic Wars*, Book IV, translated by John Warrington (Everyman, 1965).

Charles Callwell: from *Small Wars: Their Principles and Practice* (Tales End Press, 2012).

Barney Campbell: from *Rain* (Penguin, 2015).

Philip Caputo: from *A Rumor of War* (New York: Holt Rinehart, 1977).

Les Carlyon: from *Gallipoli*, Bantam, 2002, copyright © Les Carlyon 2001. Reproduced by permission of The Random House Group Limited.

Maurice Carpenter: from *The Indifferent Horseman – The Divine Comedy of Samuel Taylor Coleridge* (Elek, 1954).

Samuel Carter III: from *Blaze of Glory* (St Martin's Press, 1971).

Dio Cassius, Tacitus and Leonard Cottrell: from *The Great Invasion* by Leonard Cottrell (1958).

Stanley Casson: from *Steady Drummer* (1935).

Bruce Catton: from *Grant Takes Command* (1969). From *This Hallowed Ground: The Story of the Union Side of the American Civil War*. Copyright © 1955, 1956 by Bruce Catton.

Lord Chalfont: from *Montgomery of Alamein*, Weidenfeld & Nicolson, 1976. Reproduced by permission of The Orion Publishing Group, London.

John Chaney: quoted in *The Fire of Liberty* by Esmond Wright, published by The Folio Society for its Members in 1984.

Guy Chapman: from *A Passionate Prodigality*, Pen & Sword, 1983, copyright © The Estate of Guy Chapman, 1933, 1965, 1985, 2019. Reproduced by permission of the publisher.

Winston S. Churchill: from *My Early Life*, 1941, copyright © The Estate of Winston S. Churchill. Reproduced by permission of Curtis Brown, London on behalf of The Estate of Winston S. Churchill.

A. F. N. Clarke: from *Contact*, Pan Macmillan, 1984. Reproduced by permission of the Licensor through PLSclear.

Peter Coats: from *Of Generals and Gardens* (1976).

Arthur Conan Doyle: from *The Great Boer War* (Smith Elder, 1902).

Edwin Cook: from *Letters* (London, 1855).

Sir Edward Creasy: from *Fifteen Decisive Battles of the Western World* (Everyman, 1908).

René Cutforth: from *The Listener*, 19 December 1968.

Saul David: from *Operation Thunderbolt*, Hodder and Stoughton, 2015, copyright © Saul David, 2015. Reproduced by permission of Hodder and Stoughton Limited; and Peters Fraser & Dunlop (www.petersfraserdunlop.com) on behalf of Saul David.

C. P. Dawnay: from *Montgomery at Close Quarters*, edited by T. E. B. Howarth (Leo Cooper, 1985).

Daniel Defoe: from *The Life and Adventures of Mrs Christian Davies, Commonly Called Mother Ross* (1740).

David Divine: from *Dunkirk*, Faber & Faber Ltd, 1945. Reproduced by permission of David Higham Associates Ltd.

Norman Dixon: from *On the Psychology of Military Incompetence* (1976).

Christopher Duffy: from *The Army of Maria Theresa*, 1974. Reproduced by permission of Peters Fraser & Dunlop (www.petersfraserdunlop.com) on behalf of the Estate of Christopher Duffy.

Henri Dunant: from *Born in Battle: The Origins of the Red Cross, A Memory of Solferino* (JR Publishing, 1862).

Nadezhda Durova: from *The Cavalry Maiden*, translated by Mary Fleming Zirin, University of Indiana Press, 1988. Permission conveyed through the Copyright Clearance Center.

Sir James Edmonds: quoted from *Montgomery of Alamein* by Lord Chalfont (1976).

Bernard Fall: from *Street Without Joy*, Stackpole Books, 1961, copyright © 1961, 1963 and 1964 by Bernard B. Fall. Renewed © 1989 by Dorothy Fall. Reproduced by permission of Globe Pequot and Tantor Media.

Ladislas Farago: from *Patton: Ordeal and Triumph*, Weidenfeld & Nicolson, 1966, copyright © 1963, 1966 by Faracom Ltd. Reproduced by permission of The Orion Publishing Group, London.

Anthony Farrar-Hockley: from *The Edge of the Sword*, Pen & Sword, 2007, copyright © Anthony Farrar-Hockley, 1954, 1993, copyright © Estate of Sir Anthony Farrar-Hockley, 2007. Reproduced by permission of the publisher.

Byron Farwell: from *Queen Victoria's Little Wars*, Pen & Sword Books, 2009, copyright © Byron Farwell, 1973. Reproduced by permission of Pen & Sword Books.

Dorothie Feilding: from *Lady Under Fire on the Western Front: The Great War Letters of Lady Dorothie Fielding MM*, edited by Andrew & Nicola Hallam, Pen & Sword, 2010, copyright © Andrew & Nicola Hallam, 2010. Reproduced by permission of the publisher.

Nathaniel Fick: from *One Bullet Away*, Weidenfeld & Nicolson, 2006, copyright © Nathaniel Fick, 2005. Reproduced by permission of The Orion Publishing Group, London.

Gordon Fleming: from *The Young Whistler*, George Allen & Unwin Publishers, 1964, 1978. Reproduced by kind permission of the Estate of the author.

Florus: from *Epitome of Roman History*, Book II, translated by Edward Forster (1927).

Denis Forman and Max Hastings: from *To Reason Why*, Pen & Sword, 2008. Reproduced by permission of the publisher.

Robert Fox: from *The Listener*, 17 September 1964.

Joseph Frank: from *Dostoevsky: The Seeds of Revolt, 1821–1849* (1976).

John Frost: from *A Drop Too Many* (Cassell, 1980).

Sir John Froissart: from *Chronicles of England, France, Spain and the Adjoining Countries*, Books I and II, translated by Thomas Jones (William Smith, 1839).

J. F. C. Fuller: from *Memoirs of an Unconventional Soldier* (Nicolson, 1936).

Roy Fullick and Geoffrey Powell: from *Suez: The Double War* (Hamish Hamilton, 1979).

Edward Gibbon: from *Autobiography* (London, 1796). From *The Decline and Fall of the Roman Empire*, vols. I and II (Everyman, edn., 1938).

Vishnu Bhatt Godse: from traditional accounts.

G. N. Godwin: from *Memorials of Old Hampshire* (London, 1906).

John B. Gordon: from *Reminiscences of the Civil War* (Constable, 1904).

Ulysses S. Grant: from *Personal Memoirs*, vol. I (Low, 1885).

William Grattan: from *Adventures with the Connaught Rangers, 1808–1814*, edited by Charles Oman (Edward Arnold, 1902).

Robert Graves: from *Goodbye to All That*, Cape, 1929, copyright © Robert Graves Trust, 1929. Reproduced by permission of United Agents LLP (www.unitedagents.co.uk) on behalf of Accuro Trustees (Jersey) Ltd as trustees of the Robert Graves Copyright Trust.

Fulke Greville: from *The Life of the Renowned Sir Philip Sidney* (London, 1652).

Llewelyn Wyn Griffith: from *Up to Mametz* (London, 1931).

Rees Howell Gronow: from *Last Recollections* (Selwyn & Blount, 1934). From *Reminiscences and Recollections*, abridged by John Raymond (Bodley Head, 1964).

Vasily Grossman: from *A Writer at War* (Harvill, 2005).

Major H. R. Hadow: quoted in *Sandhurst* by Sir John Smyth, Weidenfeld & Nicolson, 1961. Reproduced by permission of The Orion Publishing Group, London.

Alan Hankinson: from *Man of Wars: William Howard Russell and 'The Times', 1820–1907* (Heinemann Educational, 1982, distributed by Gower Publishing Company).

Dr Hare: from *Journal*. Quoted from *Marlborough: His Life and Times*, vol II (Harrap, 1934).

Henry Harford: from *Zulu War Journal* (Shuter & Shhoter Pietermaritzburg, 1978).

Charles Yale Harrison: from *Generals Die in Bed* (Noel Douglas, 1930).

Toby Harnden: from *Dead Men Risen*, Quercus, 2011, copyright © Toby Harnden, 2011. Reproduced by kind permission of the author.

Lewis Hastings: from *Dragons are Extra* (Penguin Books 1947). Copyright 1947 by Lewis Hastings.

Max Hastings: from *Montrose: The King's Champion*. (Gollancz, 1977). Reprinted by permission of the author. From the *Spectator*, 26 June 1982 and 3 April 2005. © Max Hastings 1982, 2005. From the *Standard*, 6 October 1983. © Max Hastings 1983. By permission of the author.

G. F. R. Henderson: from *Stonewall Jackson and the American Civil War*, vol. I (Longmans Green & Co., 1898).

Johnny Henderson: from *Montgomery at Close Quarters*, edited by T. E. B. Howarth, Leo Cooper, 1985. Reproduced by permission of Peters Fraser & Dunlop (www.petersfraserdunlop.com) on behalf of the Estate of T. E. B. Howarth.

Henrici Quinti Angliae Regis Gesta, edited by B. W. Williams. Quoted from *English Historical Documents, 1327–1485*, vol. IV, edited by A. R. Myers (Eyre & Spottiswoode, 1969).

Robert Henriques: from *A Biography of Myself: A Posthumous Selection of the Autobiographical Writings*, Secker & Warburg, 1969. Reproduced by kind permission of the Estate of Robert Henriques.

Aubrey Herbert: from *Mons, Anzac and Kut* (Edward Arnold, 1919).

Herodotus: from *The Histories*, Books III, IV and VII, translated by Aubrey de Sélincourt (Penguin Classics, 1954).

Chaim Herzog: from *The War of Atonement*, Weidenfeld & Nicolson, 1975. Reproduced by permission of Greenhill Books.

Christopher Hibbert: from *A Brief History of the Battle of Agincourt*, Robinson, copyright © 1964. Reproduced by permission of David Higham Associates.

Raymond Horricks: from *Marshal Ney: The Romance and the Real* (Midas Books, 1982).

Michael Howard: from *The Franco-Prussian War*, Hart-Davis, 1961, and *Captain Professor*, Continuum 2006, reproduced by kind permission of the Estate of the author; from *Clausewitz: A Very Short Introduction*, Oxford University Press, 2002. Reproduced by permission of David Higham Associates.

David Howarth: from *A Near Run Thing* (1971).

Sir Edward Hulse: from *Letters from the English Front in France* (Leopold Classic Library, 2015).

Josephus: from *Josephus: The Jewish War* translated by G. A. Williamson, Penguin Classics, 1970, copyright © G. A. Williamson, 1959, 1969. Reproduced by permission of Penguin Books Limited and David Higham Associates.

E. J. Kahn: from the *New Yorker*, 24 April 1951.

John Keegan: from *The Face of Battle*, Jonathan Cape, copyright © The Estate of John Keegan, 2014. Reproduced by permission of The Random House Group Limited and Aitken Alexander Associates.

Jack Kelly: from *Gunpowder*, Atlantic, 2004, copyright © Jack Kelly, 2004. Reproduced by permission of Basic Books, an imprint of Hachette Book Group, Inc. and the author.

Peter Kemp: from *Mine Were of Trouble: A Nationalist Account of the Spanish Civil War*, Mystery Grove Publishing, 2020, copyright © Peter Kemp, 1957. Reproduced by permission of the publisher.

David Kilcullen: from *The Accidental Guerrilla*, Hurst & Co, 2009, copyright © David Kilcullen, 2017, reproduced by permission of Hurst Publishers.

Robin Lane Fox: from *Alexander the Great*, Penguin Books, 1973.

F. S. Larpent: from *Private Journal of F. S. Larpent During the Peninsular War*, edited by G. Larpent (Bentley, 1854).

Blaise de Lasseran-Massencôme, Seigneur de Monluc: from *Commentaries* (1574).

T. E. Lawrence: from *Seven Pillars of Wisdom* (1922).

James Lees-Milne: from *Another Self*, Hamish Hamilton, 1970. Reproduced by permission of David Higham Associates Ltd.

Pierre Leulliette: from *St Michael and the Dragon: A Paratrooper in the Algerian War* (Heinemann, 1964).

Ronald Lewin: from his own Commonplace Book.

B. H. Liddell Hart: from *Alchemist of War: The Life of B. H. Liddell Hart* by Alex Danchev, Weidenfeld, 1998. Reproduced by permission of David Higham Associates.

Eric Linklater: from *Fanfare for a Tin Hat*, Macmillan, 1970, and *The Man on My Back*, Macmillan 1941. Reproduced by permission of Peters Fraser & Dunlop (www.petersfraserdunlop.com) on behalf of the Estate of Eric Linklater.

Lives of the Two Illustrious Generals, John, Duke of Marlborough, and Francis Eugene, Prince of Savoy, vol. I, anon. (A. Bell, 1713).

Livy: from *History of Rome*, Books, II, V, VIII, XXI, XXII and XXV, translated by Church and Broadrill (Macmillan, 1890).

Elizabeth Longford: from *Wellington: The Years of the Sword*, Weidenfeld & Nicolson, 1969, copyright © Estate of Countess of Longford, 1992. Reproduced by permission of Curtis Brown Group Ltd, London on behalf of the Estate of Countess of Longford.

Sister Katherine Luard: quoted in Lyn Macdonald, *The Roses of No Man's Land* (Michael Joseph, 1980).

Franklin Lushington: from *The Gambardier* (Ernest Benn, 1930).

George MacDonald Fraser: from *Quartered Safe Out Here*, HarperCollins, 2000, copyright © George MacDonald Fraser, 1992. Reproduced by permission of HarperCollins Publishers Ltd; Skyhorse Publishing and Curtis Brown Ltd, London on behalf of the Trustees of the Estate of George MacDonald Fraser.

A. G. Macdonell: from *Napoleon and His Marshals* (1936).

Margaret MacMillan: from *War: How Conflict Shaped Us*, Profile Books, 2020, copyright © Margaret MacMillan, 2020. Reproduced by permission of the publisher.

Philip Magnus: from *Kitchener*, John Murray, 1958. Reproduced by kind permission of Sir Laurence Magnus for the Estate of the author.

Baron Marcellin de Marbot: from *Memoirs*, translated by A. J. Butler (Longmans Green & Co., 1893).

George Monck: from *Observations upon Political and Military Affairs* (1645).

Viscount Montgomery: quoted in *Out of Step* by Michael Carver, Hutchinson, 1989. Reproduced by permission of United Agents on behalf of the Estate of the author.

James Carrick Moore: from *The Life of Lieutenant-General Sir John Moore*, vol. II (London, 1834).

W. Stanley Moss: from *Ill Met by Moonlight*, Orion, 2014. First published by George G. Harrap & Co., Ltd, 1950, copyright © *The Estate of William Stanley Moss* and Reproduced by kind permission.

Simon Murray: from *Legionnaire: An Englishman in the French Foreign Legion*, Pan Macmillan, 1978. Reproduced by permission of the Licensor through PLSclear.

A Book of Naval and Military Anecdotes, anon. (London, 1824).

Nigel Nicolson: from *Alex*, 1973, copyright © Nigel Nicolson, 1973. Reproduced by permission of Curtis Brown Group, Ltd, London on behalf of the Estate of Nigel Nicolson.

David Niven: from *The Moon's a Balloon* (Hamish Hamilton, 1971).

MacCarthy O'Moore: from *The Romance of the Boer War: Humour and Chivalry of the Campaign* (Elliot Stock, 1901).

Hiroo Onoda: from *No Surrender: My Thirty Year War*, translated by Charles S. Terry, Kodansha, 1975. Reproduced by permission of Naval Institute Press.

George Orwell: from review published in *The Wound and The Bow*. From *Homage to Catalonia* (1938).

Arthur Osburn: from *Unwilling Passenger* (Faber & Faber, 1932).

J. Outram: quoted from *A Season in Hell*, by Michael Edwardes (Hamish Hamilton, 1973).

George Painter: from *Marcel Proust* (Pimlico, 1996).

Roger Parkinson: from *The Fox of the North* (Peter Davies, 1980). © Roger Parkinson 1976.

Francis Parkman: from *Montcalm and Wolfe*, vol. II (Macmillan, 1906).

Léonce Patry: from *The Reality of War*, translated by Douglas Fermer (Cassell, 2001).

Heather Paxton: from *Powder: Writing by Women in the Ranks* edited by Shannon Cain and Lisa Bowden, Kore Press, 2008, copyright © Heather Paxton. Reproduced by permission of Kore Press.

Miss Payne: quoted from *Memoirs of Sir Harry Smith*, vol. II, by G. C. M. Smith (Murray, 1901).

Hesketh Pearson: from *The Hero of Delhi* (Penguin Books, 1948).

The Percy Anecdotes, edited by S. and R. Percy (London, 1823).

Jocelyn Pereira: from *A Distant Drum* (Uniform, 2020).

Plutarch: from *Lives*, translated by B. Perrin (Loeb Classical Library, 1917).

The Poem of the Cid, translated by Hamilton and Perry, edited by I. D. Michael (Manchester University Press, 1975), pp. 73–7.

Anthony Powell: from *Marcel Proust, 1871–1922: A Centenary Volume*, edited by Peter Quennell. Weidenfeld & Nicolson, 1971. Reproduced by kind permission of the Estate of the author.

Lynette Powell: quoted in Lyn Macdonald, *The Roses of No Man's Land* (Michael Joseph, 1980).

Sednham Poynts: from *The Relation of Sydnam Poyntz, 1624–1636* (Camden Third Series, vol. xiv, 1908).

John Prebble: from *Mutiny: Highland Regiments in Revolt, 1743–1804*, Pimlico, copyright © John Prebble, 1957. Reproduced by permission of The Random House Group Limited; and Curtis Brown Ltd, London, on behalf of The Literary Estate of John Prebble.

William H. Prescott: from *History of the Conquest of Mexico*, vol. I (Bentley, 1843).

Christopher Pulling: from *They Were Singing* (1952).

Ludwig Reiners: from *Frederick the Great*, translated by Lawrence Wilson (1960).

Frank Richards: from *Old Soldier Sahib* (Anthony Mott edn, 1983).

Lord Roberts: from *Forty-One Years in India* (Bentley, 1898).

Manfred Rommel: from *The Rommel Papers*, translated by Paul Findlay, edited by B. H. Liddell-Hart, copyright © 1953 by B. H. Liddell-Hart, renewed 1981 by Lady Kathleen Liddell-Hart, Fritz Bayerlein-Dittmar and Manfred Rommel. Reproduced by

permission of Mariner Books, an imprint of HarperCollins
Publishers; and HarperCollins Publishers Ltd. All rights
reserved.

Gabriel Ronay: from *The Tartar Khan's Englishman* (Cassell, 1978).

Lord de Ros: from *The Young Officer's Companion* (Murray, 1868).

Theodore Roosevelt: from *An Autobiography* (1913).

Steven Runciman: from *A History of the Crusades, Volume II*,
Cambridge University Press, 1952. Reproduced by permission of
the publisher through PLSClear.

Captain H. Sadler: from *The History of the Seventh (Service)
Battalion, The Royal Sussex Regiment 1914–1919*, ed. Owen Rutter
(London, 1934).

Guy Sajer: from *Forgotten Soldier* (London: Weidenfeld & Nicolson
Ltd., 1971/New York: Harper & Row, 1971).

Siegfried Sassoon: from *Memoirs of an Infantry Officer*, Faber &
Faber Ltd, 1931, copyright © Siegfried Sassoon. Reproduced by
permission of the publisher and the Estate of George Sassoon.

James Settle: from *Anecdotes of Soldiers in War and Peace* (Methuen,
1905).

'S. H.': quoted from *History of the British Army*, vol. vi, by Sir John
Fortescue (Macmillan, 1899–1930).

William T. Sherman: from *From Atlanta to the Sea*, vol II (New York:
Charles Webster, 1891).

William Siborne: from *History of the War in France and Belgium*, vol.
II (Boone, 1844). From *Waterloo Letters: Hitherto Unpublished,
Bearing on 16–18 June 1815*, vol. II, edited by H. T. Siborne
(Cassell, 1891).

Emile Simpson: from *War from the Ground Up*, Hurst & Co, 2012,
copyright © Emile Simpson, 2012, reproduced by permission of
Hurst Publishers.

Osbert Sitwell: from *Great Morning*, Macmillan, 1951. Reproduced
by permission of David Higham Associates Ltd.

Emma Sky: from *The Unravelling*, Atlantic Books, 2015, copyright
© Emma Sky, 2015. Reproduced by permission of Aitken
Alexander Associates and PublicAffairs, an imprint of Hachette
Book Group, Inc.

Viscount Slim: from *Defeat Into Victory*, Cassell, 1956. Reproduced by permission of David Higham Associates Ltd.

David Smiley: from *Albanian Assignment* (1984).

Rupert Smith: from *The Utility of Force: The Art of War in the Modern World*, Penguin Books, 2005, copyright © Rupert Smith and Ilana Bet-El, 2005, 2007, 2019. Reproduced by permission of Penguin Books Limited, and Alfred A. Knopf, an imprint of the Knopf Doubleday Publishing Group, a division of Penguin Random House LLC. All rights reserved.

The Soldier's Companion, anon.

Sir Edward Spears: from *Liaison 1914* (Heinemann, 1930).

Earl Stanhope: from *Conversations with Wellington* (1888).

Richard Steele: from *Tatler*, no. 164, 27 April 1710.

William B. Stevens: quoted in *Storm Over Savannah: The Story of Count d'Estaing and the Siege of the Town in 1779*, by Alexander A. Lawrence (University of Georgia Press, 1951).

Rory Stewart: from *Occupational Hazards: My Time Governing in Iraq*, Picador, 2006, copyright © Rory Stewart, 2006, 2007. Reproduced by permission of Aitken Alexander Associates and the author.

Michael Strachan, 'M.L.S.', *Blackwood's*, May 1949.

Suetonius: from *Suetonius: The Twelve Caesars*, translated by Robert Graves, Allen Lane, 1979, copyright © Robert Graves Trust, 1979. Reproduced by permission of United Agents LLP (www.unitedagents.co.uk) on behalf of Accuro Trustees (Jersey) Ltd as trustees of the Robert Graves Copyright Trust.

Douglas Sutherland: from *Sutherland's War* (1984).

Duchess of Sutherland: from *Six Weeks at the War*, 1915. Reproduced by kind permission of the Estate of the author.

Wilfred Thesiger: from *The Life of My Choice*, W.W. Norton, 1987, copyright © 1980, 1987 by Wilfred Thesiger. Reproduced by permission of HarperCollins Publishers Ltd, W. W. Norton & Company, Inc. and Curtis Brown Group Ltd, London on behalf of the Estate of Wilfred Thesiger.

Le General Baron Paul François Charles Thiébault: from *Memoirs*, vol. I, translated by A. J. Butler (Smith Elder, 1896).

Hugh Thomas: from *The Spanish Civil War* (1961).

Lee Thornton: from 'Last Words of Love from Killed Soldier' by Nigel Bunyan, *Daily Telegraph*, 23/09/2006, copyright © Nigel Bunyan/Telegraph Media Group Limited 2006.

Claire Tisdall: quoted in Lyn Macdonald, *The Roses of No Man's Land* (Michael Joseph, 1980).

Pamela Toler: from *Women Warriors: Unexpected History*, Beacon Press, 2019, copyright © Pamela Toler, 2019. Reproduced by permission of Beacon Press, Boston.

Leo Tolstoy: from *Sebastopol Sketches* (1855).

Nicholas Tomalin: from 'Zapping Charlie Kong', *Sunday Times*, 5 June 1966, copyright © The Sunday Times/News Licensing.

Geoffrey Trease: from *The Condottieri*, copyright © 1970. Reproduced by kind permission of Thames & Hudson Ltd., London.

Anthony Trollope: from *Complete Works of Anthony Trollope*, Delphi.

Barbara Tuchman: from *August 1914*, Constable/*The Guns of August*, Macmillan, copyright © Barbara W. Tuchman, 1962. Reproduced by permission of Little, Brown Book Group Ltd; and the A. M. Heath & Co. Ltd. Authors' Agents.

E. S. Turner: from *Dear Old Blighty* (1980).

Sun Tzu: from *On The Art of War*, translated by Lionel Giles, Kegan Paul, 2002, copyright © Kegan Paul, 2002. Reproduced by permission of Taylor and Francis Group, LLC, a division of Informa plc.

Unnamed British medical officer: quoted from *Incidents in the China War* (London, 1862).

John Verney: from *Going to the Wars*, 1955, reprinted 2019. Reproduced by permission of Paul Dry Books, Inc.

Voltaire: from *Ancient and Modern History*, vol. XIV, translated by William Fleming (Dingwall-Rock, New York, 1927). From *Siécle de Louis XIV*, translated by Max Hastings.

Sir William Waller: quoted from *Cornwall in the Great Civil War and Interregnum, 1642–1660* (Barton, 1977).

Auberon Waugh: from *Will This Do?* Ebury, 1980, copyright
© Auberon Waugh. Reproduced by permission of Peters Fraser &
Dunlop (www.petersfraserdunlop.com) on behalf of the Estate of
Auberon Waugh.

Evelyn Waugh: from *The Diaries of Evelyn Waugh*, edited by M.
Davie (1976). From *The Letters of Evelyn Waugh*, edited by Mark
Amory (1980).

C. V. Wedgwood: from *The King's War*, copyright © 1958, C. V.
Wedgwood. Reproduced by permission of The Wylie Agency
(UK) Limited.

James Wellard: from *The French Foreign Legion* (1981).

Nehemiah Wharton: quoted from *The Blessed Trade*, by Marjorie
Ward (1971).

William Wheeler: from *The Letters of Private Wheeler*, edited by Basil
Liddell Hart (1951).

Sir John Wheeler-Bennett: from *The Nemesis of Power: The German
Army in Politics 1918–1945*, Palgrave Macmillan, copyright
© John Wheeler-Bennett, 1954. Reproduced by permission of the
publisher through PLSClear.

Adrian Carton de Wiart: from *Happy Odyssey*, Pen & Sword, 2007,
copyright © Adrian Carton de Wiart, 1950, 2007, 2011.
Reproduced by permission of Pen & Sword Books.

Frank Wilkeson: from *The Soldier in Battle* (Redway, 1896).

Kayla Williams: from *Love My Rifle More Than You: Young and
Female in the U.S. Army* with Michael E. Staub, W. W. Norton,
2006, copyright © Kayla Williams and Michael E. Staub, 2005.
Used by permission of W. W. Norton & Company, Inc. and The
Orion Publishing Group, London.

Henry Williamson: from *The Wet Flanders Plain*, Faber & Faber
Ltd, 1929, copyright © Henry Williamson, 1929. Reproduced
by permission of the A. M. Heath & Co. Ltd. Authors'
Agents.

Pete Winner: from *Soldier 'I': The Story of an SAS Hero* with Michael
Paul Kennedy, Osprey Publishing, copyright © 2010. Reproduced
by permission of the publisher.

A. D. Wintle: from *The Last Englishman* (1962).

Cecil Woodham-Smith: from *Florence Nightingale* (1950). From *The Reason Why* (1953).

Alexis Wrangel: from *The End of Chivalry: Last Great Cavalry Battles, 1914–1918* (New York: Hippocrene Books, 1982; London: Secker & Warburg, 1984).

F. Yeats-Brown: from *The Bengal Lancer* (Gollancz, 1930).

A. Donovan Young: from *A Subaltern in Serbia, and Some Letters from the Struma Valley* (1922).

Zeno: from *The Cauldron*, Macmillan, 1966. Reproduced by permission of the Licensor through PLSclear.

Index